DEVELOPING ECOLOGICAL CONSCIOUSNESS

DEVELOPING ECOLOGICAL CONSCIOUSNESS

Path to a Sustainable World

Christopher Uhl

ROWMAN & LITTLEFIELD PUBLISHERS, INC.

Lanham • Boulder • New York • Toronto • Oxford

ROWMAN & LITTLEFIELD PUBLISHERS, INC.

Published in the United States of America
by Rowman & Littlefield Publishers, Inc.
A wholly owned subsidiary of The Rowman & Littlefield Publishing Group, Inc.
4501 Forbes Boulevard, Suite 200, Lanham, Maryland 20706
www.rowmanlittlefield.com

PO Box 317
Oxford
OX2 9RU, UK

British Library Cataloguing in Publication Information Available

Library of Congress Cataloging-in-Publication Data

Uhl, Christopher, 1949–
   Developing Ecological Consciousness: Path to a Sustainable World /
   Christopher Uhl.
      p. cm.
Includes bibliographical references.
   ISBN 0-7425-3290-9 (hardcover: alk. paper)—ISBN 0-7425-3291-7
   (pbk. : alk. paper)
   1. Ecology. 2. Sustainable development. I. Title.
   QH541.U44 2003
   333.7'2—dc21   2003010298

Printed in the United States of America

♾™ The paper used in this publication meets the minimum requirements of
American National Standard for Information Sciences—Permanence of Paper
for Printed Library Materials, ANSI/NISO Z39.48-1992.

For Jake and Genny

CREDITS

The following sources have given permission for quoted material:

"At the Cliff Edge of Life," in *Crossing the Unknown Sea,* by David Whyte (Riverhead Books, 2001). Reprinted with permission of Riverhead Books, an imprint of Penguin Group.

Buddha's Nature, by Wes "Scoop" Nisker (Bantam Books, 2001). Reprinted with permission of Bantam Books, a division of Random House.

"Catastrophe on Camels Hump," by Herbert Vogelmann (*Natural History Magazine,* 1982). Reprinted with permission of *Natural History Magazine.*

Coming Back to Life, by Joanna Macy (New Society Publishers, 1998). Reprinted with permission of New Society Publishers.

The Courage to Lead, by R. Brian Stanfield (New Society Publishers, 2000). Reprinted with permission of New Society Publishers.

Creating a World That Works for All, by Sharif Abdullah (Berrett Koehler Publishers, 1999). Reprinted with permission of Berrett Koehler Publishers.

Earth Education: A New Beginning, by Steve VanMatre (Institute for Earth Education, 1990). Reprinted with permission of The Institute for Earth Education.

Leadership and the New Science, by Margaret J. Wheatley (Berrett Koehler Publishers, 1999). Reprinted with permission of Berrett Koehler Publishers.

Mindfulness, by Ellen J. Langer (Perseus Books, 1989). Reprinted with permission of Perseus Books.

Mindfulness in Plain English: Updated and Expanded Edition, by Bhante Henepola Gunaratana (Wisdom Publications, 2002). Reprinted with permission of Wisdom Publications.

Natural Prayers, by Chet Raymo (Ruminator Books, 1999). Reprinted with permission of Ruminator Books.

Navigating the Tides of Change, by David La Chappele (New Society Publishers, 2001). Reprinted with permission of New Society Publishers.

Nonscience and Nonsense: Approaching Environmental Literacy, by Michael Zimmerman (Johns Hopkins University Press, 1995). Reprinted with permission of The Johns Hopkins University Press.

Nonviolent Communication: A Language of Compassion, by Marshall Rosenberg. (PuddleDancer Press, 1999). Reprinted with permission of PuddleDancer Press.

Peace Is Every Step, by Thich Nhat Hanh (Bantam Books, 1991). Reprinted with permission of Bantam Books, a division of Random House.

The Sacred Balance, by David Suzuki (Prometheus Books, 1997). Reprinted with permission of Prometheus Books.

Sharing Nature's Interest, by Nicky Chambers, Craig Simmons, and Mathis Wackernagel (Earthscan Publications, 2000). Reprinted with permission of Earthscan Publications.

"Small Change," by Bill McKibben (*Orion,* January/February 2003). Reprinted with permission of *Orion Magazine.*

"A Special Moment in History," by Bill McKibben (*The Atlantic Monthly,* 1998). Reprinted with permission of *The Atlantic Monthly.*

Stepping Lightly, by Mark A. Burch (New Society Publishers, 2000). Reprinted with permission of New Society Publishers.

"Strategic Questioning," by Fran Peavy. Reprinted with permission of author.

CONTENTS

FIGURES

PREFACE

I was born in 1949, turned sixteen in 1965, twenty-one in 1970. In other words, I came of age when Americans were slowly coming to see themselves not as separate from, but as part of, a living planet. The shift was barely perceptible at first. Some peg its beginning to 1962, the year Rachel Carson's book *Silent Spring* was published. Carson told the story of familiar creatures, such as backyard songbirds, dying in grotesque ways—writhing in the grass, their bodies contaminated with pesticides. She described how an American symbol, the bald eagle, was failing to reproduce, their eggshells rendered thin and brittle because of pesticides.

Carson was not alone in sounding an alarm. Donella Meadows and colleagues made headlines in 1972 with the release of *The Limits to Growth*, a book that advanced the thesis that, if human population and consumption—and their associated pollution—were to continue to grow at then-current rates, humanity was headed for catastrophe. I remember clearly the day a fellow graduate student at Michigan State passed me a copy of *Limits*. I read it in a single sitting. Meanwhile, reporters described not only rivers so polluted that they actually caught fire but also U.S. cities so encumbered by smog that older residents were advised to stay indoors.

And it was at about this time (1969) that several human beings left Earth and traveled to the moon. It took leaving Earth in a spaceship to realize that Earth itself was a "spaceship." Soon, ecologically attuned progressives began to call for "spaceship ethics" and "spaceship economics" (i.e., economics cognizant of Earth's finite resources).

These often subtle perceptual shifts were reflected in the political and social movements of the time. In 1970, Earth Day was inaugurated. Many environmental organizations that have now grown to national prominence got their start in the 1960s and 1970s, and comprehensive national laws designed to protect the environment such as the Endangered Species Act and the Clean Air Act were enacted at this time.

It was also in the sixties and seventies that a remarkable cadre of writers—Edward Abbey, Gary Snyder, Peter Mathiesen, Wendell Berry, Barry Lopez, and Annie Dillard, among others—began to soulfully describe ecological wonders and ecological tragedies. I remember receiving Annie Dillard's *Pilgrim at Tinker Creek* as a gift from my parents in 1971. Dillard described in stark language the encounters I had had—or longed to have—with wild creatures. A bit later I picked up Wendell Berry's *The Unsettling of America*, a brutally honest book about our loss of connection to "place" and the dissolution of agrarian culture in America. And then I stood at attention reading Edward Abbey and, in the process, discovered the now legendary Aldo Leopold. I was inspired, as were others at that time, by the courageousness and authenticity of these writers.

ECOLOGICAL LITERACY REACHES OUR SCHOOLS

The spread of environmental consciousness eventually became manifest in our universities. Courses on the environment, a rarity in the 1950s, were commonplace by the mid-1970s. At many schools these courses were directed to non–science majors, the idea being that all students should receive some exposure to basic ecological principles and should have an understanding of the environmental problems facing humanity, such as population growth, acid rain, and natural resource depletion.

In the 1970s, Penn State (where I now work) initiated such a "general education" course on the environment (BiSci 3). In 1983, when I joined the faculty of the biology department, I was asked to teach this course. I had never taught a college course before, and—like many freshly minted, research-oriented Ph.D.'s in the sciences—I had never thought much about the characteristics of effective teaching, much less received training in this area.

My department chair told me to expect between four and five hundred students (all non–science majors) in BiSci 3 and wished me well. I set to work developing a course outline, reading environmental science and general ecology textbooks, and writing lectures. My goal was to survive.

Initially, I taught BiSci 3 in the only way I thought possible—by embracing "the system." I picked a fat textbook, delivered lectures, and administered multiple-choice exams. When colleagues visited my class to evaluate my teaching, they offered bits of advice (e.g., write bigger on the overhead projector; leave your transparencies up longer; check to see if students have questions), but they never questioned the content of my course or my overall approach to teaching. I tinkered with the course from one year to the next—inviting in outside speakers to spice things up, switching to a snazzier textbook, offering lectures on local environmental controversies—but none of this seemed to help very much. Students didn't "perform" well on my tests, and they seldom asked probing questions about the material.

My teaching was floundering, but my research was going well. This mattered because I was at a "research university," and my superiors had made it clear that promotion and tenure would be based primarily on my effectiveness as a researcher and not on my teaching accomplishments. Specifically, this meant that I was being evaluated based on the number of scientific papers I published each year (three or four per year is a standard expectation) and my ability to bring grant money into the university to support my research and contribute to university operations.

This scenario was just fine by me; I loved doing research and I was good at it. My field work was centered in the Amazon Basin. I was studying how rain forest ecosystems responded to different sorts of disturbances, like logging and slash-and-burn farming. This was a hot topic. Scientific journals quickly published my research papers; organizations, such as the National Geographic Society, supplied funding for my field work; I was awarded Guggenheim and Fulbright Fellowships; and the Pew Foundation named me a Conservation Scholar. Penn State, meanwhile, was delighted with my performance. When I was offered a position at Yale, Penn State successfully enticed me to stay by offering me tenure and a significant salary increase.

Years came and went. All was well with one exception: my performance in the classroom was lackluster, which was especially true for BiSci 3, the earth-literacy course I had been teaching since my early days at Penn State. BiSci 3 was supposed to be a course about life, but it seemed lifeless—ossified. The nadir came at the end of the fall 1995 semester. While reading student evaluations, it occurred to me that many students were probably leaving BiSci 3 feeling more alienated from the environment than when they came in. Take the case of Jason, a student who spent the whole semester reading about and listening to a depressing litany of environmental woes that he didn't cause and that he feels he can't do anything about. To add insult to injury, Jason ends up receiving a D in my course. A year later, when someone says "environment," Jason thinks "Ugh, no thanks." Rather than engendering a sense of care for the natural world in themselves, along with a clear-headed awareness of environmental problems and a sense of hope and personal efficacy, many students, like Jason, were probably leaving my course feeling depressed and disempowered.

It dawned on me that I myself had been feeling depressed about the condition of the world for some time and that I had been transferring my own despair on to my students. As education specialist Parker Palmer points out, "We teach who we are."[1] At the time, I was literally in mourning over the condition of Earth. It was clear that I needed to do some work on myself. So, rather than sit in my office and stew, I decided to put on my boots and take a week-long hike through Penn's woods. During my walkabout, I remembered what had attracted me to ecology in the first place. It was my love for the natural world, the diversity of life-forms on the land and in the sea, the fascinating life cycles of Earth's creatures, the spaciousness and splendor of the night sky. . . .

TEACHINGS TO AWAKEN THE HEAD

The course (and my life) needed to be grounded in awe and empowerment, not doom and gloom. I needed to say "yes" to life. In redesigning the course, I thought of the tens of thousands of college students who take general ecology/environment-type courses each year, as well as the many upper-level high school students who are offered courses in the realm of environmental studies. And I asked myself, "What do these students— indeed, what do all of us—need to know to become more environmentally literate and ecologically conscious?" As I pondered this question, the word *awaken* kept popping into my head. So, I rephrased the question, "What do we need to *awaken* to?" The answers came quickly: first, to the awe and wonder of the living earth; second, to the dreadful beating we are inflicting on Earth and one another; and, third, to our collective capacity to reverse present trends and to create a life-sustaining and just world.

This book is an elaboration of these three answers. Part I, "Earth, Our Home," offers an ecology-based, wonder-filled initiation to the universe and planet Earth. Part II, "Assessing the Health of Earth," examines the alarming ways in which we humans are damaging the living body of Earth and, in the process, our own bodies and spirits. Part III, "Healing Ourselves, Healing Earth," presents the paradigms, values, tools, and consciousness that are an essential part of both planetary and personal transformation.

The book's three parts work synergistically. Indeed, it is only when we are physically, mentally, and emotionally grounded in the web of life (part I) that we can unflinchingly face the depth and severity of the current ecological crisis (part II); and then we can be empowered to respond (part III), provided we are fortified with knowledge and nurtured with hope and a belief in our own power to effect change.

Each of the book's ten chapters focuses on a major theme. For example, the theme for chapter 1 is "Earth's Origins and History," chapter 2, "A Living Planet," chapter 3, "Earth's Web of Life," and so forth. Each chapter's theme is then divided into three "foundations"—that is, three core ideas or concepts that contribute to ecological literacy. In addition, a short "reflection" is associated with each foundation, where I offer personal thoughts on the material in the foundation, along with some "questions for reflection."

PRACTICES TO AWAKEN HEART

After completing the first draft of this book, it was clear to me that something important was missing. In fact, I was ready to scrap the whole project because it seemed to lack vitality. Feeling frustrated, I went on a retreat. I spent my days in sitting and walking meditation. One evening, as I was reflecting on the role of meditation practices in cultivating awareness, I realized that the book was only half written. The book's foundations were indeed ecological teachings that engaged the intellect, but to be fully absorbed—fully embodied— these teachings needed to be accompanied by practices that awakened the heart.

After the retreat, I went back to work with renewed energy, determined to instill heart into the book by offering "practices" alongside each of the book's thirty foundations. In developing the practices for each individual foundation, I asked myself, "What could someone actually do, as a regular practice, to fully and deeply explore the essence of the ecological teaching contained within the foundation?" Collectively, the practices are invitations to experience the full spectrum of what it means to be alive. As such, they are invitations to care—to care for ourselves, to care for each other, to care for the entire community of life.

Implicit in these practices is the recognition that humans are much more than mere brains on the end of a stick. We are spirit beings—with the capacity for love, forgiveness, compassion, grief, hope, wonder, intuition, despair, insight, joy, celebration—searching for meaning in a universe imbued with mystery. "We are," as Pierre Teilhard de Chardin once observed, "not human beings having a spiritual experience; we are spiritual beings having a human experience."

In the end, it is not new laws or more efficient solar cells that will play the leading role in solving humankind's environmental and social problems; it is our awakened and caring hearts. When our hearts awaken, our resolve quickens; our courage grows; our compassion stirs; and our imagination expands. It is because an awakened heart can do all this that I have tried to imbue this book's practices with heart.

WHO IS THIS BOOK WRITTEN FOR?

When I shared early drafts of some of this book's chapters with friends, they invariably asked me, "Who is the audience?" At first, I thought the book would be for other teachers. Indeed, the first draft was written almost as a "letter" to fellow teachers. But after a time I rejected this idea and decided instead to address the book to those interested in developing and expanding ecological consciousness. The book's primary audience is students and teachers, but it should also be of interest to the general reader interested in ecology, social change, earth spirituality, sustainability, and the evolution of human consciousness.

The struggle to define my audience was intertwined with a struggle to find my voice. I remember describing this book to an acquaintance when I was still in the early writing stages. He said, "You know, Chris, when you describe this book, it seems as if your are talking from a place very high up in your chest." Although his comment made me uncomfortable, I knew instantly what he was saying. I had the book clearly worked out in my head. I knew the story that I wanted to tell, but I didn't know how to anchor the story in my gut. As a research scientist, all the writing I had ever done was from the neck up—the language, terse; the tone, objective and nonpersonal. But in recent years the themes of this book—for example, story, despair, deep ecology, vision, citizenship, healing, listening, sustainability, empowerment, activism—have weaved their way through both my professional and private life. Hence, the "voice" I have sought in the telling of this "story" isn't strictly professional or private; it is a blending of the two.

The image I held before me as I wrote was that of a spirited young person searching for truth and authenticity in a world crippled by suffering and cynicism. Someone not afraid to fully explore life's great questions:

- What do I love about being alive?
- What do I believe to be true?
- What are my most deeply cherished values?
- What do I trust most within myself?
- What does it mean to be fully human?
- Where does my passion meet the world's deepest yearnings?[2]

By openly tangling with these kinds of juicy questions, we respond to the soul's deep longing for meaning and purpose.

The book is autobiographical insofar as its contents mirror my efforts to grow in self-understanding and make sense of the times in which we live. In my search I have found science, and especially the science of ecology, to be a helpful tool; but all tools have their limitations. Indeed, I understand Ken Wilber when he writes:

> I had spent my entire life studying science only to be met with the wretched realization that science was, not wrong, but brutally limited and narrow in scope. If human beings are composed of matter, body, mind, soul and spirit, then science deals handsomely with matter and body, but poorly with mind and not at all with soul and spirit.[3]

We grow in understanding, I believe, by exercising both our powers of reason and discrimination (our "left-brain" attributes), as well as by nurturing our visual, imaginative, metaphoric, and intuitive sides (our "right-brain" attributes). It is through the marriage of the left and the right—science and spirit—that wisdom and transformation most readily emerge.

NOTES

1. Parker Palmer, *Courage to Teach* (San Francisco: Jossey-Bass, 1998), 1.
2. Questions inspired by Sharon D. Parks, *Big Questions, Worthy Dreams* (San Francisco: Jossey-Bass, 2000), 137–138.
3. Ken Wilber, *Grace and Grit* (Boston: Shambhala, 1991), 12.

ACKNOWLEDGMENTS

I confess at the outset that I have never written a book before and didn't consciously choose to write this one. It feels more like I became pregnant with it. Slowly, over the years, the book's various facets have gathered and grown in my head and heart. Now, the weight of it has become too much to hold any longer, and so it comes forth. As with a biological birth, this book has benefited from the caring presence of midwives.

First, I am grateful to my dear Brazilian friends Tatiana Correa and Adalberto Verissimo, who offered friendship and support during this project's formative years.

Next, I am deeply indebted to all the magnificent undergraduate students who have served as my teaching assistants over the years in BiSci 3; we have done such good work together and have had such fun doing it! Next, I take my hat off to all those who made special contributions over the years to the design and content of BiSci 3: Jeff Gerwing, Jim Lockman, Eric Sheffer, Dominik Kulakowski, Christine Coughlin, Sara Eisenfeld, Jen Foight, Frank Frey, Austin Mandryk, and Melissa Rihl. Then there is Sally Maud Robertson who joined me in the classroom as a "process" guru and who has been both a mentor and friend.

My gratitude goes to Jimmy Davidson, who combined his artistic talents and ecological sensitivities to produce this book's illustrations and cover; and to Scott Jerard, for his careful and sensitive editing.

A warm thanks goes as well to all colleagues and friends who were kind enough to read and comment on pieces of the text: Kim Corrigan, Karen Litfin, Laurie Mulvey, Alice Skipton, Brian Swimme, Mary Evelyn Tucker, Eric Feigelson, Richard Fox, Jim Minesky, Chet Raymo, Lucky Yappa, Warren Abrahamson, Lee Kump, Jack Schultz, Alan Durning, David Orr, Robert Brooks, Mark Abrams, Mathis Wackernagel, Joshua Pearce, Sandra Steingraber, Theo Colburn, Jackson Spielvogel, Andy Lau, Denis Hayes, David Korten, Larry Schardt, Jim Shortle, David Morris, Lisa Miller, Dean Snow, Dana Stuchul, Eric Post, Alan AtKisson, Lincoln Brower, Neil Johnson, Steve Lachman, Madhu Prakash, Bill Sharpe, and Nancy Wright. The views and interpretations that I have presented in this book are not necessarily those of these reviewers.

A special bow to Jon Schach and Jeff Gerwing, who spent days reading an early draft of the entire book; and to Richard Chadek, who provided just the right nudges and guidance during the later phases of the writing.

My thanks also goes to Brian Romer and Lori Pierelli at Rowman & Littlefield, for their skillful means in moving this book through the publication process.

The book also benefited from a cadre of conscientious student reviewers: Julie Dunn, Lindsey Alexander, Aaron Angert, Garen Jenco, Derek Heine, Court VanTassell, Kerri Bielski, Amy Early, Pete Weinberg, Jenn Mulhausen, Shelly Laczynski, Julia Hynes, Brett Lester, Cheryl Bailey, Kathryn Willaman, Michelle Grasser, Jennifer Lane, and Erica Wechsler.

I am grateful to Penn State for granting me a sabbatical so that I could write this book. Thanks, too, to my colleagues in the Department of Biology, especially Robert Mitchell, Richard Cyr, and Doug Cavener, who have supported my experimentation with BiSci 3 over the years.

My appreciation goes as well to Jim Adams and Heidi Loomis for allowing me use of their cabin in the woods for writing retreats; and to Pierre MacKay, whose home in Seattle was a wonderful sabbatical haven.

I wish to include in my thanks the teachers and mentors who have inspired me over the years—Lucille Bowen, Stacy Jackson, Francis Craddock, Peter Murphy, Carl Jordan—along with those shining souls I encountered during my sabbatical year: Joanna Macy, Marshall Rosenberg, Vicki Robin, Sharon Parks, Larry Daloz, Rodney Smith, Daeja Napier, Fran Peavey, Richard Chadek, Karen Litfin, and Claudia McNeil.

And, at last, comes those who are near and dear. So it is that I thank my mother, Fran Uhl, for what she has taught me about life and for her abiding faith in me; Vince, Ann Marie, and Monica for their feistiness and enduring good will; and Nan for her generosity and friendship.

Part I: Introduction

EARTH, OUR HOME

In reality there is a single integral community of the Earth that includes all its component members whether human or other than human. In this community every being has its own role to fulfill, its own dignity, its inner spontaneity. . . . Every being enters into communion with other beings. This capacity for relatedness, for presence to other beings, for spontaneity in action, is a capacity possessed by every mode of being throughout the entire universe.

—Thomas Berry[1]

All of us—through our lifestyle choices, political dispositions, and general sensibilities—will influence to a significant degree, and for better or worse, the future ecological health of planet Earth. The goal of part I (chapters 1–4) is to tune in to the mind-boggling physical and ecological relationships knitting each of us and our world together. The goal, in a word, is to connect—connect to the cosmos, to the earth, to the life around us, and to ourselves.

In connecting, we open ourselves to life-altering experiences such as those described by some of the early astronauts. For example, when astronaut Rusty Schweikert was released from his space capsule on an umbilical cord during an early Apollo mission, he looked back to the earth, "a shining green gem against a totally black backdrop," and he realized that all that he loved was on that "gem": his family; the land and rivers of his home place; art, history, culture. . . . This jet-fighter pilot was so overcome with emotion that he wanted to "hug and kiss that 'gem' like a mother does her firstborn child."[2]

Schweikert had another breakthrough: As he observed the earth from outer space, he saw that clouds did not stop at country borders to check for political ideology; and he saw that ocean currents, rivers, and mountain chains took no heed of nation-states. This red-white-and-blue American came to understand in a profound way that national boundaries are political constructs devoid of biological meaning. The natural world, built on the principle of interdependence, ignores these artificial boundaries. Schweikert's "awakening" cost American taxpayers tens of millions of dollars, but I believe it is possible to expand ecological consciousness without leaving Earth.

This call to awaken to the real world of air, soil, water, and wild beings is bewildering to many people, young and old. You yourself might be thinking: "I am awake; I am alive; I know what I am doing." If this is the case, consider for a moment: What is it that supports life here on our planet? When I ask people this question, they often respond that it is taxpayers or our local businesses or our industries that support life on the planet. Often they forget to mention the Sun, the great furnace that powers Earth; water, the giver of life; and soil, the foundation of all life.

Their forgetting is really not so surprising. As humankind has become more specialized and dependent on technology, our connections to the natural world have been weakened. In countries such as the United States, citizens spend more and more time indoors each year, inhabiting "boxes"— apartments, dormitories, classrooms, cafeterias, buses, cars. Indeed, every day, hundreds of millions of people, in cities and towns all over the world, get up in the morning inside a box (apartment building). Then, they rush off to work or school in a small horizontally moving box (car). They leave their moving box in an underground box (parking garage) and quickly enter another rectangular box (office building, school, business, factory). Then they proceed to spend their entire day inside miniboxes (cubicles, classrooms, offices). In the evening, they reverse the process. Amazingly, they may pass their entire day without any direct contact with the elements of life on planet Earth: unfiltered sunlight, clean air, free-running water, soil, wild plants, and animals.[3] At night they may read a science fiction book or view a movie about aliens visiting Earth from a distant planet, all the while failing to recognize that they have become, to a significant extent, alienated from their home planet.

Living our lives in boxes, as many of us now do, disconnects us from the living world. We become indoor people with indoor concerns, "connected" to our televisions and computers, substituting virtual reality for the real thing. Our alienation from place often extends to our very bodies. When asked to identify ourselves, we point to our head because this is, more and more, where we live. We imagine, mistakenly, that our body and mind are separate. Michael Lerner in his book *Spirit Matters* challenges us to take a new approach:

> Why not let awe and wonder be the first goals of education? Why not let our teachers be judged on how successful they are at generating students who can respond to the universe, each other and their own bodies with awe, wonder, and radical amazement at the miracles that are daily with us? . . . Why not declare the person who is most grateful and most awe-filled as the person most likely to succeed in building a fulfilling life and most likely to have the qualities of soul desired by the institutions that eventually do the hiring and promotion at every level of society?[4]

Part I of this book is a response to Lerner's challenge. It is an invitation to see everything with new eyes—the universe (chapter 1), planet Earth (chapter 2), Earth's biota (chapter 3), and ourselves (chapter 4). Lerner knows that the human being is not simply a brain attached to a stick. To be fully human is to be awake in body, mind, and spirit.

NOTES

1. Thomas Berry, *The Great Work* (New York: Bell Tower, 1999), 4.
2. Matthew Fox, *Coming of the Cosmic Christ* (San Francisco: Harper San Francisco, 1988), 32.
3. Steve VanMatre, *Earth Education* (Greenville, W.Va.: Institute for Earth Education, 1990). Contact information: The Institute for Earth Education, Cedar Cove, Greenville, WV 24945.
4. Adapted from Michael Lerner, *Spirit Matters* (Charlottesville, Va.: Hampton Roads, 2000), 243.

1

DISCOVERING HOME: EARTH'S ORIGIN AND HISTORY

Indigenous peoples . . . live in a universe, in a cosmological order, whereas we, the people of the industrial world, no longer live in a universe. We in North America live in a political world, a nation, a business world, an economic order, a cultural tradition, a Disney dreamland.

—Thomas Berry[1]

Humans come into the world gifted with curiosity. At an early age, we began to ask the great cosmic questions: Where did the earth come from? What are stars? How did I come to be? It is in the nature of the human to ask probing questions in the search for meaning. Yet, I confess that I taught courses in the realm of environmental studies for fifteen years without helping my students engage with the granddaddy of all environmental questions: Where did the "environment"—everything around us—come from in the first place?

The theme of this chapter is "origins." Today, by joining discoveries in the sciences with humanity's great wisdom traditions, we are able as never before to fathom the origins of the universe (foundation 1.1), the emergence of the solar system (foundation 1.2), and the origins and history of life on Earth (foundation 1.3).

The practices that accompany this chapter's three foundations center on "discovering our home" in the universe. We are all part of the community of life on Earth, members of the Milky Way galaxy, and participants in an expanding universe. The practices invite us into this more expansive consciousness.

FOUNDATION 1.1: THE ORIGINS OF THE UNIVERSE

For thousands of years, shamans, sages, philosophers, theologians, and just plain folk have mused about the origins of the universe. This same question has occupied the intellects and imaginations of modern scientists.

A big question calls for a big answer. So it is that human cultures, both past and present, have created countless creation myths. "Myth is the ancestor of science: it's our first fumbling attempt to explain *how* things happen. It is also the forbearer of philosophy and religion in that it tries to explain *why* things happen."[2]

The Old Paradigm

A paradigm is a worldview or belief system that explains why things are as they are. Paradigms are often transmitted from generation to generation through story. Some paradigms last for millennia; others have a shorter tenure.

The early Greek astronomers, puzzling over the question of origins, saw the Sun "circling" the earth each day; and at night, they saw the moon and stars "move" across the sky. Based on their limited perspective, their conclusion, quite sensibly, was that the earth must be nested at the very center of a series of concentric celestial spheres. The Sun, they thought, was embedded in one sphere, the moon in another, stars in another, and so forth. Think of a basketball as defining the Sun sphere, with the earth as a grape located at the center of this ball. Now imagine a larger ball with all the stars embedded in its surface, enveloping the basketball (Sun sphere). Early astronomers reasoned that the difference in brightness among the stars was the result of star's different sizes, not their distance from Earth. This was the prevailing view up to the sixteenth century.

These early astronomers knew that the greater the distance between the earth and the boundaries of their hypothesized "celestial spheres," the faster the elements embedded in the various spheres would have to be moving in order to be able to rotate around the earth once each day. Here on Earth, this simple fact can be appreciated by examining a merry-go-round. If you were to stand close to the axis when the merry-go-round spins, you would move slower than a rider standing out at the edge. Using this principle, early mathematicians could calculate the speed that a "celestial sphere" would have to spin to complete its daily rotation around the earth. For example, if the sphere containing the stars were 10,000 miles away from Earth, it would have to move approximately 2,500 miles per hour.

These early investigations occurred at a time when some of the fastest known objects were horses and arrows. Thus, it was probably hard for most people to imagine speeds of hundreds much less thousands of miles per hour. Indeed, early astronomers may have preferred to think that the "celestial sphere" containing the stars rotated at a dignified ten miles per hour, in which case the stars would only be forty miles away from Earth. Living in this mental construct, the only object of significant size in the universe would be Earth, which, of course, would conveniently explain why Earth stood at the center, with everything else circling obediently around it.[3]

But, then, in the sixteenth and seventeen centuries, along came Copernicus, Galileo, and Kepler. Using empirical observation of the heavens based on telescopes and more sophisticated mathematics, these investigators blew the lid off the "celestial spheres" hypothesis, positing that, first, the Sun, not Earth, was at the center of the universe; and, second, the Sun was not close by, but millions of miles away. For us this all seems so obvious, but put yourself back five hundred years. One evening:

> you go out under the stars to consider these ideas. You know the stars well for they are your calendar and nighttime clock. You stand at the center of your familiar universe, looking up at the well-known, nearby stars, when suddenly— it can happen no other way—the celestial sphere shatters and the stars are hurled at varying, unknown distances.[4]

Suddenly, you consider that the brighter stars might be closer stars and that the fainter stars might be those located farther away. You stand awestruck as you consider that there might be stars so far away that they are too faint to see. Your world is no longer small: the universe may be too big to see, and you are no longer at the center of this world. Feel the dissonance, as the old worldview collapses, but also feel the excitement as you are beckoned to construct a new paradigm.

A New Paradigm

Just as people in the 1600s resisted the new knowledge presented to them by science, so it is that today we sometimes find it difficult to embrace new scientific discoveries. For example, it is easy to accept old truths, like the fact that the earth revolves around the Sun; but many struggle to comprehend that our sun is just one of billions of stars in the Milky Way galaxy and that the Milky Way is just one of an estimated one hundred billion galaxies (more than ten galaxies for each person on Earth) that make up the known universe. It has been especially humbling for us to acknowledge that

Earth—instead of being at the center of "creation"—is a little planet circling a relatively small star out toward the edge of a galaxy containing a multitude of other "suns" in a universe seething with galaxies.

Yet, in a very real and fascinating way, the earth really is at the center of things because we now know that we live in an "omnicentric" universe. There is no one center; rather, each galaxy is at the "center." How could this be? In the twentieth century, astronomers discovered, to their astonishment, that all the known galaxies (or galaxy clusters) are moving away from each other. One way to visualize this is to think of a loaf of raisin bread baking in the oven. Let each raisin be a galaxy. Place yourself on one of the raisins. As the bread bakes and expands and as you look around, you will see all the other raisins moving away from you. No matter which raisin you are on, you will see it the same way.[5]

Not only are the galaxies expanding away from each other, but the farther apart they are, the faster they are rushing apart. So it is that galaxies that are twice as far apart are separating two times faster, and those that are ten times farther apart are racing away ten times faster. Space, literally, rushes into existence as the galaxies separate from one another.

Scientists have been able to trace this process of expansion backward to a hypothesized beginning point—the birth of the universe—estimated to have occurred some thirteen billion years ago. Imagine running the "universe movie" backward and using the laws of physics to determine when things began. As the myriad galaxies of today's universe draw together, they merge into a hot, high-density gas. As the volume of space decreases, the temperature steadily rises. Atoms eventually begin to dissolve as atomic mass is converted to pure energy. Ultimately, the entire universe collapses into an infinitely small, infinitely hot mathematical point. The singularity. The progenitor of the "big bang" (see box).[6]

Scientists don't simply "run the movie backward" and then accept that the universe began from a big bang thirteen billion years ago. They check for other evidence. For example, here on Earth they do experiments using high-energy particle accelerators to understand what happens to matter at very high temperatures; they use this information to calculate the kind of universe that would have come forth from the hypothesized big bang. These calculations predict that the matter formed from the big bang should have consisted largely of hydrogen and helium in a ratio of three to one, which is in fact the makeup of the universe as determined by astrophysicists.[8] Furthermore, if there really were a big bang, it stands to reason that the faint light from this cosmic bursting forth should still be detectable—a kind of "calling card" from the beginning of time in the form of infrared ra-

THE BIG BANG

Hearing the phrase "big bang," we have a tendency to picture a Fourth of July sort of event, an explosion that occurs in already existing space; but this image is misleading. Current scientific understanding suggests that not only was all matter created in those first moments of the universe, but so, too, were time and space.

Because the big bang metaphor—a ball of "something" exploding into empty space—is inherently faulty, science writer Chet Raymo has whimsically suggested referring to the big bang as the "big sneeze," where everything including space and time come into being in one spontaneous act. Or perhaps, muses Raymo, the "big ha," where space, time, energy, and matter are brought into existence by successive chuckles of the Creator. Raymo points out that "an ancient Jewish creation text has God create the world with seven laughs. The first laugh is light. A blaringly luminous hoot of laughter. A side-splitting guffaw of gamma rays, X-rays and a rainbow of colors. The 'Big Ha' puts a little fun back into creation."[7]

diation and short radio waves. Although this radiation wouldn't be visible to the naked eye, scientists reasoned that it could be measured with infrared and radio telescopes. So it was that in 1965 two scientists at Bell Telephone Laboratories, Arno Penzias and Robert Wilson, detected a uniform background microwave radiation pervading the sky—all that is now left of the big bang's fury. Later in 1989 the satellite dubbed *COBE* (cosmic background explorer) measured this radiation with great precision and found that its temperature and behavior matched, very closely, to that predicted by the big bang theory. Here's the punch line: The big bang isn't lost far back in time; today planet Earth receives light that was emitted when the universe was first forming. The big bang is all around us. [9]

A huge question remains: Where did the stuff that caused the big bang originate from? When this question is posed to physicists, they explain that it is theoretically possible for things to materialize out of a vacuum—out of nothingness. According to MIT physicist Alan Guth, a speck of matter, one billionth the size of a proton, may have bubbled into being and created a repulsive gravitational field—a "false vacuum"—so strong that it exploded into our universe. This is theoretically possible. Science writer Brad Lemley explains:

According to Einstein's theory of relativity, the energy of a gravitational field is negative. The energy of matter, however, is positive. . . . Calculations totaling up all the matter and all the gravity in the observable universe indicate that the two values seem to precisely counterbalance. All matter plus all gravity equals zero. So the universe could come from [essentially] nothing because it is fundamentally, nothing.[10]

If you think this scenario sounds weird, you are not alone. Even the scientists involved in this research acknowledge that their findings are disorientating and difficult to fathom.

Guth's model for the origin of the universe, called inflation theory, has gained wide acceptance. According to Guth, as the repulsive forces of the incipient universe compounded, vast quantities of ever-doubling energy were created. Upon cooling, this energy transformed to a soup of particles—electrons, positrons, and neutrinos—which eventually gave rise to the first simple elements, then the stars and galaxies.[11]

The question of how the universe will end has also occupied the scientific community in recent decades. There appear to be two possibilities: the universe will continue expanding forever; or it will eventually reverse direction and collapse back on itself, ending as it began in a blaze of heat and light. It all depends on the interaction of two forces—the speed at which the galaxies are rushing apart and the mutual attraction of gravity that counteracts their outward rush. Raymo compares the situation to that of a rocket ship leaving the earth. The rocket must achieve a velocity of at least twenty-five thousand miles per hour at the earth's surface to escape the earth's gravitational pull. "Like a rocket engine, the big bang hurled the galaxies outward; gravity is pulling them back together. Do the galaxies have sufficient speed to overcome their mutual attraction? Do they possess the escape velocity?"[12]

If the universe eventually begins to contract, it will end just as it began, with a bang—the "big crunch." If it continues to expand (as most scientific evidence now indicates), it will end with a whimper tens of billions of years in the future, when there is no more energy for the birthing of new stars. Either way, the universe has a finite lifetime . . . or does it? Cosmologists Andrei Linde and Alexander Vilenkin, operating on the assumptions of Guth's inflation theory, remind us that it is theoretically possible to have an infinite number of inflating universes.[13] How science delights us with its mysteries!

Reflection

Recently I took a backpacking trip through the forests of central Pennsylvania. One evening I awoke because (in the spirit of the outdoors) "nature called." I rose from my hammock, which was slung between two pine trees, and I stumbled off into the darkness. When I turned around, I couldn't see my hammock. I wandered about—hands out in front—embracing nearby trees, feeling for hammock strings. I sampled four likely trees but found no hammock. I began to feel disoriented and distressed, but then I took a deep breath and considered the delicious absurdity of the moment. The worst that could happen was that I would spend a few hours on a star-strewn night propped up against a tree, free to contemplate the heavens.

Eventually I found my hammock; and, as I laid back down, I gazed out through the trees toward the Big Dipper and recalled the stunning Hubble Space Telescope "Deep Field Photograph." Remembering advice from Raymo, I reached down and picked up two pine needles; I held them in a crossed position, defining a point within the Big Dipper's bowl. For ten days, the Hubble camera did just this—that is, it focused on a sky speck equivalent to that defined by the intersection of two pine needles held at arms length, all the while soaking up the faint light from distant galaxies. The result: an extraordinary photograph showing nearly two thousand galaxies in that speck.

It would take twenty-five thousand such photographs to survey just the bowl of the Big Dipper. Such a survey would reveal an estimated forty million galaxies in the Dipper's bowl alone; a survey of the entire sky, in this manner, would probably turn up one hundred billion galaxies—each galaxy with billions of stars, many of those with families of planets.

Raymo calls on us to

Go outside and hold those crossed [needles] against the night sky. Let your imagination drift away from the Earth into those yawning depths where galaxies whirl like snowflakes in a storm. From somewhere out there among the myriad galaxies, imagine looking back to the one dancing flake that is the [billions of] stars of our Milky Way. Galaxies as numerous as snowflakes in a storm.[14]

Questions for Reflection

- What does it mean to live in a universe?
- How does this new cosmological understanding of the age and origin of the universe affect your understanding of yourself?

Practice: Discovering Our Home in the Night Sky

Spending most of our time indoors, as so many of us now do, we easily forget that we are part of the Milky Way galaxy. Fortunately, we can engage in an age-old practice—stargazing—to befriend our home galaxy. It is not hard to do. On a moonless summer night, pack a blanket and a thermos; and with some friends, head out to a quiet spot in the country or the mountains, far from the glare of city lights. Spread out your blanket, and then lay down and gaze up at the night sky. You are looking at the Milky Way galaxy from within.

If you could somehow remove yourself from the Milky Way galaxy and look down on it from outside, it would appear like a gigantic frisbee—with a bulge at the center (see figure 1.1).

The Milky Way galaxy is not out there separate from us; we live within a great wheel of stars. Our Sun is one star among over a hundred billion stars that cluster together to form the Milky Way. It is worth repeating: We are citizens of the Milky Way galaxy.

As you gaze skyward, you will see stars in all directions; but when your gaze is straight across the plane of the Milky Way's body, the commingling of the light from myriad distant stars will cause you to see a shimmering milky path.

Just how big is this home galaxy of ours? You will need to think in terms of "light-years" to grasp its immensity. One light-year—the distance that light travels in one year—is almost six million million miles:

(186,000 miles/second) × (60 seconds/minute) × (60 minutes/hour)
× (24 hours/day) × (365 days/year) = 5,865,696,000,000 miles/year

Yes, one light year is a colossal distance, and the Milky Way is approximately *one hundred thousand* light-years across!

Figure 1.1.   Representation of the Milky Way galaxy[15]

Now, as you lay there gazing into your home galaxy, imagine yourself whirling on your multiple journeys. First, recall that the Milky Way is still racing outward from the big bang—that is, you yourself are currently part of the Milky Way's racing out from the big bang! Next, consider that Earth, in addition to spinning on its axis each day and orbiting around the Sun each year, is involved in a grand galactic rotation—traveling at 150 miles per second, whereby it spirals around the Milky Way galaxy once every 225 million years.

As you lie on your back, it is natural to assume that you are looking "up" at the stars. But cosmologist Brian Swimme reminds us that "up" is just a cultural construct. Neither the earth nor the Milky Way have an "up" or "down." Indeed, when you stand on Earth, you are not standing "up"; rather, you are sticking "out" into space. So, as you lie on your back, instead of thinking of yourself as looking up, picture it so that you are on the "underside" of the earth looking down into the inky night sky. It may take a while, but eventually you will experience all the stars as "way down there below you"; and you will be surprised that you are not falling down there to join them.

You don't fall because the earth's gravitational pull holds you. It is not your weight, but the Earth's hold that suspends you above the stars. If this power were to suddenly vanish, you would fall down into the dark chasm of stars below. Swimme concludes:

> As you lie there feeling yourself hovering within this gravitational bond while peering down at the billions of stars drifting in the infinite chasm of space, you will have entered an experience of the universe that is not just human and not just biological. You will have entered a relationship from a galactic perspective, becoming for a moment a part of the Milky Way Galaxy experiencing what it is like to be the Milky Way Galaxy.[16]

Skywatching can be a "movie" that you go to once and say "been there"; but it can also be a practice. Skywatchers know that no two nights are the same. Atmospheric conditions, hour of the night, time of year, and physical location all affect what can be seen. Binoculars or a telescope can enhance the experience. But why stop there? Raymo recommends combining stargazing with music and imagination. In addition to a thermos and blanket, he suggests that the next time you go stargazing you take along a CD player and a recording of Joseph Haydn's *Creation Oratorio*. Once you are comfortably settled, place your finger on the "play" button, then close your eyes and breathe yourself into a relaxed state. Then, with your eyes still closed, press "play":

> Silence. A C-minor chord, somber, out of nowhere. Followed by fragments of music. Clarinet. Oboe. A trumpet note. A stroke of timpani. A prelude of

shadowy notes and thrusting chords, by which Haydn meant to represent the darkness and chaos that preceded the creation of the world. . . . Open your eyes! A brilliant fortissimo C-major chord! A sunburst of sound. Radiant. Dispelling darkness. A universe blazes into existence, arching from horizon to horizon. Stars. Planets. The luminous river of the Milky Way. As you open your eyes to Haydn's fortissimo chord and to the (almost) forgotten glory of a truly dark starry night, you will feel that you have been a witness to the big bang.[17]

Practices that befriend the night sky help us put our lives into a larger context. At present, most of us are much more likely to be sitting in front of the television at night than under a star-strewn sky. The messages from television and the night sky are strikingly different: One often promotes mindlessness and isolation; the other invites a sense of presence and belonging.

FOUNDATION 1.2: THE ORIGINS OF THE SOLAR SYSTEM

The famous astronomer and cosmologist Carl Sagan observed that we are all stardust. It is only by studying how stars (such as our Sun) and planets (such as Earth) are formed that we can appreciate the literal truth in Sagan's statement.

Formation of Stars and Planets

Our Sun came into being 4.55 billion years ago, and it owes its existence to the death of earlier generations of stars. As early stars burned up all the hydrogen fuel in their cores, they developed into "red giants." These red giants made heavier elements like carbon, nitrogen, oxygen, and silicon. These elements were then released as interstellar gases (forming red giant winds). Some red giants created monstrous explosions called "supernovae," which released very heavy elements like iron, nickel, and gold into the interstellar medium. It was from these gaseous releases—by-products of dying suns—that later stars and planets, like those in our solar system, were formed. So it is that the iron in your blood, the calcium in your fingernails, the carbon and oxygen that make up much of your body mass came from past stellar events.

Imagine the gas and dust from these red giant winds and supernovae as they first gather along weak lines of magnetic force and then gradually collect in whirlpools and eddies; watch as the atoms begin to clump and form a common center. Sense the enormous attractions and pressures that lead the hydrogen atoms to fuse into helium atoms, provoking massive outbursts of energy that cause the core to ignite, forming a star. First there was a cloud of hydrogen atoms; then, because of the immensity of the undertaking and the forces that come into play, there was our Sun (see box).

The everyday phenomena we know as gravity and rotation contribute to planetary stability. If gravity alone controlled the universe, all the galaxies and solar systems would collapse in a pile. It is inertia that keeps things from falling inward. This principle can be demonstrated by placing a bucket of water above your head and turning it upside down. Instantly, the force of gravity brings the water down on your head. But now suppose you fill the bucket and begin swinging it in a vertical arc from ground to sky. The bucket still spends an instant upside down in the same position above your head as before; but now the rotational movement counteracts gravity, and you stay dry. So it is that Earth's rotation keeps it from literally falling into the Sun. Likewise, it is the force of gravity that keeps the earth in orbit around the Sun. If the Sun's gravitational pull were somehow extinguished, the earth would fly off along a tangent.[19]

SPECKS IN SPACE

Here is a mental model of the solar system: Start by visualizing a grapefruit—the Sun—placed on the goal line of a football field. The planets extending out from the goal line are tiny compared to the grapefruit-sized Sun. For example, imagine a grain of sugar on the four-yard line—that's Mercury; a grain of salt on the seven-yard line—Venus; Earth, another salt grain, on the ten-yard line; Mars, a grain of sugar, on the fifteen-yard line; a pea at the fifty—that's Jupiter; and Saturn, another pea, at the opposite goal, a hundred yards out from the grapefruit (Sun). Tack on a second football field and we encounter Uranus, the size of a peppercorn; add on a third field and we find Neptune, also of peppercorn proportions. It would take four fields laid end to end (almost a quarter-mile) to reach the speck that is Pluto. Now, let the fields, the bleachers, and everything else dissolve; and set those nine isolated flecks with vast empty spaces between them in motion, rotating around the Sun. Mercury zips around completing an orbit in eighty-eight days; Earth makes the trip in one year; Pluto takes 250 years to complete its loop. This is the solar system in which we live.[18]

In sum, this solar system that we call home has come into being as the result of the forces of gravity and rotation. But do not forget that the substance of our solar system—the elemental composition of our very bodies—is the stuff of stars. We are stardust through and through. It was through the bestowal of the gifts of a myriad of dying stars that blue whales, prairies, clouds, strawberry sundaes, and the whole dazzling earth have become possible.

The Workings of the Sun

It is easy to take the Sun for granted. It has been burning brightly for almost five billion years, and astrophysicists calculate that it still has five billion years of fuel left to burn. Only in the twentieth century did scientists come to understand how the Sun "works." It is an extraordinary story; and in the hearing of it, we can gain a fuller understanding of Einstein's famous equation, $E = mc^2$, which describes how mass is converted to energy. According to Einstein's equation, the amount of energy (E) being produced by the Sun is related to some mass (m) multiplied by the speed of light squared.

What a remarkably simple equation! Look at it:

$$\text{energy} = \text{mass}(\text{speed of light})^2$$

Two things multiplied together to give an answer! And one of those two things, the speed of light (c), is already known—186,000 miles per second. The only thing missing to solve the equation is mass (m).

Where does the mass come from to produce the Sun's energy? In the center of stars, like our Sun, horrendous temperatures cause protons to fuse together. When two protons fuse they lose a small amount—about 1 percent— of their total mass. This quantity of lost mass is the m in $E = mc^2$. This 1 percent m multiplied by the c^2 (186,000 × 186,000 = 34,596,000,000) results in the creation of a lot of energy from a small amount of mass.

Normally, it is hard to get two protons to stick together because their like charges repel each other ferociously. Here on Earth, for example, this union only happens either in enormously expensive particle accelerators or in the fury of atomic bomb detonations.[20] But the immense size of stars creates the massive gravitational forces and associated temperatures that allow for nuclear fusion.

The energy of proton fusion at the Sun's core percolates upward for a half million miles through the Sun's seething interior until it reaches the Sun's roiling surface and is hurled into space as heat-light. The Sun converts four million tons of its proton mass (m) as energy each second (see box).

A SNIPPET THAT MAKES ALL THE DIFFERENCE

During a recent class, when a student asked how much of the Sun's energy the earth intercepted, I responded by saying I didn't know but I was sure we could figure it out. After class, I gathered the necessary background information, and the next time the class met, I told the students to take out a piece of scrap paper. I then asked them to figure out the percentage of the Sun's total energy that Earth intercepts. I told them I was available to help, if there were pieces of information they needed.

They set to work and after a bit of struggle managed to figure it out. One thing they needed to know (which I provided) was the radius of the Earth (approximately four thousand miles); with it, they could calculate the area of the disk that the earth presents to the Sun at any given moment (area of Earth disk = $3.14 \times 4{,}000$ miles^2 = $50{,}240{,}000$ miles^2). Next, they needed to know the distance from Sun to Earth (approximately 92,700,000 miles) and the formula to calculate the area of a sphere (Area = $12.57 \times r^2$). Why? Because the Sun's energy is going out all directions, and they needed to know the total receiving surface before they could calculate the percentage of that "receiving surface" that Earth commands. This calculation revealed that 1.1×10^{17} miles^2 ($12.57 \times 92{,}700{,}000^2$) is the area of the sphere receiving the Sun's energy. The fraction of this sphere occupied by the earth disk (i.e., 50,240,000 miles^2; see above) turns out to be a mere 0.00000008 percent of the total energy emitted by the Sun.

This small snippet of the Sun's energy makes "all the difference in the world" for us. If it weren't for the Sun's warming radiation, temperatures on Earth would be hundreds of degrees below zero—far too cold for life. And without the Sun's rays, there would be no evaporation of water from oceans and leaf surfaces (i.e., no rainfall, no hydrologic cycle). Nor would there be photosynthesis. The Sun is, in effect, a giant engine that powers the winds, the water cycle, photosynthesis, and, ultimately, our very bodies.

So, our Sun releases four million tons of its proton mass (m) as energy each second; and eight minutes later, a half of a billionth of that radiant energy is intercepted by Earth. Think of it as five pounds of vanished solar mass arriving to Earth as radiant energy each second—lighting up and powering Earth. Five pounds of solar mass doesn't sound like much until we understand $E = mc^2$.

Recalling the entirety of this epic story, we see that our solar system, coalescing into existence from cosmic dust generated by solar winds and star explosions, has now given birth to us. Incredibly, our very bodies, our cells, the air we breathe, are the work of the heavens—gifts of the stars.

Reflection

I am struck by how often I use and hear the word "awesome" in the course of daily conversation. I talk about the "awesome" movie, the "awesome" musician, the "awesome" team, and so on. But lately I have been trying to reserve the word "awesome" for things that put me on my knees and truly leave me breathless. I am awestruck when I learn that the universe emerged thirteen billion years ago from a tiny seed of energy smaller than a pinhead; I am awestruck when I hear that there are ten times as many galaxies in the universe as there are human beings on Earth. I am awestruck when I realize that Earth—composed almost entirely of iron, silica, sulfur, nickel, aluminum, oxygen, magnesium, and calcium—has brought forth tree frogs and orchids and soaring eagles, as well as babbling brooks and salmon.

As I think about "awe," I am reminded of a story told about Saint Francis of Assisi. In the middle of the night, when everyone was asleep, Francis climbed the church tower in the center of Assisi and began to ring the bells. He rang and rang until the townspeople came out into the streets.

"What is it, Francis? Why are you ringing the church bells like a madman?" they asked.

It was the night sky, of course—the glories of the heavens; Francis couldn't contain his awe.

Questions for Reflection

- How might the realization that your lineage goes back thirteen billion years give added meaning to your existence?
- What might you do on a daily basis to acknowledge your utter dependence on the Sun's benevolence?

Practice: Discovering Our Home in the Solar System

In grade school I learned that day and night occur because the earth spins on its axis—alternately turning toward the Sun, then turning away from the Sun. However, it wasn't until I was in my early twenties that I finally experienced the earth as the pirouetting partner in this solar dance. It happened when I was on a freighter headed for Japan. Each evening I went out to the foredeck to watch the Sun "go down." One night a crew member joined me, and he asked if the Sun was actually "going down." I had always perceived the Sun as setting on the horizon, even though in my head I knew that it was the earth, not the Sun, that was moving. Challenged by this Japanese crewman to align my perceptual experience with my book knowledge, I was able to begin to experience, in a bodily way, the earth's rolling away from the Sun.

In addition to "rolling" with the Earth, other practices can give us a physical sense of being part of the solar system. For example, cosmologist Brian Swimme suggests going out a half hour before sunset to a place offering a good view of the western horizon. Once you are settled, bring the simple model of the solar system to mind. Then, as you look at the setting Sun, recall that it is a million times the size of Earth. Focus your attention on the horizon and locate Venus, which is located sixty-three million miles from the Sun. (On Earth, your distance from the Sun is ninety-three million miles.) Next, extend your view "up" off the horizon and locate Jupiter—480 million miles from the Sun. Now, put it all together and experience it: Venus, Earth, and Jupiter are orbiting in a plane around the Sun; you are on the middle planet. All three planets are being whipped around by the Sun's gravitational power. Be patient with yourself. There should come a moment when a door opens and you experience yourself "inside" the solar system.[21]

Implicit in such practices is the belief that our cognitive understanding of Earth's movements needs to be joined with an emotional and bodily experience of these movements. As Swimme observes:

> to contemplate the solar system until you feel the great Earth turning away from the Sun and until you feel this immense planet being swung around its massive cosmic partner is to touch an ocean of wonder as you take a first step into inhabiting the actual universe and solar system and Earth.[22]

Given our utter dependence on the Sun and our growing understanding of solar science, we have a remarkable opportunity to develop modern ceremonies to initiate our children into the solar system. Swimme asks us to imagine awakening our children a few minutes before dawn and walking with them out into the gray light:

> As they're yawning away the last of their sleep and as the Earth slowly rotates back into the great cone of light from the Sun, they will hear the story of the Sun's gift. How five billion years ago the hydrogen atoms, created at the birth of the universe, came together to form our great Sun that now pours out this same primordial energy and has done so from the beginning of time. How some of this sunlight is gathered up by Earth to swim in the oceans and to sing in the forests. And how some of this had been drawn into the human venture, so that human beings themselves are able to stand there, are able to yawn, are able to think only because coursing through their blood lines are molecules energized by the Sun.[23]

Each year, children and adults alike have the opportunity to consciously celebrate the earth's journey around the Sun in bringing us the seasons. As the fall equinox approached last year, I asked the students in one of my classes if they would like to participate in an outdoor ceremony. They agreed, and so on the evening of the equinox we gathered in a field under a starry sky. I explained that the equinox is a time of balance and harmony. On this one day in fall, day and night, light and dark, are of equal length.

After sitting quietly for a time, I suggested that we all rise and face north. I then encouraged those present to think about the essence of "north" and, if they felt so inclined, to speak of the images, values, and feelings evoked by north. Several individuals spoke. In a similar fashion, we placed our attention to south, east, and west. Then, we paused and turned our gaze and invocations to the heavens above and the earth below. Finally, returning to our sitting positions, we turned our gaze inward on ourselves—inhabitants of the middle realm between Sky and Earth—and we spoke of our place in the annual dance of Earth and Sun.

We concluded our ceremony by using our imaginations to go back to the time of the big bang and the formation of the galaxies and early generations of stars; we recognized that it was these primordial events that ultimately gave rise to all the elements that now compose our solar system and, by extension, our very bodies. We visualized the wondrous journeys of these elements— from molten rock to organic compound to bacterial colonies to algal mat to horseshoe crab to turtle to dragonfly to dinosaur to sequoia to hummingbird to deer mouse to whale, and now, for a short time, to us. Understanding our origins—from the big bang through the stars to Earth to us—can bring both meaning and mystery to our existence. Celebrations like this don't require any significant planning. All we have to do is create a space for them to happen and trust that we will know what to do. We surely do.

FOUNDATION 1.3: THE ORIGINS AND HISTORY OF LIFE ON EARTH

By remarkable good fortune, Earth offers conditions propitious for life as a result of its distance from the Sun and its size. For example, Earth is not so close to the Sun that its atmosphere is lost and its surface scalded, as is the case with Mercury; nor is Earth so far from the Sun that its life-giving water is perpetually frozen, as is the case with Jupiter. Likewise, Earth is not so large that its gravity holds a dense blanket of atmospheric gases that filter out light, nor is it so small that it fails to hold onto a life-sustaining atmosphere.[24]

Although Earth is well situated, its environment at birth—a fiery magma ball—was hostile to life. With time, however, a crust formed over Earth's molten core, much like the thin surface layer does on pudding as it cools. During the cooling process, gases were released, primarily carbon dioxide, nitrogen, and water vapor. Earth's early atmosphere then acted as a blanket, creating a severe "greenhouse" effect. This was a steamy time for Earth; but as it cooled, water vapor condensed into clouds, then fell to the ground as rain.[25]

The Origin of Life on Earth

Water served as a kind of planetary womb, providing the necessary medium to nurture life; and it was in Earth's primeval seas that life probably first appeared, but nobody knows how. Was it simply by chance? Was there some, as yet undiscovered, self-organizing principle involved? We simply don't know.

One theory of life's origins is known as "panspermia," and it holds that life didn't arise on Earth at all, but somewhere else—out in space. According to this view, the primitive seeds of life are adrift throughout our galaxy; and, by chance, they occasionally colonize a planet like ours, where conditions are favorable for life, by arriving via meteorites or comets.

Another theory proposes that the first complex organic molecules on Earth were produced with energy contributed by lightning and/or by the release of heat from deep-sea vents. To test the theory, scientists have placed water (to represent the early oceans) in closed glass containers and added the gases that were presumably present in Earth's primeval atmosphere. Then, they applied heat and electric sparks (to simulate lightning) and in so doing were able to generate amino acids (building blocks of proteins), fatty acids, and urea (a molecule common to many life processes); but, of course, a collection of organic molecules still isn't life.[26]

Somehow, the molecules for life must have become encased in a cell. But how? It is likely that among the many compounds in Earth's early chemical soup, there were greasy compounds (i.e., lipids) that formed bubbles (cells) when mixed with water. Microbiologist Lynn Margulis sketches out how life's first cells might have come into existence:

> cell-like membrane enclosures form as naturally as bubbles when oil is shaken with water. In the earliest days of the still lifeless Earth, such bubble enclosures separated inside from outside. Prelife, with a suitable source of energy inside a greasy membrane, grew chemically complex. . . . These lipid bags grew and developed self maintenance. . . . After a great deal of metabolic evolution, which I believe occurred inside the self-maintaining greasy membrane, some, those containing phosphate and nucleotides with phosphate attached to them, acquired the ability to replicate more or less accurately.[27]

Of course, this is pure speculation, but it forms the basis of a hypothesis that scientists are investigating. All we know for sure is that, through a process that will perhaps forever be shrouded in mystery, "that first organism, from which we are all descended was sparked into being, full of a life force that has so far persisted tenaciously for close to four billion years."[28] Earth's first organisms were bacteria-like; they probably scrounged energy from hydrogen sulfide spewing forth from deep-sea vents or from complex molecules dispersed in the sea water.

Earth's first "sun eaters" appeared about 3.6 billion years ago; they were photosynthetic bacteria that formed dense mats, the fossilized remains of which exist today in structures called "stromatolites." With the spread of photosynthetic life, the earth's atmosphere was gradually enriched with oxygen, a by-product of photosynthesis. Earth's original atmosphere had almost no oxygen. This atmosphere "was as different from ours as that of an alien planet. But purple and green photosynthetic microbes, frantic for hydrogen, discovered the ultimate resource, water, and its use led to the ultimate toxic waste, oxygen. Our precious oxygen was originally a gaseous poison dumped into the atmosphere."[29] Eventually, microbes discovered ways of detoxifying and using what originally was a dangerous pollutant. Today, oxygen is essential to the aerobic metabolism of all the plants and animals that occupy Earth.

The first four billion years of Earth history were a long prelude to the magnificent bursting forth of life that has occurred over the last six hundred million years. It has just been in this recent period that Earth's many life forms have appeared—crustaceans, mollusks, insects, fish, mammals, birds, fungi, mosses, ferns, flowering plants. The first mammals broke onto the scene two hundred million years ago. The primate line from which we trace our own ancestry began roughly two million years ago with *Homo habili.*

At present, the diversity of life on Earth is extraordinary. The total number of species is estimated at between ten and fifteen million, although only one and a half million species have been formally classified. Many of Earth's species are too small to be seen. For example, using just the naked eye and searching very carefully, one could locate perhaps a hundred different species—worms, snakes, frogs, caterpillars, grasses, trees, and so forth—in a forest, grassland, or wetland ecosystem in North America. Several thousand additional species—mites, tiny crustaceans, fleas, microscopic worms, algae, bacteria, and fungi—could be located with the assistance of a more sophisticated microscope. Indeed, an invisible world of algae, bacteria, and fungi is located in the soil, on the bark of trees, and on the surfaces of leaves. Beattie and Ehrlich compare this process of searching for species in ecosystems to the search for stars in the night sky:

The situation is not unlike astronomy; most of the stars in the sky cannot be seen with the naked eye, but with the help of sophisticated technology, millions of stars appear in the cosmos. In biology, it is microscopes that reveal the otherwise invisible objects. In this case they are species and the 'cosmos' is the Earth's myriad environments and habitats.[30]

Life on Other Planets?

Given the abundance of life on Earth and the immensity of the known universe, what is the possibility of life on other planets? This question is fun to think about. Consider the possibility just for the Milky Way, one of an estimated hundred billion galaxies that make up the known universe. Within this galaxy, our Sun is one of approximately one hundred billion stars. That's one hundred billion other suns, each with the potential to have solar systems of their own; but, to be conservative, let's imagine that only one half of those suns actually have planets. So, this leaves us with roughly fifty billion suns, each with planets in the Milky Way galaxy.

However, planets with the characteristics that make them candidates for life—just the right distance from their sun and just the right size for the formation of a healthy atmosphere—might be relatively rare. Thus, let's be conservative again and imagine that only one in every five hundred of those fifty billion suns in the Milky Way has a planet with basic characteristics needed for life. This still leaves us with roughly a hundred million planets with the potential for life support in just our galaxy.

With the possibility of extraterrestrial life seemingly all around us, perhaps a more appropriate question is: Why haven't we had any visitors? Or if we have had visitors, why aren't they more common? The reason, in large part, is the extreme size of our galaxy. The average distance between stars, considering all stars, is five hundred light-years, which is ten billion times the distance between the earth and the moon! The average distance between that subset of stars with potentially life-supporting planets would be even greater. In addition to these vast distances in space, we also have "distance" in time. Possible civilizations would be scattered in time. To grasp this concept, consider our solar system. It has been around for almost five billion years, but intelligent life (i.e., us), capable of reaching out into space, has been in existence for only a very, very tiny fraction of our solar system's total lifetime. Thus, at any one time, we may only have a few thousand stars simultaneously supporting a planet with life; and the average distance between these stars, given the size of our galaxy, would be likely to be on the order of two thousand light-years (i.e., forty billion times greater than the distance between the earth and the moon). And to top it all off, if life did develop on another planet, we have no guarantee that it would be anything like us or that it would even be interested in us.

Reflection

Cosmologists take our existing scientific knowledge and attempt to imbue it with meaning. We can all be amateur cosmologists. For example, when I consider that our universe, including planet Earth, is structured by a force that we call "gravity" but that it could have been structured in many other ways, I am led to conclude that the universe is grounded in attraction, or allurement. Likewise, when I learn that our Sun, in a sense, "gives" five pounds of itself each second to the earth—so that we can all live—I am inclined to bow in gratitude, acknowledging that a forerunner of human generosity is stellar generosity.[31]

On the other hand, when I learn that cosmic explosions (e.g., supernovae) have played a part in the creation of our solar system, I realize that I live in a dynamic universe. Stars and life are snuffed out in fits of destruction and ushered forth in creative outbursts. This conception helps me to understand my own life. I, too, have times of stasis and times of creativity, and often my creativity arises in moments of turmoil.

For me, the most remarkable thing about the universe's story is that it is not over. We live in the midst of an unfolding universe; we are participants in an epic story. To understand the "story line," we have to understand our past, because what happened in the past preconditions, to a significant extent, what can happen today. For example, at one specific time in our story line, all the hydrogen that makes up much of the universe came into being. This was a one-time event; the conditions were just right. It is unlikely that there will ever be another period when the universe can create hydrogen again.

It is the same with the galaxies. They could only come into existence once hydrogen was present. It was then that gravitational attractions came into play, allowing the galaxies to begin to form and creating the conditions that to this day allow for the birth of stars and planets. It appears that the time for galaxy making in the universe is over; that time has passed. And the same is true for Earth's oceans. Early in the earth's history, a set of specific conditions existed that allowed the oceans to come forth; those conditions no longer exist. And so it is for life: The conditions for creating life—the chemical-rich soup from which life emerged—no longer exist. Life, in effect, ate the womb that brought it forth. The universe's unfolding has now led to the emergence of human consciousness. In sum, the story of the universe reveals an inexorable process toward differentiation and complexification.[32]

Questions for Reflection

- How does this new understanding of the universe and the earth—not as static entities, but as dynamic "happenings"—change your understanding of yourself? Your past? Your future?

Practice: Discovering Our Home in Time

Most people experience time as a linear progression of moments—what the Greeks called "chronos." Our time horizon is usually in terms of days, weeks, or years; seldom do we think in terms of millions or billions of years. An appreciation of "deep time" is required to conceive of our origins and place in the universe (table 1.1).

George Fisher, a geologist at Johns Hopkins, helps his students visualize the thirteen-billion-year period from the big bang to the present by linking time with distance. Specifically, he asks them to imagine that one millimeter equals one year. Thus, one meter would equal one thousand years, and one kilometer, a million years. Using this scale, the entire thirteen-billion-year history of the universe would span a distance from the center of Tokyo (the beginning of the universe) to the center of Washington, D.C. (the present moment). Imagine yourself standing at the base of the Washington monument in downtown D.C., preparing to walk back to the beginning of time. Your first step (about a half-meter) would take you back five hundred years, to the time when the Spanish discovered America. Four more steps and you would be back to the time of Socrates. Trekking through northern Virginia you would be among the dinosaurs. To arrive at the time when the earth and solar system were formed, you would have to walk all the way to the California coast. And if you were truly earnest in traveling to the beginning of time, you would still have to make your way across the immense expanse of the Pacific separating you from Tokyo.

I don't imagine that any of us is ready to set out on this trek, but smaller deep-time journeys are possible. For example, I recently invited fifteen students to join me in a meadow on a star-strewn night. At the far end of the meadow, 130 feet away, was a small pine tree. We could barely make it out. I told the students that each foot between us and the pine tree represented one hundred million years of universe history. Thus, ten feet represented one billion years, and the entire 130-foot distance represented the entire thirteen billion years of the universe's history.

We lit a candle and set it on the ground at "Time Zero," the moment of the great flaring forth. Then, we walked forward three feet to a time three

Table 1.1.   The History of the Universe and Life on Earth from Thirteen Billion Years Ago to the Present[33]

Years before Present*	Cosmic/Evolutionary Event
13 billion years	Big Bang—the primordial flaring forth of space, time, and matter.
12.9 billion years	Hydrogen and helium, the first atomic "life," coming into existence.
12 billion years	Galaxies and stars forming from the hydrogen and helium issuing from the big bang.
4.6 billion years	Stardust and gases from the dissolution of earlier stars offering "raw material" for the formation of our solar system.
3.9 billion years	Earth coming to life; bacteria appearing in Earth's primitive seas.
3.6 billion years	Bacteria "inventing" photosynthesis whereby they capture light energy from the Sun.
2 billion years	Two distinct forms of life merging to form the first nucleated cells.
1 billion years	Single-celled organisms combining genetic material (sex) thereby opening up immense opportunities for creativity.
600 million years	The first "multicellular" organisms appearing.
500 million years	The first organisms with backbones moving through the oceans.
400 million years	Life forms leaving water; worms, mollusks, and crustaceans breathing air for the first time; algae and fungi venturing ashore.
300 million years	The first forests appearing; reptiles and amphibians flourishing on land.
200 million years	Dinosaurs diversifying in size, form, and habits; mammals appearing and birds budding off from dinosaurs.
100 million years	Flowers evolving and insects diversifying to feed on nectar and pollen.
65 million years	Dinosaurs disappearing; mammals flourishing: whales, rodents, antelopes, cats, horses, elephants, camels, bears, baboons appearing.
4 million years	Human ancestors leaving the forest and standing up to walk on two legs.
2 million years	Early humans using their hands and heads to shape tools.
100,000 years	Modern humans emerging; language, clothing, shelter, fire, musical instruments, art, and religion appearing.
10,000 years	Humans learning to cultivate plants and domesticate animals.
75 years	Humans discovering that they live in an expanding universe.
35 years	Humans traveling to outer space and looking back to see their Earth home for the first time.
Today	Humans beginning to tell the story of the universe; the "primordial flaring forth" continuing in this moment as us.

*All dates are approximate.

hundred million years following the big bang, when hydrogen and helium, the first atomic "life," came into existence. Here, too, we placed a candle. Next, we moved ahead seven feet to the time when the universe was one billion years old. This was the time of immense creativity, when the galaxies, including our Milky Way, were coming into being. Then, we walked slowly for seventy feet (through seven billion years of universe time) to a period of high adventure—roughly 4.6 billion years ago—when our solar system came into being.

We proceeded likewise through twenty-one stops. At each stop a student described the events of that time. For example, for the step that represented 3.9 billion years ago, when life first arose on Earth, the student said:

> Three billion nine hundred million years ago, this vibrant and fertile womb (Earth) brought forth the first living cells. These primal beings had the power to organize themselves, as did the stars and galaxies, and they had stunning new gifts as well. They could remember significant information, including the patterns necessary to knit together new living cells.[34]

We placed the final candle by the pine tree in the present moment, and students offered personal statements of presence, such as "I, Sarah, stand here with the full knowledge that my story extends back thirteen billion years to the primal flaring forth of space and time." Finally, we walked meditatively back through time to the first candle—that mysterious singularity—at the far end of the field.

A physical "walk through time" under the night sky is a wonderful deep-time practice. It provides an opportunity for the awe and mystery of the universe to seep into our bones and insinuate itself into our psyche (see box).

A second dimension of time is worthy of our attention—what the Greeks called "kairos." Kairos has nothing to do with clocks and calendars, the past and future, but it has everything to do with the abysslike nature of "now."[35]

You can cultivate kairos—that is, living in the "now"—through simple daily observances. For example, go outside first thing in the morning and simply experience the day: the temperature and movement of the air; the condition of the sky; the movements and sounds of the creatures that are co-inhabitants of your home place. And as your day unfolds, pay attention to the changing position of the Sun and the changing qualities of light as the earth turns on its axis. Note the first signs of a change in the weather. When night comes, again, step outside and pause to pay attention to the air, its humidity and movement. Take a moment to behold the heavens. Don't "think" the heavens, just experience them. In this way, you can resynchronize your "clock" to the rhythms of the earth.

As we cultivate the habit of experiencing time's different textures and dimensions, we become less anxious about our lives. Entering the "now," we set aside our fears and worries; and entering into the fullness of deep time, we come to see that our lives are part of a grand cosmic unfolding.

UNIVERSE STORY BEADS

To create a concrete reminder of deep time, the "walk through time" practice can be combined with the crafting of a "universe story necklace"—an idea my friend Paula Hendrick introduced me to. All that is necessary to make one is a string twenty-six inches long (two inches for each of the thirteen billion years in the universe's history) and an assortment of beads, charms, and trinkets to symbolize major deep-time events. The first item on the necklace symbolizes the singularity that gave rise to the universe. The second bead symbolizes the birth of the galaxies. Next come the beads or charms to symbolize the formation of the solar system, and the Sun and the earth born of this event. These are followed by symbols representing the first bacteria; the first photosynthetic organisms; the evolution of sex; and the appearance of mollusks, insects, dinosaurs, birds, mammals, flowering plants, and humans. One may also secure a bead to symbolize the date that one was born into the universe. Remember, it is the temporal sequencing of the beads that is important, not the precise physical spacing. The final bead on Paula's necklace, a small concave mirror, is about two inches from the initial big-bang bead. This mirror bead symbolized the Hubble Space Telescope, which now gives humankind the ability to look back to the beginning of time and, in so doing, to craft our story of origins.

CONCLUSION

So now in our modern scientific age, in a manner never known before, we have created our own sacred story, the epic of evolution, telling us, from empirical observation and critical analysis, how the universe came to be, the sequence of its transformations down through some billions of years, how our solar system came into being, then how the Earth took shape and brought us into existence. . . . This is our sacred story.

—Thomas Berry[36]

Humans, since our earliest days on Earth, have puzzled over and tried to make sense of the cosmos. This is part of what makes us human. In our time, we now continue this process. The story that has emerged from science explaining the origins of the universe and life is as fantastic and awe-inspiring as any creation story ever conjured up by humans in the past.

This science-based creation story, however, need not be seen as disproving the creation stories coming from the world's different cultures and religious traditions. All these stories represent a yearning for meaning and understanding; they are different ways of fathoming the same mysteries, and at their core they share remarkable commonalties. For example, the Christian creation story ("First there was light . . .") offers an intuitive vision of how the universe came to be while science attempts to discern the underlying mechanisms that ushered forth the universe. Similarly, Genesis (e.g., "God formed us from the dust of the ground and breathed into us the breath of life . . .") offers a poetic way of describing the origins of life, which science is now articulating in detail.

While science can tell us some of what happened and can offer us powerful explanations for how it happened, first-order mysteries may always lie outside its realm. As Albert Einstein said, "Science without religion is lame. Religion without science is blind."

Chet Raymo provides a concrete example of how the merging of heart and intellect can carry us forward:

> Two things are required to truly see: love and knowledge. Without love, we don't look. Without knowledge, we don't know what it is we are seeing. With love, the attentive observer will recognize that the middle star in Orion's sword is smudged, slightly different from its neighbors. With knowledge, the smudge becomes a window on the birth of stars.[37]

As a result of the advances in modern cosmology, the universe itself has become a profound source of revelation.[38] Our consciousness expands as we develop the capacity to see our lives in the context of a wondrous, unfolding universe. But this expansion doesn't come by simply reading a book chapter; it requires that we engage in practices. We actually have to fill that thermos and pack that blanket and leave the lights of the city behind; we have to befriend the night sky; we have to celebrate the solstices and equinoxes; we have to take that "walk through time" and don those "universe beads." In short, we have to actively cultivate awareness. Only in this way will we experience a genuine homecoming.

NOTES

1. Thomas Berry, *The Great Work* (New York: Bell Tower, 1999), 14–15.
2. Gregg Levoy, *Callings: Finding and Following an Authentic Life* (New York: Three Rivers Press, 1997), 139.
3. Paul Krafel, *Seeing Nature* (White River Junction, Vt.: Chelsea Green , 1999).
4. Krafel, *Seeing Nature,* 16.

5. Brian Swimme, *The Hidden Heart of the Cosmos* (Maryknoll, N.Y.: Orbis Books, 1996).

6. Michael Seeds, *Horizons: Exploring the Universe* (Pacific Grove, Calif.: Brooks/Cole, 2002); Chet Raymo, *An Intimate Look at the Night Sky* (New York: Walker, 2001).

7. Chet Raymo, *Natural Prayers* (St. Paul, Minn.: Hungry Mind Press, 1999), 21.

8. Raymo, *An Intimate Look.*

9. Phillip Ball, *A Biography of Water: Life's Matrix* (Berkeley: University of California Press, 2001); Seeds, *Horizons.*

10. Brad Lemley, "The Last Great Mystery," *Discover* 23, no. 4 (2002): 32–39.

11. Lemley, "The Last Great Mystery."

12. Raymo, *An Intimate Look,* 208.

13. Lemley, "The Last Great Mystery."

14. Chet Raymo, *Skeptics and True Believers* (New York: Walker, 1998), 243.

15. This figure was inspired by a sketch in Seeds, *Horizons,* 231.

16. Swimme, *The Hidden Heart of the Cosmos,* 52–53.

17. Raymo, *An Intimate Look,* 6–8.

18. Raymo, *An Intimate Look.*

19. Raymo, *An Intimate Look.*

20. Raymo, *Natural Prayers.*

21. Swimme, *The Hidden Heart of the Cosmos.*

22. Swimme, *The Hidden Heart of the Cosmos,* 30.

23. Swimme, *The Hidden Heart of the Cosmos,* 43.

24. Lynn Margulis and Dorion Sagan, *Microcosmos* (Berkeley: University of California Press, 1997).

25. David Suzuki, *The Sacred Balance* (Amherst, N.Y.: Prometheus Books, 1998).

26. Suzuki, *The Sacred Balance;* Seeds, *Horizons.*

27. Lynn Margulis, *Symbiotic Planet* (New York: Basic Books, 1998), 71–72.

28. Suzuki, *The Sacred Balance,* 114.

29. Margulis and Sagan, *Microcosmos,* 99.

30. Andrew Beattie and Paul Ehrlich, *Wild Solutions* (New Haven, Conn.: Yale University Press, 2001), 32–33.

31. Brian Swimme, *The Universe Is a Green Dragon* (Santa Fe, N.Mex.: Bear, 1985).

32. Brian Swimme, *Canticle of the Cosmos Study Guide* (New York: Tides Foundation, 1990). Contact information: Tides Foundation, 40 Exchange Place, Suite 1111, New York, NY, 10005.

33. Sidney Liebes, Elisabet Sahtouris, and Brian Swimme, *A Walk through Time* (New York: John Wiley and Sons, 1998); Brian Swimme and Thomas Berry, *The Universe Story* (San Francisco: Harper San Francisco, 1992).

34. Adapted from Swimme and Berry, *The Universe Story.*

35. Thomas Stella, *The God Instinct* (Notre Dame, Ind.: Sorin Books, 2001).

36. Berry, *The Great Work*, 31.

37. Raymo, *Natural Prayers*, 36.

38. Diarmuid O'Murchu, *Our World in Transition* (New York: Crossword, 1992).

②

CULTIVATING AWARENESS:
A LIVING PLANET

Imagine [the earth] is your biological mother—because, in a very real sense, she is. Imagine the Sun is your biological father—because, in equally real, life giving ways, he is. Imagine that after the spirit of God touched them, your distant but brilliant father and 70-million-square-mile mother not only fell in love, but began making love: imagine Ocean and Sun in coitus for eternity—because they are. Imagine your ocean mother's wombs are countless, that her fecundity is infinitely varied, and that her endless slow lovemaking with Sun brings about countless gestations and births and an infinity of beings: great blue whales and great white sharks; endless living castles of coral; vast phalanxes of fishes; incalculable flocks of birds; gigantic typhoons; weather patterns the size of continents—because it does.

—David James Duncan[1]

The language in David James Duncan's passage—"Earth is your biological mother. . . . Earth and Sun in coitus . . . slow lovemaking"—may sound peculiar because we have grown up in a culture that tends to view the earth as an object, something to be acted upon. But as Duncan reminds us, the earth is not an object; it is a living system.

This chapter provides the foundational material necessary to experience Earth as a living system. We will bring our attention and awareness, first, to Earth's respiration—the planetary "breath" (foundation 2.1); next to Earth's energy transfers—the planetary "metabolism" (foundation 2.2); and, finally, to Earth's cycles—the planetary "circulatory system" (foundation 2.3). The practices accompanying each foundation are invitations to consciously participate in Earth's breath, Earth's metabolism, and Earth's cycles.

FOUNDATION 2.1: EARTH'S BREATH

The earth's plants, animals, and microbes are constantly breathing, taking in and releasing gases. The collective result of this breathing creates a "planetary breath." The record of the earth's atmospheric carbon dioxide concentrations since 1960 (figure 2.1) is, arguably, the most important graph produced in the history of environmental science. This graph reveals a steady upward trend in atmospheric carbon dioxide concentrations. The steady rise is primarily the result of growing fossil-fuel consumption, although deforestation, worldwide, has been a factor in the increase as well.

A fascinating and sometimes overlooked part of this graph is the annual oscillations in atmospheric carbon dioxide concentrations: seven parts per million up, and then seven parts per million down, each year. These oscillations chart, in a sense, "Earth's breathing." To understand this concept more fully, imagine that you are able to create a self-contained forest ecosystem— make it an oak forest—with its own private atmosphere. Now, imagine this forest in winter. The leafless oaks continue to respire, breaking down stored sugars in their stems and roots, and in the process releasing carbon dioxide. All the animal, bacterial, and fungal life in this forest also continues to respire. Hence, carbon dioxide concentrations in the atmosphere of this self-contained forest ecosystem would rise during the winter—this is the season of the planetary out-breath. Now, fast-forward to May: The oaks are flushing out with new leaves. Since carbon dioxide is a foodstuff for photosynthesis, these new leaves begin removing carbon dioxide from the atmosphere. The forest vegetation as well as the associated animals and microbes also continue to release carbon dioxide via respiration, but the carbon dioxide uptake by the photosynthesizing oaks begins to exceed the release of carbon dioxide to the atmosphere through respiration. Hence, carbon dioxide concentrations in the atmosphere go down from late spring to early fall—this is the "season" of the planetary in-breath.[3]

Of course, if the amount and distribution of land were equal in the Northern and Southern Hemispheres and if the atmosphere were perfectly mixed, we would have no annual oscillation in atmospheric carbon dioxide concentrations because the out-breath in the North's winter would be balanced by the in-breath of the South's summer, and vice versa. However, most of the earth's

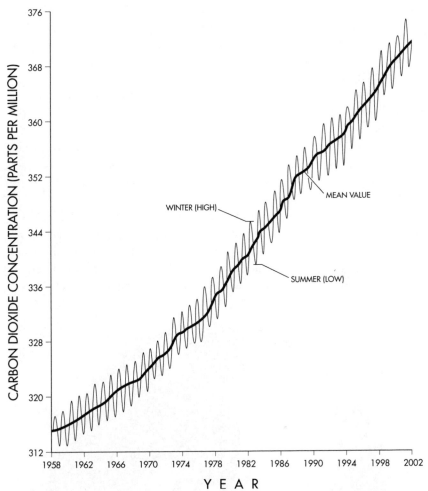

Figure 2.1. **Increases in atmospheric carbon dioxide concentrations on Earth since 1960[2]**

breathing occurs on land, and the Northern Hemisphere has much more land than the Southern Hemisphere. So it is that we see a rise in atmospheric carbon dioxide of seven parts per million during the Northern Hemisphere's winter and a dip of seven parts per million during the North's summer.

Breathing Surfaces

If you wished to discover the location of this planetary breathing, the place to look would be on life's surfaces, where the exchange of life's gases

and materials occurs. Much of the world, as biogeochemist Tyler Volk reminds us, is covered by and made of surfaces (figure 2.2).

It is natural to suppose that most of life's gaseous exchanges would occur
on the leaf surfaces of land plants since this surface type is most visible and
abundant, from a human perspective. Indeed, if all the leaves of Earth's
land plants were stitched together, they would make a complete covering
for the Earth (i.e., the surface area of plant leaves is about the same as the
Earth's total surface area).

Just as leaves capture sunlight, roots are organs designed to capture the
water and nutrients that plants need to grow. It turns out that the surface
area of plant roots is thirty-five times greater than the surface of the earth
(see box).[5]

Because humans are land-bound, it is not surprising that we think of life's
surfaces as concentrated on land; but what about the oceans, which cover
roughly 70 percent of the earth's surface? The surface area of the small
planktonic organisms that dominate the world's oceans is approximately six
times greater than the surface area of leaves on land (figure 2.2). This fact is

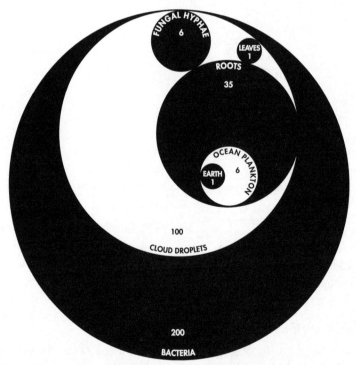

Figure 2.2.   The surface area of Earth relative to other surfaces[4]

COULD THIS REALLY BE TRUE?

It seems hard to believe that the surface area of the roots of a tree would be greater than the tree's leaf area. After all, plant leaves are so conspicuous and roots are so fine and inconspicuous. As I was pulling out my hair wondering how this could be, it occurred to me that the hair on my own head is like a fine network of roots.

Although I think of myself as mostly skin, could it be that most of my exterior surface area is hair (roots) and not skin? I decided to explore this hypothesis by doing a few back-of-the-envelope calculations. Here are the "givens" that I worked with:

- Average density of hair on human scalp: 1,500 hairs/square inch
- Surface area of the human scalp: 120 square inches
- Average diameter of human hair: 0.0030 inches
- Average hair length: 9 inches

Given these parameters, I calculated that there are 180,000 individual hairs on the human scalp (1,500 hairs/square inches × 120 square inches of scalp). Next, using the formula

$$A = \pi \times \text{diameter} \times \text{length},$$

I determined that the surface area of a single hair is 0.085 square inches (3.14 × 0.0030 × 9). Knowing this, I estimated that the total surface area for all hairs on my model human head was 15,300 square inches (0.085 × 180,000 hairs). Then, dividing by 144, I determined that the total surface area of my model's hair was 106 square feet. This area is about ten times greater than the surface area of the human body. After completing this exercise, the idea that the surface area of a plant's root system might greatly exceed that of its leaves no longer seemed like an impossibility.

paradoxical because the actual mass of oceanic plankton is tiny compared to the mass of leaves on land. But as we saw in the hair exercise (see box), appearances can be deceiving. Oceanic organisms usually don't live as multicellular leaves but as unitary, free-living cells. Imagine dispersing the millions of cells in a single oak leaf, one by one, into the ocean; to do so would magnify the total surface area of the oak "leaf" dramatically, which is precisely the strategy of the ocean plankton—to be small and numerous. Volk explains this tactic: "One can show mathematically that for a given mass divided into smaller and smaller spherical packets, the increases in surface area are proportional to the decreases in diameter."[6]

But the surface areas of all the foregoing categories pale in comparison to the surfaces of the most abundant of all organisms, the bacteria. Lest you imagine that bacteria are insignificant creatures, think again: Bacterial surfaces exceed the earth's surface area by two hundred (figure 2.2). Again, bacteria achieve their enormous surface area by being small (only about one-tenth the diameter of algal cells) and by being very numerous.

Given the importance of surfaces for "life's breath," it should come as no surprise that the air, water, and soil in which life is embedded are also rich in surfaces. Any way you cut air, it is all surface; ditto for water. Soil, too, is a world rich in surfaces. Volk illustrates this point for soil with the following calculation:

> Let's assume that the entire land area is covered with a typical loam of forty percent sand, forty percent silt, and twenty percent clay. Computing over a depth of just 10 centimeters (four inches) and for a moment neglecting the clay, I derive a surface area index equivalent to 2,000 Earth areas. Adding the bacteria-sized clay particles boosts this number to 11,000 Earth areas.[7]

Life probes and plants itself into the matrixes formed by soil and air and water. "These matrixes bathe life, harbor life, surround life, nourish life."[8] Seeing the locus of "life's breath" as surfaces helps us remember that the essence of life is exchange.

Earth's Physiology: The Gaia Hypothesis

This discussion of Earth's breathing surfaces suggests that the earth, as a whole, has its own "physiology"; and, indeed, many scientists are now engaged in studying Earth's capacity to self-regulate. British scientist and inventor James Lovelock was the first to propose that Earth's species, acting in concert, might be able to regulate Earth's atmosphere and temperature, thereby creating conditions that ensure the continuation of life. Lovelock first advanced this idea, now known as the Gaia hypothesis, in the early 1970s. He was working for NASA at the time, trying to figure out a way to check for the presence of life on Mars by studying its atmosphere. But before looking to Mars, Lovelock examined the atmosphere of planet Earth. What he found—an atmosphere shockingly out of chemical equilibrium—intrigued him greatly. Using the rules of chemistry and physics, one wouldn't expect any methane, ammonia, or sulfur gases in Earth's atmosphere because these gases are highly reactive, especially in an oxygen-rich environment like that of Earth. However, these and other reactive gases are present in Earth's atmosphere at concentrations often billions of times higher than expected, based on the expectations of equilibrium chemistry. Lovelock concluded that Earth's unusual atmosphere was the result of Earth's biota (Earth's breathing surfaces), which in any given moment are producing prodigious amounts of reactive gases. "The atmosphere is not merely a biological product," wrote Lovelock, "but more probably a biological construction: not living, but like a cat's fur, a bird's feathers, or the paper of a wasp's nest, an extension of a living system."[9]

Lovelock reasoned that if other planets, like Mars, harbored living organisms, they would have the signature of life's gases in their atmosphere. Because each gas type has a unique spectral fingerprint, it was possible to check for life on Mars without going there by mounting a spectroscope on a telescope and aiming it at Mars. The result: The Martian atmosphere was wholly predictable based on energy, chemistry, and physics alone, suggesting that the planet was without life. Lovelock's prediction that there would be no life on Mars was confirmed when the Viking spacecraft visited the Red Planet in 1976.[10]

Later, Lovelock and colleagues began using computer models to further explore the notion that the Earth's biota can actually regulate Earth's temperature, maintaining it within a narrow range favorable for life. He knew from the fossil record that Earth's temperature has remained relatively stable during Earth's history. However, this temperature stability was unexpected since the laws of physics make it virtually inescapable that the Sun's total luminosity—that is, its output of energy as light—has increased by an estimated 33 percent over the past four billion years. Lovelock posited (in accord with the Gaia hypothesis) that it is the Earth's biota that has regulated the Earth's temperature in the face of this rising solar luminosity.[11]

Although it has been challenging to offer definitive proof for the Gaia hypothesis, Lovelock has developed some elegant mathematical models that explain how Earth's biota could act collectively, but without conscious intent, to optimize the environment for their own use. His best-known model, developed with Andy Watson, is called "Daisyworld." The model starts with a planet like Earth illuminated by a star that slowly grows hotter over millions of years. There are two species of plants that can grow on this hypothetical planet: white and black daisies. The daisies act as a giant thermostat, maintaining the temperature of the planet within a narrow range favorable for life. Biologist Lynn Margulis in her book *Symbiotic Planet* describes Daisyworld:

> The assumptions are straightforward. Black daisies tend to absorb heat and white daisies to reflect it. Neither flower grows below 10 degrees C and they all die above 45 degrees C. Within this range, black daisies tend to absorb local heat

and therefore grow faster in colder conditions. White daisies, since they reflect and lose more heat in warmer conditions, thrive to produce more offspring. Let us begin with the black daisy world. As the sun increases in luminosity, the black daisies grow, expanding their surface area, absorbing heat, and heating up their surroundings. As the black daisies heat up more of the surrounding land surface, the surface itself warms, permitting even more population growth. This positive feedback continues until daisy growth has so heated the surroundings that white daisies begin to crowd out the black ones. Being less absorbent and more reflective, the white daisies begin to cool down the planet. The cumulative result of these actions is to heat the planetary surface when it is cooler during the early evolution of the sun, and then to keep the planet relatively cool as solar luminosity increases. Despite the ever-hotter sun, the planet maintains a long plateau of stable temperature.[12]

Stephan Harding, a professor at Schumacher College in England, has made a more complex version of Daisyworld containing more than twenty different-colored "species" of daisies, along with herbivores feeding on the various daisies and carnivores that eat the herbivores. His model reveals that it is at least theoretically possible for the earth's biota to regulate the earth's temperature while still behaving in accord with the principles of evolution and natural selection.

In the real world, "it is the growth, metabolism, and gas-exchanging properties of microbes, rather than daisies, that form the complex physical and chemical feedback systems which modulate the biosphere in which we live."[13]

Reflection

I am struck by how difficult it is to see life's myriad surfaces. The bacterial, fungal, and algal exchange surfaces, where so much of Earth's "breathing" is transacted, are invisible to me. Meanwhile, the things that I do see are in a sense illusions. For example, when I look in the mirror, what I see is a reflection that is the result of light reflected off the atoms that make up my skin. Although I look solid, I know that I am mostly emptiness—visible emptiness. Cosmologist Brian Swimme explains it this way:

> You are more fecund emptiness than you are created particles. We can see this by examining one of your atoms. If you take a single atom and make it as large as Yankee stadium, it would consist almost entirely of empty space. The center of the atom, the nucleus, would be smaller than a baseball sitting out in center field. The outer parts of the atom would be tiny gnats [electrons] buzzing about at an altitude higher than any pop fly Babe Ruth ever hit. And between the baseball and the gnats? Nothingness. All empty. You are more emptiness than anything else.[14]

The reason that the chair we sit in or the cup we drink from seems to be solid, even though it is almost entirely empty space, is because electric forces operate to hold the specklike nuclei of atoms together. The more I learn about the world, the more I appreciate that things are not as they appear to be.

Questions for Reflection

- Consider that you are not so much breathing as you are being breathed by your body: How does this change your notion of selfhood?
- Consider that you are mostly empty space: How does this change your notion of material existence?

Practice: Cultivating Awareness of Earth's Breath

The words *breath* and *air* come from the Latin *spiritus* with connections to *soul, spirit, liveliness, emotional vigor,* and *essence.* Our breath is our lifeline. The following practice is a simple breath-centered exercise that cultivates an awareness of the primordial elements—earth, water, fire, and air—that we owe our life to.

As you read over this breathing practice, you may find it weird and reject it. In fact, I remember making a joke about this practice the first time someone described it to me. We all tend to resist things that we are unfamiliar with. Later, though, I gave this practice a try and was surprised to discover great power in it—which has led me to refrain from rejecting practices until I at least allow myself to experience them once.

Breathing with the Primordial Elements

Earth element. Start by bringing your attention to the earth element, characterized by solidity and mass. Note these same "earth" qualities in the seemingly compact mass of matter that is your body. Next, bring you attention to the earth beneath you. You are not floating around free from Earth's surface; rather you are partially attached to Earth. The mysterious force of gravity embraces you, tightly connecting you to Earth like a magnet. Next, register that the elements making up your body are "earth" elements. You trace your origins to the earth's primordial seas, where the first life—your ancient ancestors—sprang forth. Earth is truly your birth mother.

Now, breathe in through your nose; and as you do so, imagine yourself pulling the power of the earth up through the base of your spine to the level of your heart; on the out-breath (through the nose), release any bodily impurities to the abiding earth. Slowly repeat this breath—in and out—five times.

Water element. Next, bring you attention to the water element, with its qualities of flow and cohesion. Experience the water element by tasting the liquid in your mouth. Your body owes its softness to water; pat your belly and poke your muscles—notice the watery nature of your flesh. Now, breathe in through your nose; and as you do so, imagine air flowing in through the crown of your head; on the out-breath (out the mouth), imagine cleansing water rushing down through your mind and body. Slowly repeat this breath five times.

Fire element. Next, bring your attention to the fire element, characterized by heat and ripening. The great fire of the Sun warms the earth and the air you breathe. Feel the heat inside you and the heat radiating from the surface of your skin—generated by the "burning" of food produced from Sun's "fire." All that lives "burns."

Now, take in air through your mouth; and as you do so, feel the breath entering through your solar plexus, fanning the flame at the base of your heart and burning away fear. Then, as you breathe out (through your nose) experience light emanating from your heart in all directions. Slowly repeat this breath five times.

Air element. Finally, bring your attention to the air element, characterized by motion. As you breathe, air is meeting the interior surfaces of your lungs, which, if spread out, would cover half a tennis court. Experience the air element inside you as your chest expands on your in-breath.

Conclude, now, by taking in and exhaling air through your mouth. As you do so, experience it as a wind that first approaches from the front (in-breath) and then leaves from behind (out-breath). This wind rushes through the mostly empty spaces of your body, leaving you cleansed and refreshed. Slowly repeat this breath five times.[15]

Breathing with Plants

The following is a complementary breathing practice for forging a direct link between your body and the breathing body of Earth. Start by taking hold of a tree branch. Behold the leaves, then bring your attention to your breath. As you breathe out, carbon (as carbon dioxide) is passing from your body into the leaves on the branch. Fired by solar rays, this carbon—your carbon—is being forged into sugars in each leaf's interior. As you breathe in, the by-product of the tree's solar forging, oxygen, is finding its way into your lungs and thence, via blood, to the billions of cells of your body.

Breathe in and then out with full consciousness of the exchange taking place. Now, slowly let go of the branch and, as you do so, hold your breath. Sense the discomfort as your body yearns for connection. Grasp the tree branch again, then breathe in and then out. Establish and break this life connection as you "conspire" with the tree's leaves. Next, slowly release your hold on the tree branch and cup your hands around a single leaf. Continue to breathe, giving your breath to and receiving breath from this leaf. Finally, open your arms to embrace the whole tree, then breathe in and out, feeling your breath going out to the tree and experiencing the tree's "breath" entering your body. As you engage in this practice, you give yourself the opportunity to develop a bodily sense of your participation in Earth's breathing. It is a vital exchange that connects you to the greater cycles of the planet.[16]

FOUNDATION 2.2: THE EARTH'S METABOLISM

Animals, ourselves included, rely on the food-manufacturing skills of plants for sustenance. Ever since grade school, we have heard it: *Plants use solar energy to create complex sugars from the simple building blocks, carbon dioxide and water.* Perhaps because we have heard it so often, the wonder has ebbed away; but consider anew the magnificence of photosynthesis: Two formless things, water and air, forged into a solid: a molecule of sugar. Look around at the vegetation close at hand—the wildflower meadows, cattail marshes, or pine forests that are part of your home place. Remind yourself that all this plant life was sparked into being through the alchemy of photosynthesis. Photosynthesis is what shapes the formless into form.[17] The star performer in photosynthesis is the 120-atom chlorophyll molecule, with its platelike head and long tail.

Chlorophyll is designed to snag photons—very tiny and very fast-moving light "particles." Normally, when photons strike a surface, they are converted into heat; but chlorophyll grabs photons in just the right way so that their energy can be used to initiate photosynthesis. Fortunately, the world is awash in chlorophyll. Volk offers this perspective:

> Imagine the human population (six billion) increased by a few ten-million fold. This multitude, reduced to chlorophyll molecules, would comfortably find niches latched onto internal cellular membranes within a single square centimeter of leaf area. That's the same scale as the density of photons striking the same leaf surface. It enables each chlorophyll molecule to absorb a couple of photons per second from a shower of bright sunlight. Each absorbing plate

(i.e., chlorophyll molecule) thus flickers into excited states at about the pace of a heartbeat during brisk aerobic exercise.[18]

Volk has suggested that chlorophyll

would be a perfect icon of a science-based, earth-centered religion. In the form of a molecular model, its head and tail might easily replace, for example, the Catholic chalice [see figure 2.3]. A nature priestess could hold the iconic molecule by its tail and lift its illuminated, green head, truly the bringer of light to life, glittering high in the air before a reverent congregation.[19]

In spite of chlorophyll's effectiveness at capturing light, less than 1 percent of the solar radiation striking the earth's surface is used in photosynthesis. The reason, in part, is that, first, plants can only utilize certain wavelengths of solar energy; and, second, they can only thrive in areas where temperature, water, and soil conditions permit photosynthesis. But the rest of the Sun's energy is by no means wasted. A significant fraction is used to heat the earth's atmosphere as well as land and water surfaces. This differential heating of Earth's surfaces creates winds that, in turn, power Earth's weather systems. Another major chunk of the Sun's energy is used in the evaporation of moisture from water, soil, and leaf surfaces—that is, the energy activates the water cycle. Hence, the Sun powers the winds, the water cycle, and all photosynthesis on Earth (see box on page 46).

Divvying Up the Bounty of Photosynthesis: The 10 Percent Rule

It bears repeating that the ability of green plants to "feed" on solar energy, air, and soil creates food for almost all the other creatures on Earth—millions of species. These myriad species are able to coexist, in part, because they feed on plants in specialized ways. For example, imagine that you are in the middle of a corn field. It is the end of the growing season; and, as you look around, you see mature corn plants—food—everywhere. Examining individual corn plants, you note that certain insect species have chewed out holes in the leaves while others have bored their way into the corn stalks. Yanking corn plants from the soil and inspecting their roots with the aid of a microscope, you discover lots of protozoans, nematodes, and insect larvae feeding on these roots. Insect herbivores might consume 10 percent of the corn plant's production. Meanwhile, humans harvest about 40 percent of the field's yield as corn kernels. So, all told, perhaps 50 percent of a living corn plant is piped to other organisms.

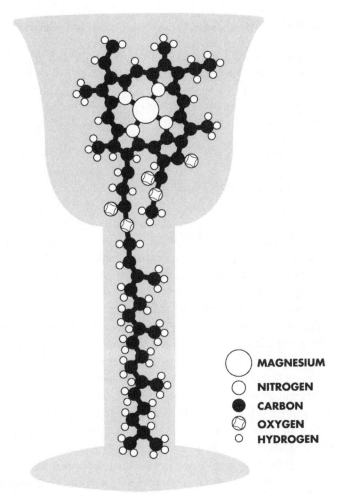

MAGNESIUM
NITROGEN
CARBON
OXYGEN
HYDROGEN

Figure 2.3.   The chlorophyll molecule[20]

Now, what happens when animals, like insects or humans, eat corn leaves or corn kernels? Does the material they eat show up as new animal growth? Not to the degree that you might expect. In fact, on average only about 10 percent of the plant material consumed by animals results in new growth, which makes sense if you think about it from a human perspective. When humans act as herbivores and eat leaves (lettuce), or flowers (broccoli), or grains (corn kernels), about a quarter of this plant food passes through the gut undigested, and most of the rest is "burned" in respiration to maintain normal body functioning. At most, only 10 percent goes to growth (i.e., weight gain or tissue replacement). The same is true for the blackbird eating the corn

EXPRESSING SOLAR INCOME IN HUMAN TERMS

To appreciate the earth's solar income, we may find it useful to use a familiar energy currency, the Calorie (note: a capital C calorie is also referred to as a *kilocalorie*). An adult male requires about twenty-eight hundred Calories of food energy each day to keep his body functioning. Expressed on an annual basis, this comes to about one million Calories (2,800 Cal/day × 365 days = 1,022,000 Calories/year). Remarkably, this is just about the same as the average amount of solar energy arriving to each square yard of the earth's surface each year.

kernel and the hawk feeding on the blackbird. Although the percentages may differ somewhat depending on the quality of the food and the type of organism (e.g., cold-blooded organisms can convert a higher percentage of the food they consume to new growth), 10 percent is a reasonable estimate for the overall efficiency of energy transfer in feeding relationships among organisms. In fact, ecologists refer to this biophysical constraint as the 10 percent rule. Applying the 10 percent rule, we can expect that a herbivore consuming a hundred units of plant biomass would transform them to ten units of new body mass; and a carnivore consuming these ten units of "herbivore" would transform them into one unit of new carnivore body mass.

The 10 percent rule reveals that it is more ecologically efficient for humans to eat plant foods, such as corn or soybeans, than it is for us to feed these foods to animals and then eat the animals. The reason is that it takes roughly ten pounds of grain to produce a pound of beef.

In the United States, with a per capita meat consumption of approximately 250 pounds per year, two-thirds of domestic grain consumption goes to feed poultry, swine, and livestock. The situation is further exacerbated when we consider the large amounts of fertilizers, pesticides, herbicides, and irrigation water devoted to grain-producing lands, not to mention the increased risk of heart disease and stroke associated with meat-rich diets (i.e., diets high in saturated fats).[21]

Returning to the corn field, if only 50 percent of the corn plant's production goes to herbivores, including humans, what about the other 50 percent? After the harvest, the uneaten roots, stems, and leaves of the corn plants die and settle to the ground (barring removal for silage). The same is true in a forest: Each year the trees in the forest grow bigger and eventually they grow old and, one by one, fall to the forest floor. Tons of tree leaves also drop to the ground each year. If all this dead material simply accumulated on the forest floor, there would be several feet of leaf and log debris on the ground after a few hundred years; and the trees would eventually die of "starvation" because the nutrient elements they require to grow would all be imprisoned in the accumulating debris on the forest floor. But, of course, this is not the way things work. All this dead plant material, in the case of the corn field or the forest, provides "lunch" for trillions of organisms that are largely hidden from sight.

It is not far-fetched to think of the soil as a densely settled city teeming with bacteria, fungi, ants, beetles, worms, slugs, mites, and nematodes. As writer Evan Eisenberg points out:

> These "citizens" are often in motion, hurrying along the vast expressways made by moles, the boulevards of earthworms, the alleys between particles of sand or clay, the dank canals that these alleys often become. Certain districts and certain intersections—mainly close to the roots of plants—get especially busy. The "citizens" move in the dark, sniffing at chemical trails. They are constantly doing business with one another. They traffic in molecules: minerals, organic compounds, packets of energy.[22]

In everyday terms this means that the dead plant debris in forests and fields is heavily colonized by all sorts of debris-eating organisms. The feeding relationships among these organisms allow the nitrogen and phosphorus and other plant nutrients in plant debris to return to the soil where they become available, once again, for uptake by plant roots.

Ants, worms, millipedes, beetles, and their brethren play an important role early on in this decomposition process by physically chewing apart the plant debris, which greatly increases the surface area available for the action of bacteria and fungi. Bacteria species—which specialize in easily decomposed compounds, such as sugars and amino acids—are the first to colonize: their poplulations explode in direct proportion to the food source, then they die off as their food supply is exhausted. The leftovers—lipids, alcohols, chitin, cellulose—become lunch for slower-growing decomposers such as *Arthrobacter*, the most common group of soil bacteria. Finally, only tough plant material, like lignin, is left; and these recalcitrant compounds become the substrate for certain specialized types of fungi.[23]

In the big picture, photosynthesis is the "lifting up" process of life, whereby energy is used to build new tissue; respiration, however, is life's "flowing down" process, whereby tissue is decomposed, releasing energy.[24] Everything in the natural world can be thought of as part of these "lifting up" and "flowing down" processes. The garden plants in full flower in summer are "lifting up," while the rotting leaves of late fall are "flowing down." The cow pie in the pasture "flows down," releasing to the soil nutrients that, in turn, contribute to a "lifting up" of the grasses that are subsequently nourished by these nutrients.

These paired concepts of "lifting up" (plant production) and "flowing down" (decomposition) allow us to see the land—the fields and forests where we live and walk—in a fuller, more dynamic way. For example, imagine that while walking along a stream-bottom trail, you come upon a dead doe. You crouch down to have a close look. Judging from her size, you conjecture that this deer died in her second year. During her life, her body was fueled by acorns, berries, fresh shoots, and, perhaps, corn from nearby fields—all products of photosynthesis. Now, you observe that this doe is returning to the earth, and in the process myriad organisms are finding sustenance in her body.

Educator Paul Krafel describes the remarkable parade of organisms that visit a deer carcass. First, the vultures and blowflies arrive. The adult blowflies come to mate and lay eggs. The tiny larvae hatching from these eggs burrow into the deer's flesh to feed, and a portion of the dead deer is thus transformed into a writhing mass of fly maggots. Then, predators such as the rove beetle visit the deer and feast on the maggots. Then there are the robber flies that hang out around the perimeter of the carcass, ready to catch flying insects attracted to the carcass. Looking closely, one might also discover parasitic wasps injecting their eggs into the insects feeding on the carcass. In the final stages of this "feast," new types of insects come to feed on the deer's greasy bones and fur. All told, the energy within the deer is likely to nourish more than a hundred different species. Most of the deer "flows down," rendered into gas, water, and mineral elements. Meanwhile, some 10 percent of the deer is momentarily "lifted up," contributing to the growth of new tissue in vultures, maggots, and beetles.[25]

Hence, in both life and death, plants and animals provide food (nutrients) for one another in myriad ways. One of the reasons that so many species are able to coexist in a given ecosystem is that most species are specialized in what they eat. For example, we humans have morphological and physiological adaptations that allow us to eat the corn kernel but not the corn leaf; the grasshopper can eat the corn leaf but lacks the adaptations to eat the corn root; the meadowlark can eat the grasshopper but not the corn stem; and so forth. These specializations create a multitude of feeding relationships. Meanwhile, the 10 percent rule operates, placing a biophysical upper limit on the earth's capacity to support herbivores and carnivores.

Reflection

Simply by eating food (other organisms) each day, I am involved in the earth's "lifting up" dynamic, and when I excrete wastes, I participate in the "flowing down" process. The big "flow down" will occur when I die. When that happens, my body will probably be slowly devoured by microbes, rather than predators . . . but one never knows. There was a time not long ago when I came face to face with a panther. I was walking through a forest in northern Brazil at midday, lost in thought. Suddenly, I saw the panther walking across the trail just thirty yards in front of me. I stopped; he stopped. We gazed at each other for perhaps a minute. There was nothing menacing in the panther's posture. He seemed to be curious yet cautious. As quickly as he had appeared, he quietly slipped away. This experience reminded me that I am a potential prey item for some species. In fact, I had a second visceral experience as a potential food item in life's web, when I went snorkeling in a tropical sea. Swimming in shark habitat and surrounded by large barracuda, I felt vulnerable and very alert.

My most recent experience of vulnerability in a wild setting occurred while hiking in the Appalachians. I was waking along a rocky ridge when suddenly the rock shelf on which I was standing crackled to life. The sound of one rattlesnake and then a second snapped me to attention. I abandoned the rock, leaping forward, but then a third and fourth rattler sounded to my right. Both were coiled and only a few feet from the trail. I had always wanted to see a rattlesnake; now I was in their midst and found myself gingerly tiptoeing my way to safety.

When I return to wild places, it seems that some measure of my wildness returns. But no matter how much I venture out, I doubt that I will ever be picked off by a predator. It is more likely that I will continue to participate in life's "flowing down" processes in dribs and drabs—via the wastes leaving my body, which provide nourishment for microbes. It will only be upon my death that I will offer up a big feast. Then, just like the deer, my body will be a food bonanza for many organisms.

Questions for Reflection

- In what sense are you engaged in the biological processes of "lifting up" and "flowing down" each day?
- What might you do to bring fuller consciousness to your participation in life's "lifting up" and "flowing down" processes?

Practice: Cultivating Awareness of Earth's Metabolism

Author and photographer Paul Rezendes tells a poignant story that reminds us that eating is a biological act:

> One summer morning I stepped outside and heard an unworldly, totally unfamiliar crying sound, not very loud, but strange enough to stop me in my tracks. I looked in the direction of the sound but saw nothing except an open area of grass and pine needles. I didn't dare move. Again, I heard a distressed crying, a sound neither human nor animal. I saw nothing until I noticed the grass move, which I attributed to the wind. I waited until I heard the cry again, then slowly turned my body in its direction. I was amazed to see a toad being swallowed by a garter snake!
>
> The snake's jaw had dislocated to accommodate the size of the toad; it had already swallowed the toad's left hind foot and one third of its body. The toad alternated between helplessly struggling to extricate itself from the snake's jaws and bleating its mournful cry while lying still. It seemed to be pleading for me to do something. I felt overwhelmed with pity for this toad, crying for its life.[26]

Perhaps take a moment to consider how you would have responded to Rezendes' dilemma. Might you find a "middle way" between what appear to be two polarities—either rescuing the toad or feeling "overwhelmed with pity."

> To rescue it was my first inclination, and I came very close to intervening in the drama of the snake and the toad. But there was a larger picture here. Before my very eyes the toad was becoming the snake. The toad had hunted, killed, and eaten many insects, which had become the living toad. The snake, by eating the toad, became the toad and the insects, and the snake, in turn, would be fair game for a hawk or owl.
>
> I was witness to an ancient and sacred ritual, something that could not and must not be stopped. Recognizing and accepting the interconnection of these lives allowed me to have as much compassion for the snake as I did for the toad.[27]

Eating is no less of an ecological act for humans than it is for other species. When we eat, we are like the insect feeding on the corn plant or the bird eating the insect.

We can engage in certain practices to become more conscious of our eating as an ecological act. A first step is to make conscious choices about what to eat. Too often, it seems, we simply reach for whatever is close at hand when we get hungry. Our eating can become more mindful if we figuratively place a skillful question between our mouth and our appetites. For example, ask yourself, "Do I want this [baked potato, apple, order of french fries, Coke, etc.] for me?" Doing so allows you to consider if you really want to put a certain food item into your body. The "for me" on the end of this question invites consciousness.

After choosing what to eat, we can practice bringing mindfulness to how we eat. Physician Jon Kabat-Zinn uses the following exercise to help his patients (people who have been living under chronic stress) bring mindfulness to eating.

We give everybody three raisins and we eat them one at a time, paying attention to what we are actually doing and experiencing from moment to moment. . . .

First we bring our attention to seeing [one] raisin, observing it carefully as if we had never seen one before. We feel its texture between our fingers and notice its colors and surfaces. We are also aware of any thoughts we might be having about raisins or food in general. We note any thoughts and feelings of liking or disliking raisins if they come up while we are looking at it. We then smell it for a while and finally, with awareness, we bring it to our lips, being aware of the arm moving the hand to position it correctly and of salivating as the mind and body anticipate eating. The process continues as we take it into our mouth and chew it slowly, experiencing the actual taste of one raisin. And when we feel ready to swallow, we watch the impulse to swallow as it comes up, so that even that is experienced consciously. We even imagine, or "sense," that our bodies are now one raisin heavier.[28]

This is what it means to eat mindfully. We bring our attention to the raisin, and we really see it; we pay attention as we chew it; we are present as the many different tastes of the raisin are revealed to us.

Making physical contact with our food—actually touching our rice, peas, salad, or fish—can help bring us to mindfulness as well (see box).

An important aspect of mindful eating is eating slowly. When people develop the habit of truly savoring each bite of food, from start to finish—instead of starting the next bite before the current bite is finished—they discover that they need to eat less to satisfy their bodies' cravings for flavors.[29]

An important final step in mindful eating is to use honest language to describe our food. Educator Steve VanMatre illustrates this point:

By chance, I was watching the cook post the dinner menu one day when I suddenly realized [her posting] wasn't what we were going to eat at all. It was a disguise. So after [she left], I went up and erased it. On the first line she had written, "Roast Pork." I changed that to "Roast Pig," then I decided that still

GETTING YOUR FINGERS INVOLVED

I had no choice but to eat with my fingers when I spent a month in southern India with Nitin, one of my graduate students. Our meals usually consisted of rice or other grains and a variety of curries and chutneys. My "plate" for each meal was a fresh banana leaf. At our first meal, I watched Nitin casually mix fish, lentil curry, and yogurt with rice; and then deftly guide tidy packets of food to his mouth. The grace and artistry in his movements reminded me that eating with fingers is what humans have always done. I hesitated, hoping that a fork or perhaps a pair of chopsticks would be made available for me, but I was wrong. Finally, Nitin nodded for me to eat. My feelings of self consciousness quickly dissolved as I experienced the enhanced sensual pleasure of carefully eating food with my fingers.

wasn't clear enough and changed it again to "Roast Hoofed Mammal." The next line became "Boiled Orange Roots" (carrots), followed by "Mashed White Tubers" (potatoes). And the last line read simply: "Tossed Leaves" (salad). That is what we were really going to eat that evening—roots and tubers and mammals and leaves. [Let's call our] food by that which it is instead of by that which it is not. By peeling away such disguises we can begin to see our connections again with the systems of life around us.[30]

Using honest language to describe the food we are about to eat helps ground us in the web of life. In sum, cultivating an awareness of the earth's metabolism means paying attention to our own metabolism—what we eat and how we eat.

FOUNDATION 2.3: THE EARTH'S CYCLES

If you were to put an ink dot on a map of the earth to designate the origins of the trillions of atoms that make up your body, the map would be covered in ink. Our atoms have journeyed to us from literally everywhere on the planet. We are part of their cycles.

Planetary Cycles

The hydrologic cycle is particularly important to life because most organisms, ourselves included, are more than half water. In this cycle, salty ocean waters are transformed to pure water through the Sun-driven "lifting up" process of evaporation.

On an annual basis, approximately 950 billion cubic yards of rain fall to Earth. If evenly distributed, this deluge would cover Earth to a depth of approximately 2.5 feet.[31] The water falling to land seeps into the soil where it may be quickly "lifted up" by tree roots; there, either it contributes to the manufacture of sugars via photosynthesis, or it passes up through tree leaves and evaporates back to the atmosphere via transpiration. Alternatively, this rainwater might find its way into nearby rivers (and thence to the ocean), or it might slowly percolate down through the latticework of soil particles and thus contribute to water reservoirs (aquifers) located far below the soil surface.

All of life's important elements have cycles, not just water. Sometimes, as in the case with nitrogen, the cycles have a gaseous phase. In other cases (e.g., calcium, magnesium, and phosphorus), the cycles are limited to solid and liquid phases. Such cycles are referred to as "nutrient cycles" when the elements involved are the basic building blocks of life.

One way to begin to grasp the remarkable journeys of life's "nutrients" is to accompany the movements of a single element over time. For example, imagine for a moment that you *are* a calcium atom that has just been weathered loose from bedrock in a New England forest. Initially, you find yourself dissolved in a drop of water lodged between two grains of sand, but soon you are sucked up by a tree root, incorporated into a leaf; and then, as winter sets in, you are shed to the ground. In the early spring, and still in your guise as a calcium atom, you are absorbed by a bacterial cell, which is then eaten by an earthworm, which then deposits you as part of one of its castings. Once again, you find yourself being lifted up by a tree root. The cycle repeats many times—soil to tree and back to soil—but a time comes when you find yourself encased in an acorn, which is carried off by a squirrel. On a frigid winter afternoon, the squirrel consumes you. This same squirrel later dies by the edge of a stream and is washed into the water by a spring flood. So now you float down the stream, entombed in the dead squirrel's flesh. Soon microbes set to work on the squirrel, and you are set free—but not for long. An algal cell absorbs you into its tiny body; and then a mayfly larva eats you; and then this wriggling larva is snatched up by a trout. When the trout dies, you are again set free by microbes that dismantle the trout's body.

Decade by decade, you join one organism after another. Meanwhile, the inexorable force of gravity is taking you further downstream until one day you reach the ocean. There you continue to be passed around from organism, to water, to organism until one day you find yourself lodged in the bone of a sperm whale. When this whale dies, its bones settle to the ocean's depths and with time become buried in ocean sediment. At this moment, it might seem that your journey as a calcium atom is over. It is true that you may remain locked in ocean sediments for millions or even billions of years. However, an epoch eventually arrives when the clashing of Earth's tectonic plates thrusts the sedimentary deposits, of which you are a part, up into the sky in a fit of mountain building. Then, weathering processes act on the exposed rock, inviting you, that entombed calcium atom, once again back into the biotic cycle. If it weren't for these cycling processes, Earth's diverse ecosystems would have never come into existence.

Ecosystem Cycles

If we had the eyes to see it, we would find that the quiet and stately forests of Earth are actually buzzing with movement. On a summer day, water vapor and oxygen are wafting out of tree leaves while carbon dioxide is streaming into these same leaves; the billions of chlorophyll molecules in each leaf are flickering with excitation as they snag solar photons. Meanwhile, down in the soil, potassium atoms are being set free while fungi "digest" dead leaves and while tiny bacteria are busy snatching nitrogen from the air and incorporating it into plant roots. During each second in each square foot of forest, trillions of microscopic movements and transformations are taking place, all of which are part of the forest's nutrient cycles.

Scientists have spent a lot of time studying nutrient cycles because these cycles are essential for a healthy planet. The scientists conduct their studies by charting the movements of life-giving nutrients, such as nitrogen, phosphorus, and calcium. This scenario sounds somewhat abstract, so let's look at a real-life example. In 1963, two ecologists, Herbert Bormann and Gene Likens, initiated an ambitious "nutrient cycling" study in the White Mountains of New Hampshire, at a place called Hubbard Brook.

Bormann and Likens knew that mountain watersheds (the land drained by a particular stream) are, in effect, self-defined ecosystems. They conceptualized a watershed ecosystem as an assembly of boxes or compartments—for example, trees, soil, and bedrock—each containing a stock of nutrients. The nutrients in any given ecosystem "box" might move back and forth into other ecosystem compartments as shown by the arrows in figure 2.4.

The starting point for their work was to determine the amount of nutrients in the various ecosystem "boxes." Imagine the challenge: You walk out into the watershed to determine the weight of the forest tree compartment.

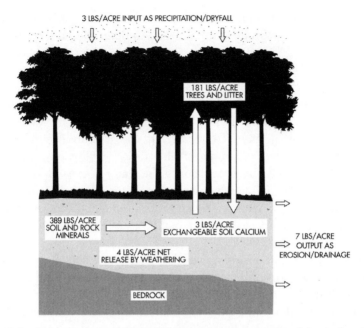

Figure 2.4. Schematic map, Hubbard Brook ecosystem, showing stocks of calcium in various forest compartments (boxes) and fluxes of calcium between compartments and into and out of the ecosystem[32]

But weighing the whole forest, even if it could be done, would leave you with a huge pile of logs, branches, and leaves; and, of course, you would end up destroying the very forest you had hoped to study. The solution was to cut down a limited number of trees of different sizes—from small to large, outside the study area—and then to carefully measure the dimensions (trunk diameter and height) and weight of each tree's components (leaves, bark, wood). These data were then used to develop equations that related tree dimensions to tree weight. Thus, rather than cutting down each tree in the watershed, the researchers merely needed to measure the dimensions of the living trees in the watershed and plug these numbers into their equations to get the tree weights. This approach yielded a good estimate of the weight of the entire forest compartment. Next, they collected samples of all the tissue types in the forest—leaves, bark, trunks, roots—and they determined the amount of nitrogen, phosphorus, magnesium, and so forth in a known weight of each tissue type. Then, by scaling up, they were able to estimate the total amount of each nutrient element in the entire vegetation compartment. Similar approaches were taken to estimate the nutrient stocks in other ecosystem compartments.

If you think this kind of work is tedious—believe me, it is. And there is more. The Hubbard Brook team also had to determine the quantities of nutrients entering (inputs) and leaving (outputs) their watershed. Small amounts of nutrients enter ecosystems in rain and snow, and as dust particulates that are carried from place to place. The Hubbard Brook team used precipitation collectors to measure these atmospheric inputs, and they employed stream monitoring stations at the base of the watershed to measure the total volume of water leaving the system. The bedrock underlying the watershed was impermeable to water, thus ensuring that all water (and associated nutrients) leaving the watershed would have to exit via stream flow. By conducting chemical analyses of both the precipitation coming into the forest and the stream water leaving the ecosystem, the researchers were able to calculate the total nutrient movements into and out of their study system.

Figure 2.4 provides a summary of the results for calcium, and it reveals that almost two-thirds of the calcium in the watershed was in the soil pool, with approximately one-third in plants and animals. Of course, nutrients such as calcium don't just sit in the soil or tree compartments. Tons of leaves and fine roots were being produced and were dying each year during the Hubbard Brook study. The "flowing down" of life's energy via decomposition of dead leaves and roots ultimately resulted in the release of nutrients to the soil.

In the end, though, the following conclusions were drawn based on per acre, per year averages. First, only seven pounds of calcium exited the watershed via stream flow, whereas three pounds of calcium entered the watershed via precipitation; hence, the net loss of calcium was only about four pounds. This was a tiny amount compared to the roughly 570 pounds of calcium per acre in the soil and vegetation. Thus, this forest watershed, in its natural state, was extremely efficient, retaining 99 percent of its calcium from year to year. The little bit of calcium that was lost each year was presumably made up through natural weathering of the bedrock underlying the soil.

In sum, natural ecosystems can serve as teachers; they are models. Materials don't rush into and out of natural ecosystems on linear trajectories. Instead, nutrients move from partner to partner—sometimes flowing down, sometimes being lifted up, seldom slipping out of the ecosystem. However, the research of Bormann and Likens (along with that of many other scientists) reveals that when forests are clearcut, they begin to hemorrhage, leaking nutrients. The reason is that normal nutrient-retention mechanisms—such as well-developed root systems and thick, organic soil coverings—are temporarily destroyed.

Reflection

As a beginning student of ecology, the earth made more sense to me after I learned about nutrient cycling; but I didn't have the opportunity to fully appreciate the efficiency of this cycling process until I confronted a paradox in the tropical forests of the Amazon Basin. I first began to conduct research in the rain forest in the mid-1970s; and like many before me, I was impressed by the lushness of the vegetation. It was easy to see why most people had assumed that Amazon rain forests were growing on highly fertile soils. After all, what else could explain their seeming exuberance. Imagine, then, the consternation of early researchers when they discovered that the soils underlying most of Amazonia's verdant rain forests were ancient, highly weathered, and extremely infertile.

For me this paradox began to resolve itself soon after I entered the Amazon forest along the Rio Negro in southern Venezuela. There were strange sounds and riotous vegetation; the air was thick with humidity. I expected all of these things, but what I didn't anticipate was that I would be walking on a sponge of roots. Most of the fine "feeder" roots of the forest trees were growing on top of the soil in a dense mat, which was sometimes a foot or more thick. These roots grew over and around the dead leaves and the woody debris that had fallen down from the canopy. I remember picking up a dead branch and discovering that it wasn't a dead branch at all; the interior was entirely colonized by living roots!

What was, at first, a paradox, now made sense. These upper Rio Negro forests were able to flourish on highly infertile soils because their roots grew right up out of the soil in search of sustenance, thereby ensuring that life-giving nutrients didn't leak away but continued cycling within the ecosystem.

Questions for Reflection

- In what ways is your life dependent on Earth's nutrient cycles?
- In what ways have you removed or isolated yourself from Earth's cycles?

Practice: Cultivating Awareness of Earth's Cycles

Strikingly, humans are the only species on Earth that actively seeks to keep its waste out of Earth's cycles. For example, in the United States, humans often gather leaves from their lawns and food scraps from their tables, and they entomb them in sanitary landfills, thereby depriving the land of the organic matter inputs necessary to maintain healthy soils.

This need not be. We can participate in Earth's cycles. One place to start is with the words we use to describe body wastes—the organic materials, water, and mineral elements that leave our body on a daily basis. Consider in our culture some of the more common words for these materials are *urine, piss, feces, poop, crap,* and *shit.* These are not very inviting words. They tend to distance us from substances we apparently don't want to have much to do with. Using such words also denies us the opportunity to see ourselves as part of Earth's cycles.

So a fundamental awakening practice might involve simply changing the words we use to refer to the earth's materials that leave our bodies. In the Orient, human organic deposits are sometimes referred to as "night soil" because these materials are spread on fields as manure (presumably at night). Using the term "night soil" allows people to see themselves as embedded in Earth's cycles.

I am not prepared to propose a new vocabulary for human "waste" in the West; but as I was mulling this over, I recalled that during high school my friends and I sometimes used the expression "shake the dew off my lily." We were likely to say this during an evening of serious beer drinking when we felt a need to urinate yet another time. It is a nice, light-hearted metaphor, and it gets at the general idea of creating vocabulary that conjures up positive images that connect us, rather than divorce us, from bodily processes and natural cycles.

A second way to participate more fully in Earth's cycles would be to overcome our fright of our own droppings. It is customary to shield ourselves from defecation by quickly flushing away our "waste." I was alerted to this "shielding" when I lived in southern Venezuela. I woke early and spent my days in the forest. Of course, there are no toilets in the forest. So, after lunch each day, I would find a nice spot and squat. I had barely finished my first "squat" when a large beetle came careening through the forest understory and landed on my steaming deposit. "Wow, this is really cool," I thought at the time. This visitor, I soon learned, was a dung beetle.

At first, I wondered what this beetle could possibly want with my organic leftovers. It was clear that he hadn't arrived by mistake; he was going to stay. I decided to sit and watch. He was about the size of a small pebble, with a hard, shiny body. Slowly he began shaping part of my deposit into a small ball, which he then proceeded to bury in the soil. Later, I learned that dung beetles lay their eggs in these dung balls. When larvae emerge from these eggs, they are surrounded by "food." All this beetle-burrowing activity helps to aerate and fertilize the soil. What I had always regarded as my "waste" appeared to be "good food" for this fellow and a benefit to the ecosystem.

With time, I have come to see squatting in the woods as a practice that allows me to give my organic leftovers back to the natural world; it is an act of generosity. When done mindfully, it helps me awaken to the fact that I am a member of Earth's ecosystems. Of course, in pursuing this practice, one should follow a certain etiquette and sense of propriety; and this I leave to your judgment and sensibilities.

Paying attention to the "resources" that leave our body ultimately leads us to consider household bathroom practices. It is more than a bit odd that in the United States we take two perfectly good resources—human manure and fresh water—and splat them together in the toilet bowl, making them both useless. Fortunately, there are now more enlightened technologies available, such as waterless urinals and composting toilets, which preserve the integrity of each "resource" while avoiding the harmful downstream effects created by our present customs. Not everyone is ready to invest in a composting toilet, but developing a simple composting system for kitchen wastes is easy and fun to do (check out www.cityfarmer.org).

A final way to participate more fully in nutrient cycling is to ensure that, upon death, our bodies are able to readily reenter Earth's cycle of life. In his book *Earth Education*, Steve VanMatre described the mutuality of life and death in a poignant story about a man who had been buried long ago in a small Massachusetts town. For some reason, the people in the town had to dig up the man's grave. When they did so, they discovered that the roots of a neighboring tree were "feeding" on the man's body and had taken his shape from head to toe.[33] Here was a man "reincarnated" as tree.

In America, it is still possible in some places to be laid down in a simple pine box and then to be slowly lifted up as tree. There is even a term for this, "green burials." As part of its "Green Burials Project," the Memorial Society of British Columbia in Canada points out:

> Green burials eliminate the negative ecological impact of cemeteries . . . and reduce the consumerism associated with current funeral practices. Graves are integrated into the existing landscape. Graves are hand dug, cement grave liners are not needed. Caskets are made of biodegradable, even recycled materials. The body is allowed to return to the earth naturally. To commemorate their dead, families may plant a memorial tree or shrub at the grave site, or install a memorial plaque on a nearby tree, a park bench, or along a trail.[34]

In Great Britain, several hundred woodland burial parks now provide a solution for people who wish to reenter nature's cycles free of frills. In the United States, a nature reserve serves as a burial park in North Carolina, and other woodland burial parks are in the planning stages.

Cremation is another option that can connect us, as well as those who mourn our passing, to Earth's elemental cycles. When my father died, our family took the elements remaining from his body to the edge of a small pond that was teeming with life. We lifted his ashes up to the sky and then let them rain down upon the water. I remember being surprised that they sparkled in the sunlight as they cascaded down. At the end, noting a blueberry bush close by, we spread some ashes around its base and, then, we each ate a blueberry, knowing that we all participate in the cycle of life and death each moment of our lives.

These three practices—using fresh vocabulary to describe our organic leftovers, bringing mindfulness to our "squatting" activities, and planning our burial—provide entry points for participating more fully in Earth's cycles.

CONCLUSION

As I look more deeply, I can see that in a former life I was a cloud. And I was a rock. This is not poetry; it is science. This is not a question of belief in reincarnation. This is the history of life on earth.

—Thich Nhat Hanh35

When I was growing up, I saw the earth as an object filled with stuff—roads, houses, stores, factories, yards, trees, animals—a planet inhabited by lots of separate things. Later, I came to see that I was not alone in this misunderstanding. It seems that most of us have inherited, to varying degrees, this fractured image of Earth. We received it from our culture, which transmits in myriad ways the message that the earth is an object composed of separate entities, with humans standing apart and above all things. Hence, it is easy to imagine ourselves as separate from life's cycles—separate from the water, air, and soil; it is easy to think that we are not part of Earth's breathing exchange surfaces, that we stand apart from the feeding relationships that interlace the earth's biota. Living this message, we miss out on the essence of what life is and what it means to be human.

This chapter is an invitation to awaken to the interconnected, not fragmented, nature of life on Earth. Planet Earth *does* breathe, and we breathe with it; Earth *does* have a metabolism, and we are part of that metabolism; Earth's elements *do* cycle, and we are part of those cycles.

In spite of appearances, life is not organized as discrete packets—the mouse, the petunia, the kelp. The tree outside your window is not restricted to the solid body that you see. Life's boundaries are not sharp, but soft. As we come to understand these things and develop practices to reinforce this understanding, we reach a point where one day the following words would no longer seem obtuse but instead would seem pregnant with meaning:

> Imagine [the earth] is your biological mother—because in a very real sense, she is. Imagine the Sun is your biological father—because in equally real, life giving ways, he is. Imagine that after the spirit of God touched them, your distant but brilliant father and 70-million-square-mile mother not only fell in love, but began making love: imagine Ocean and Sun in coitus for eternity— because they are. . . .[36] (David James Duncan, as quoted at the beginning of this chapter)

NOTES

1. David James Duncan, *My Story as Told by Water* (San Francisco: Sierra Club Books, 2001), 185.

2. These data are from the Scripps Institution of Oceanography, at the Mauna Loa Observatory in Hawaii.

3. Tyler Volk, *Gaia's Body* (New York: Copernicus/Springer-Verlag, 1998).

4. This figure was inspired by Volk, *Gaia's Body*, 120.

5. Volk, *Gaia's Body*, 120.

6. Volk, *Gaia's Body*, 121.

7. Volk, *Gaia's Body*, 122.

8. Volk, *Gaia's Body*, 124.

9. James Lovelock, *Gaia: A New Look at Life* (Oxford: Oxford University Press, 1979), 10.

10. Lynn Margulis, *Symbiotic Planet* (New York: Basic Books, 1998).

11. Lynn Margulis and Dorion Sagan, *Microcosmos* (Berkeley: University of California Press, 1997), 267.

12. Margulis, *Symbiotic Planet*, 126–127.

13. Margulis and Sagan, *Microcosmos*, 271.

14. Brian Swimme, *The Universe Is a Green Dragon* (Santa Fe, N.Mex.: Bear, 1984), 37.

15. This breathing practice has been attributed to Jazrat Inayat Khan and is described in the following: Casey Blood, *Science, Sense and Soul* (Los Angeles: Renaissance Books, 2001); Wes Nisker, *Buddha's Nature* (New York: Bantam Books, 1998).

16. I was first introduced to this practice through Project NatureConnect (www.ecopsych.com).

17. Paul Krafel, *Seeing Nature: Deliberate Encounters with the Visible World* (White River Junction, Vt.: Chelsea Green, 1999).

18. Volk, *Gaia's Body*, 128–129.

19. Volk, *Gaia's Body*, 128.

20. This figure was inspired by Volk, *Gaia's Body*, 134.

21. John Robbins, *Diet for a New America* (Walpole, N.H.: Stillpoint, 1987).

22. Evan Eisenberg, *The Ecology of Eden* (New York: Alfred A. Knopf, 1998), 23.

23. Eisenberg, *The Ecology of Eden*.

24. Krafel, *Seeing Nature*.

25. Krafel, *Seeing Nature*.

26. Paul Rezendes, *The Wild Within* (New York: Penguin Putnam, 1998), 131.

27. Rezendes, *The Wild Within*, 132.

28. Jon Kabat-Zinn, *Full Catastrophe Living* (New York: Dell, 1990), 27–28.

29. Charles Eisenstein, *The Yoga of Eating* (Washington, D.C.: New Trends, forthcoming).

30. Steve VanMatre, *Earth Education* (Greenville, W.Va.: Institute for Earth Education, 1990), 135. Contact information: Institute for Earth Education, Cedar Cove, Greenville, WV 24945

31. David Suzuki, *The Sacred Balance* (Amherst, N.Y.: Prometheus Books, 1998).

32. Frank H. Bormann and Gene. E. Likens, "Nutrient Cycling," *Science* 155 (1967): 424–429.

33. Van Matre, *Earth Education*.

34. http://greenburials.ca/homepage.html

35. Thich Nhat Hanh, *The Heart of Understanding* (Berkeley: Paralax Press, 1988), 21.

36. Duncan, *My Story as Told by Water*, 185.

⓷

SEEKING UNDERSTANDING:
EARTH'S WEB OF LIFE

Together, all species make up one immense web of interconnections that binds all beings to each other and to the physical components of the planet.

—David Suzuki[1]

Driving along a desolate stretch of road in the Everglades in 1970, I encountered a young couple hitchhiking. I picked them up and, seeing their bedraggled appearance, ventured, "Where did y'all spend the night?"

The guy, Dave, responded, "We crashed back in the 'glades. Y' know, trying to get back to nature."

"Woa!" I exclaimed, "Must of been a lot of mosquitoes? How much 'Off' did you use?"

Dave's friend, Sally, said softly, "We don't use repellents."

I probed a bit more suggesting, "Y'all must have had your blood sucked dry?"

There was a long silence, and then Dave said with a sigh, "Yeah, we got bitten up pretty good but, ya' know, it's all related." Then, Sally leaned forward, fixed me in her gaze and added, "When you really think about it, everything is connected—it's all connected."

More than thirty years later, I still remember this brief encounter as if it were yesterday. I knew those two were saying something important, but I couldn't quite grasp it. Now, I flag this event as one of my early encounters with "web" thinking.

Truly understanding that "everything is connected"—and not just grasping it intellectually, but feeling it in your bones—is a major milestone in the path to developing ecological consciousness.

The aim of this chapter is to explore Earth's web of life. We start in foundation 3.1 with Earth's species—the strings of life's web. As we will see, each species has a fascinating story to tell. In foundation 3.2, the focus is on the interactions among Earth's species. Sometimes these interactions are aggressive, as when one species hunts down another in order to survive; but, as we will discover, alliances of cooperation between species are surprisingly common. Finally, in foundation 3.3, we acknowledge that humans are also part of life's web. In fact, no other species on Earth is as dependent on life's biodiversity as *Homo sapiens*. The practices associated with these three foundations are invitations, first to observe, then to study, and finally to enter life's web.

FOUNDATION 3.1: EACH SPECIES IS A STRING IN THE WEB OF LIFE

Each species has its own way of being in the world—its own life story. Ecologists refer to this as the species' "natural history." It takes patience to figure out the natural history of a particular species of plant, animal, bacteria, fungus; but it is gratifying to do so nonetheless. The process begins when our attention, for whatever reason, is drawn to a particular organism, and we become curious about it. Our curiosity might then lead to our observing the organism's habits—what it eats, when it is active, where it sleeps, its mating behaviors, and so forth. Often these observations trigger questions; and in the process of answering these initial questions, new questions frequently arise.

Using Questions to Unravel the Monarch Butterfly's Life History

To illustrate how this discovery process works, let's examine the natural history of the monarch butterfly. We will begin by shrinking ourselves down to the size of a sand grain—small enough to fit on the head of a monarch, right between her two antennae. This idea may sound a bit ridiculous, but field biologists often exercise their imaginations in just this way, knowing that it takes creativity to begin to understand the lives of other organisms.

So, in our exercise, it is a summer day, and our monarch is flying about through a beautiful meadow. The monarch is laying eggs, with each egg no bigger than a pinpoint. Although dozens of plant species exist in the meadow, we note that she only lays her eggs on a particular plant type, the milkweeds. This observation prompts the following questions: How does the monarch distinguish the milkweeds from all the other plants in the field? Does she use sight? Smell? By removing certain sense receptors from the monarch and by observing changes in behavior, researchers have learned that monarchs have scent receptors on their antennae that allow them to zero in on milkweeds. Once a female monarch lands on a plant, she uses taste receptors on her feet to verify that it is indeed a milkweed.[2]

As we continue to fly around with our monarch, we make another observation: She lays her eggs one at a time, often laying just one egg per milkweed plant. This observation prompts a second question: Why does she spread her eggs around? It certainly would be easier to simply lay all her eggs in one place and be done with it; but, as it turns out, each caterpillar that hatches from a monarch egg gobbles up many leaves, and the supply of leaves on a milkweed plant is limited. Hence, spreading eggs around over many milkweed plants helps to ensure that individual monarch caterpillars have enough food to enter the pupal stage and then metamorphose into adult butterflies.

Riding with our monarch, we observe something else peculiar about her egg laying—eggs are generally placed on the underside of leaves, secured in place with a gummy secretion, rather than on the leaf's upper surface. Observing this tendency, we ask: Why do monarchs lay their eggs on the underside of leaves? A field scientist addressing this question might hypothesize that laying eggs on the underside of leaves improves egg survivorship, perhaps by keeping eggs from drying out. If I were doing this research, I would try to test this hypothesis by locating a hundred milkweed leaves with eggs on the undersurface. Next, I would tag half the leaves and leave them as is; then I would rotate the other fifty leaves 180 degrees so that the eggs on the undersurface now faced up. By comparing the fate—survival versus death—of the eggs on rotated versus nonrotated leaves, I would be gathering some of the necessary data to help test this hypothesis.

More observations from our perch on the monarch's head bring more questions. For example, if we gently cut the edge of a milkweed leaf, we see that it is filled with a sticky milky latex (thus, the name, milkweed). Biochemists have determined that this milky sap contains cardiac glycosides—toxins that cause heart paralysis and a severe vomiting response in vertebrates. You and I would sicken and vomit if we ate a milkweed leaf, yet monarch caterpillars eat milkweed with impunity. They have evolved the biochemical capacity to excrete the milkweed's toxins—something that very few other species are able to do. This observation prompts a fourth question: Given all the other plant species in the field, why do monarch caterpillars choose to feed exclusively on milkweeds, plants laced with cardiac glycosides? In other words, what might be the evolutionary advantage of this feeding specialization? It turns out that monarch caterpillars and monarch adults are not only able to excrete some of the toxic milkweed glycosides, but they can also store them in their tissues. In fact, the concentration of glycosides is higher in monarch tissue than in milkweeds. Because of these toxins, birds hunting for caterpillars learn to avoid monarchs. For example, researchers have found that blue jays rapidly learn to refuse to eat monarch caterpillars. So, the same glycosides that keep most herbivores from eating milkweeds keep many birds from eating monarchs. Monarch caterpillars are brightly colored, presumably so that birds who make the mistake of sampling them will remember their coloration and avoid them in the future.

So far we have seen how field observations, straightforward questions, and simple experiments can contribute to our understanding of monarch life history. But there is more to the monarch's story. Imagine, now, that the female monarch that we have been accompanying has finished laying her eggs. Her life is almost over. We sense this; and as she lays her last egg, we slide from our spot between her antennae down beside the egg. Soon a tiny caterpillar hatches. First, it nibbles away at its egg case; then it begins to eat the fine hairs on the surface of a milkweed leaf. After it grows a bit larger, it begins to eat the leaf tissue beneath its feet. After a couple of days of feeding, our caterpillar becomes confined by its cuticle and must molt.

The caterpillar continues to feed, eventually going through five molts over a three-week period, and then the feeding stops. We ride on our caterpillar's back as she descends to the ground and affixes to the branch of a honeysuckle bush. A remarkable thing then happens. The caterpillar forms a mummy-like crucible within which her body is deconstructed and then reconstructed into an adult monarch butterfly. This metamorphosis takes about one week. When the adult monarch emerges from her chrysalis, we resume our perch between her antennae and join her as she spirals high into the sky in a courtship flight. We are still with her when she settles down in the late afternoon with a male suitor. We watch as the male places a packet of sperm, complete with nutrient supplements, into a special receptacle inside her reproductive tract. The next day we fly along with our female as she begins to lay eggs, releasing sperm to fertilize each egg just before it is laid.

This entire monarch life cycle is repeated several times during the summer, and then the weather turns cold. This weather change prompts a new question: How do monarchs survive the winter? Adult monarchs would freeze to death if they stayed in the North. Many species of insects survive winter as eggs or as encased pupa, neither of which is the case with monarchs. So what do monarchs do? For many years it was a complete mystery. Some researchers, like Robert Michael Pyle, tried to literally follow the monarchs on the ground as winter approached. For fun, put yourself in Pyle's position, a young man filled with curiosity, ambition, and boundless energy:

> I began at the northwesternmost breeding grounds, in the Okanagan Valley of British Columbia. My plan was to locate monarchs at their nectar sources and overnight roosts, then follow them as far as I was able on foot. I'd catch and tag them when I could. And when they rose and flew away, as they always would, I'd take their vanishing bearings—the direction in which they disappeared—as my running orders. I'd follow in the same direction, as much as the topography would allow, until I found more habitat, more monarchs . . . and then do it all again, as far as I could.[3]

In addition to ground-tracking monarchs, Dr. and Mrs. F. A. Urquhart of the University of Toronto began tagging monarchs in the 1950s. They did so by affixing tiny, thumbnail-size tags (like the self-adhesive price tags used in grocery stores) to the wings of adult monarchs born in the fall. Each tag had a number and a mailing address for where the tagged monarch could be sent if it was found. So, for example, if one of the monarchs tagged by the Urquhart's in Michigan was hit by a car in Arkansas and if the driver later discovered this "tagged" monarch on his car's grill and sent it back, the Urquharts would know that at least one of the Michigan monarchs went as far south as Arkansas. Using both ground-tracking and tagging, researchers have slowly pieced together the migratory route and destination of the monarchs. Their winter roosts are in a mountainous region sixty miles west of Mexico City. There the monarchs settle down for the winter in a dozen or so sites, all within thirty miles of one another; the individual roosting sites are sometimes no larger than a few acres (figure 3.1).

Finding the monarch's winter roost raised yet another question: How does this delicate creature manage to fly several thousand miles from the United States to Mexico each fall? If they had to fly nonstop, they would starve to death in the air. Instead, they feed as they go—following a trail of nectar-producing flowers—all the way to Mexico. On a typical day, a monarch might fly fifty miles.

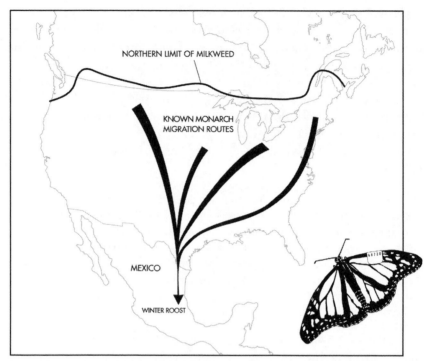

Figure 3.1. Monarch butterfly migration routes. Note monarch butterfly wing tag. These tags helped scientists determine the monarch's migratory movements.

It turns out that monarchs actually gain weight during their trip south by drinking flower nectar and then converting this nectar to fat, which they then store in their abdomen. To do so is important because they need their accumulated fat reserves to survive their three-month stay in the Mexican mountains, where little food is available for them. At their winter roosts, the monarchs cluster by the thousands on the branches of fir trees. The dense canopy of these firs protects them from freezing while keeping them cool enough to remain semidormant. Come the end of March, the monarchs make their way back to the United States.

This final part of the story presented researchers with still another puzzling question: Do the same adults that spend the winter in Mexico fly all the way back to their home meadows in the northern United States and Canada? The answer turns out to be "no." The adults flying back from Mexico settle down to mate in the southern United States as soon as they find freshly sprouted milkweeds to feed on. After laying their eggs, these monarchs finally die; and it is their offspring that continue the journey north.

It wasn't trailing or tagging that resolved this mystery; rather, it was chemical fingerprinting. By analyzing milkweed chemicals in the bodies of adult monarchs, researchers can tell where in the United States the adults lived during their larval stage. This is possible because there are scores of different milkweed species in the United States, each with its own geographic range and unique biochemical "fingerprint." Hence, a scientist capturing an adult monarch in Canada can determine where that butterfly began its life.

More recently the Internet has begun to play a role in tracking monarch movements. Through "Monarch Watch" (www.MonarchWatch.org) and Journey North (www.learner.org/jnorth) butterfly enthusiasts can communicate the migratory movements of the monarchs. For example, monarch watchers in Mexico send out the word that the monarchs are beginning to move north. Then, observers further to the north begin to post messages—for example, "monarchs just arrived here"; "monarchs mating"; "monarch caterpillars more abundant than usual"; and so forth.

The more that scientists study the life history of any particular species, the more they are presented with new questions. In the case of the monarch, a question that continues to baffle researchers is: How do the monarchs that hatch in the fall in North America find their way, year after year, to the same winter roosting sites in Mexico, given that these monarchs have never been there before? A mystery indeed!

In sum, this description of the monarch butterfly's life history illustrates the question-and-investigation approach that scientists employ to study the natural world. I chose to describe what is known about the monarch's life history—one string in life's web—but I could have just as easily selected the opossum, the spotted newt, or the fire ant.

Reflection

As I was gathering information about the monarch butterfly's life history, I encountered reports describing a surprising connection between U.S. agriculture and monarch well-being.[4] Specifically, in 1997 a team of researchers at Cornell University discovered that a genetically engineered strain of corn produces pollen that can kill monarch caterpillars. The corn in question is so-called *Bt* corn. *Bt* stands for the soil bacterium, *Bacillus thuringiensis*. This bacterium secretes a protein that wrecks havoc in the gut of many insect larvae, turning healthy caterpillars into a black, gooey mass. Now, with the advent of genetic engineering, it is possible to actually insert *Bt* genes into crop plants, like corn. The transferred genes give corn the ability to synthesize the *Bt* toxins. As the corn plant grows, all its cells—those in the stems, leaves, roots, kernels, and pollen—contain the *Bt* toxin. When corn produces flowers, its pollen blows about. If enough of this *Bt*-containing corn pollen lands on the leaves of milkweed plants that monarch caterpillars are feeding on, the caterpillars will die before becoming adults.

Reflecting on this, I am reminded that one of the fundamental axioms in ecology is that "you can't just do one thing." Everything really is "connected"; no species or person or new technology is an island unto itself.

Questions for Reflection

- It has been said that we are all "scientists"—all curious problem solvers. In what sense is this true?
- In what ways do you practice the problem-solving approaches described in this foundation in your everyday life?

Practice: Seeking Understanding by Observing Life's Web

As a beginning graduate student in ecology at Michigan State, I took a field ecology course with Dr. Patricia Werner. On the first day of class, Dr. Werner took us out to a meadow and invited us to just look around. Then she asked, "What's going on here?" We waited for her to elaborate. She did not. I no longer remember my response, but I do remember Dr. Werner's wonderful question.

When the water condenses on the inside of a window or when a ribbon of fog sits above a stream in the early morning, "What's going on here?" When earthworms appear on sidewalks after an evening rain, "What's going on here?" When you come upon a cluster of oak seedlings in a pine forest and no mature oaks are anywhere in sight, "What's going on here?"

Nonscientists often regard science as mysterious and scientists as aliens, but this is a misperception. As we saw with the monarch story, science is the art and practice of observing, asking questions, and then figuring out ways to answer those questions. Good field scientists share two qualities: They are good observers, and they are curious. Walking with them through a field is akin to walking with a child—everything is fresh and new. They stop often—here to marvel at a beautiful flower, there to note a bee pollinating a plant, further along to muse over some fresh tracks in the soil. Questions tumble forth from each observation: Why do the flowers of some plant species last only one day while those of other species persist for a week or more? Can bees tell if a flower will have nectar before they land? Can animals "read" tracks the way humans do? To exercise such curiosity requires that we be willing to slow down to nature's pace.

One of my favorite field practices for slowing down and cultivating observation skills is called Fifty Questions. All that is required is a nice spot in a field, in a patch of forest, by a stream, or even in your own backyard. Find your spot and simply sit quietly for an hour, carefully observing the life around you: the smells, the quality of the air, the sounds, the insects, and the plants. As you observe, questions will inevitably arise: What bird is that flitting around in the bushes? Does it have a nest close by? What is that ant carrying? Where is it going? Why is there moss on this rock but not on that rock over there? Why do the leaves of oaks, pines, and maples have different shapes? What will the mosquito do with the blood it is sucking from my arm?

Stay in the present moment, observing and writing down questions as they come to you. After an hour, you will have accumulated dozens of questions, perhaps even fifty. Some of your questions will have probably already been answered by scientists, but perhaps a few on your list have never been answered. Perhaps you yourself could answer them with some guidance from a mentor or on your own by simply using common sense.

This practice can be repeated again and again in different places and at different scales. For example, you might want to take a ten-foot length of string and lay it out as a circle in a meadow or forest; you can then get down on your belly and spend an hour carefully examining the life bounded within this little circle of string. If you give yourself over to the experience, you will find that questions cascade forth. You will also learn things about the way that you "sense" the world—for example, the senses that are most active for you, the colors that most attract you, and the kinds of organisms that most garner your attention.

Fifty Questions gets us in the habit of observing; it cultivates curiosity. We can carry the spirit of this practice into our daily lives. For example, the next time you sit with someone, look at the person carefully. Note their hair, the way they hold their hands, the curve of their mouth, the inflection of their voice. I don't mean to suggest that you should sit there and gawk at the person. Just try to be fully present, fully aware. Note how they hold their head, the movement of their eyes, their breathing. In other words, cultivate your interest in the other person—not in a judgmental way, but simply as an expression of your curiosity.

You can go further and cultivate the practice of observing and sensing the natural world from the interior of your being. This practice is more advanced, and as such it may seem a bit far out. If you are game to give it a try, start with a tree, trying to look at it from the center of your being. In other words, let go of the category "tree" and behold this organism as if for the first time. As you allow yourself to extend out to this new being, you might feel as if something from "tree"—a kind of "feeling-energy resonance"—is coming back and entering you:

> This is very analogous to what happens if you play a note on a guitar and there is another guitar leaning against a chair nearby. The second guitar will start to sound the same note as the one you struck on the first guitar. This is resonance. And there is actually a transfer of energy from the first guitar to the second, and then back from the second to the first. This principle of resonance runs throughout the universe.[5]

Indeed, "the world you perceive is not just there, it is a relationship between you and what is there."[6]

Whether sitting with another person or with an oak, practicing curiosity—extending our attention and interest out beyond our personal boundaries—opens us up and expands ecological consciousness.

FOUNDATION 3.2: RELATIONSHIPS AMONG SPECIES CREATE LIFE'S WEB

Somewhere along the line you were probably introduced to the concept of a "food web"—the idea that Earth's creatures are linked together in a complex web of feeding interdependencies. Once we fully grasp the concept of a food web, we are positioned to see the world differently. For example, when a fly lands on our arm, we question: What does the fly eat, and who eats it? (see box).

Relationships in a Patch of Goldenrod

A short distance from where I live, a particular field was farmed at one time but has now been abandoned. Each September, when the field is filled with flowering goldenrod, I take my ecology class there. A wealth of biology is contained in a flowering patch of goldenrod; it is a mini-ecosystem, replete with pollinators, predators, herbivores, symbionts, and parasites.[7] When we arrive at the field's edge, I give each student a hand lens. I instruct them to find a patch of goldenrod and see what they can learn, simply by observing their patch for an hour (figure 3.2).

LANDING IN LIFE'S WEB

One of my most intimate experiences with life's web of feeding relationships occurred while canoeing with a friend along Ontario's remote Spanish River. Each morning we woke early and canoed our way south. In the afternoon we bathed in the clean, cool water. One day after our swim, as we lay reading at river's edge, we noticed a large animal, five hundred yards upstream, clomping along the river shallows. It looked like a horse but we realized that it was a moose, a young female, coming straight toward us. We watched as she approached: four hundred yards, three hundred yards . . . we could hear the sound of her hooves on the river cobble . . . two hundred yards . . . at one hundred yards, she disappeared into a wet meadow and then suddenly reappeared, bursting out of the undergrowth, just thirty yards upstream from where we stood. We could see welts on her flanks and hear her labored breathing. Without pausing, she crossed the river on a gravel shoal and disappeared—this time for good—into an alder thicket.

What was that all about? Why was the moose in such a hurry? Why was she so oblivious to our presence? We sat pondering what we had just witnessed. Then, at the same river-edge spot, five hundred yards upstream where we had first spotted the moose, came a lone white wolf moving with equal purposefulness. Like the moose, the wolf also disappeared into the grassy marsh one hundred yards above us. We scampered up onto a ledge, hoping to get a view; but before we could go very far, the wolf appeared on the very same ledge, immediately in front of us. There was no showdown, however; the wolf promptly spun around and disappeared. Later, I learned that lone wolves will sometimes track a solitary moose for several days in hopes of eventually inflicting a wound and then taking the animal down.

Did the wolf pick up the moose's track again? I don't know. I do know that for a brief moment I was in a personally unfamiliar and exciting part of life's web.

Students are initially drawn to the butterflies and bumblebees on the goldenrod blossoms. They watch as the insects slurp up nectar. But the goldenrods benefit here as well. In the process of foraging for nectar, insects spread goldenrod pollen from plant to plant, thereby ensuring cross-pollination.

Looking more closely, the students discover that butterflies and bees are not the only insects at goldenrod's floral banquet. There are also predators: voracious ambush bugs that dismember their prey; and tiny yellow crab spiders that lie in wait, camouflaged amongst the blossoms, ready to immobilize an unwary pollinator and suck it dry.

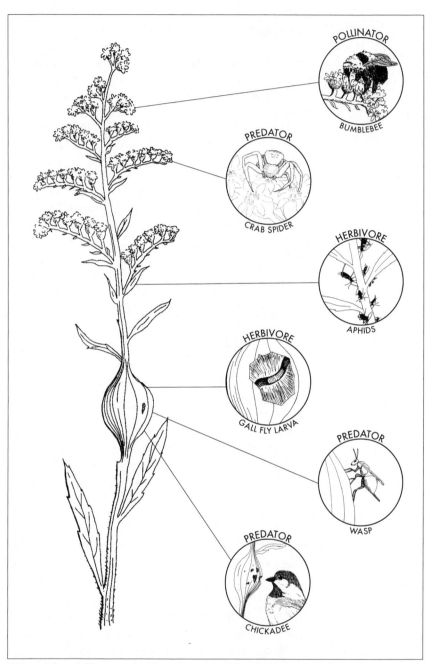

Figure 3.2. Pollinators, predators, and herbivores in a patch of goldenrod

Below the flowers, on the stem, some alert students usually discover another drama being enacted: Aphids are sticking their needle-like proboscises into goldenrod stems and extracting sugar-rich fluids. But there is a problem. This plant juice is low in the precious element, nitrogen, that aphids covet for protein synthesis. Hence, to satisfy their nitrogen needs, the aphids must suck up large quantities of the goldenrod's sugary fluid. After they extract the nitrogen, the aphids let the excess sugars drip from their anuses; but this excess doesn't go to waste. Students sit transfixed as they observe ants drinking the drops of sugary fluid leaving the aphid's anuses. Just as humans milk cows, the ants "tend" the aphids. While milking the aphids, the ants also serve as guards, warding off ladybugs and other predators. This service is important because a single ladybug, in the absence of any constraints, can easily consume fifty or more aphids a day.

Further down on the goldenrod stem, students may discover a strange, marble-sized globe of tissue—a goldenrod gall. If they cut open the gall, they sometimes find a writhing larva inside. Many people are curious about these galls. I explain that if we had come to the field the previous April, we might have observed an adult fly known as the "goldenrod gall fly" emerging from just such a ball of tissue. After emerging, the adult gall fly would have flown around laying eggs, one by one, on the tips of young goldenrod plants. Then, after the eggs of these gall flies had hatched, the larvae would have tunneled down into the growing stem tips and induced the goldenrods to produce galls, just like the ones we have been observing. Gall fly larvae feed in these protein- and carbohydrate-rich galls through the summer and early fall; then they enter a resting phase before emerging as adult flies in the early spring to repeat the cycle. But the lives of gall fly larvae are sometimes cut short by a wasp species, which deposits eggs inside galls. The wasp larvae, developing from these eggs, devour the gall flies. If the gall fly larvae manage to escape attack by wasps, they might still get gobbled up by chickadees or woodpeckers, which sometimes peck open galls in winter.

This is one classroom that students don't rush away from. Even after I tell them we are done for the day, they often linger in twos or threes by patches of goldenrod looking at this mini-ecosystem with new eyes.

Mutualistic and Symbiotic Relationships

For much of the twentieth century, biologists tended to describe nature as a "tooth-and-claw" struggle—a place of high-stakes conflict where only the strong survived. They also occasionally mentioned cooperation, such as when a plover (a bird) goes into the gaping mouth of a crocodile to rid the croc's teeth of leeches while gaining nourishment. But such stories were often presented as anecdotes or exceptions to the more general view of evolution as grounded in bloody competition. In recent decades, however, this paradigm has begun to change as scientists discover more and more instances of cooperation and mutual dependence among species. The following are four examples that showcase the ways in which organisms, sometimes very different from one another, interact in complementary ways.

Blue Jays and Oaks

I never saw a connection between oaks and blue jays until I observed a jay eating a white oak acorn one morning in late September. Grasping the acorn with its feet, the jay hammered it open with its bill and consumed the "meat." Meanwhile, higher up in the same tree, I spotted another jay swallowing entire acorns, one after the other. I learned later, reading reports in scientific journals, that jays pack whole acorns into their expandable throat and esophagus. After loading up with several acorns, they fly hundreds or even thousands of yards to their breeding territories. When they land, they disgorge their acorns and bury them, one by one, by pushing back the earth with their bills and hammering the acorns into the soil; they finish the job by covering each acorn with leaves and debris.

Jays are choosy about the acorns they pluck from oak trees, preferentially selecting the big, fully ripe ones. These, it turns out, are the very acorns that have the highest probability of producing viable oak seedlings.

An individual jay may scatter/horde four to five thousand choice acorns over a one-month period in the fall. When winter descends, jays no longer have access to insects and worms; so they rely on these buried nuts, relocating them, in part, by memorizing the location of each buried acorn in relation to nearby objects (beacons), such as rocks or logs. In fact, scientists have shown that, under experimental conditions, birds are no longer able to correctly locate their acorn caches when the positions of natural "beacons" are altered.

This relationship between oaks and jays is mutually beneficial. It helps jays survive the winter; but it also enhances oak reproduction because, over the course of the winter, a jay is likely to dig up only a fraction of its buried acorn stock. The remaining acorns are frequently "planted" in favorable conditions for germination and growth.[8]

Ruminants and Bacteria

Not all collaborations are readily apparent, like that between oaks and jays. Sometimes one of the partners is invisible. This is the case with cows and bacteria. Scientists, studying the cow–bacteria partnership, sometimes create a surgical opening (called a "fistula") in the cow's side to study the digestive processes. Imagine, if you will, reaching your gloved hand through a fistula and pulling out a handful of mush from a cow's rumen. On first glance, the "mush" would appear to be nothing more than partially digested grass. However, if you were to take a tiny pinch of it, dilute that pinch in a few drops of water, and then place it under a microscope, you would see an extraordinary community of bacteria—some large, some tiny, some ciliated, others flagellated. It is these bacteria that do the job of grass digestion in a cow's rumen. Without their bacterial symbionts, cows would be unable to digest the sugar-rich cellulose of grasses. As microbial biologist Lynn Margulis explains: "The cellulose-degrading microbes, in a very real sense, are the cow. . . . No cow would be grass-eating or cud-chewing without the microbial middlemen."[9]

Flatworms and Algae

Once, years ago, I attended a lecture where the speaker asked the audience why there isn't a mammal that can walk around like us *and* produce its own food through photosynthesis. Wouldn't it be convenient, he mused, if you and I had banana plants sprouting out of our shoulders. Then, when we became hungry, we could just reach up and pluck off some fruit.

Although such talk appears to be the stuff of fantasy, some animals—such as the flatworm, *Convoluta roscoffensis*—actually do carry their food system with them. If you were to encounter these "worms" (which look like gooey green ribbons) in their natural habitat along the beaches of northwestern France, you might mistake them for some sort of seaweed. The worms' bodies are green because their tissues are studded with cells of a photosynthesizing algae. These algae are the worm's "bananas." In fact, the worms' mouths cease to function once the algae come on board. Instead, they rely entirely on the algae's photosynthetic products for sustenance. "Algae and worm make a miniature ecosystem swimming in the sun."[10]

Fungi and Tree Roots

Walking through the woods after a wet spell, you would find that it is not uncommon to encounter mushrooms. Think of them as fungal "flowers" because, like flowers, they are the fungus' reproductive structures. If you take a mushroom cap, set it on a piece of paper, and then lift up the cap a few hours later, a print of the mushroom's spores ("seeds") will be left behind. These spores are so light that they can be carried great distances by wisps of wind. Hence, the mushroom, with its mobile spores, is the fungus' way of colonizing new ground.

Although mushrooms can be quite conspicuous, they represent only a tiny part of a fungus' body. In fact, mushrooms wouldn't exist at all if it weren't for their underground "bodies," which consist of complex networks of tubular threads known as "hyphae." Fungal hyphae absorb water from the soil and release enzymes that decompose organic debris (fungal food).

Many fungal species, known collectively as "mycorrhizae," have formed symbiotic partnerships with tree roots. For example, oaks, pines, maples, and hickories usually have mycorrhizal fungi associated with their roots. Trees in these groups "feed" their mycorrhizal associates carbohydrates, and in return the mycorrhizae act as root extensions, bringing nutrients into the tree.

All these various symbiotic relationships are like marriages with an important distinction—the partners are distinct species. We may never know the exact steps that led to the formation of the many symbiotic relationships that now exist in the world; we do know that these relationships are among the most important in life's web. As renowned biologist Lewis Thomas has pointed out, "The urge to form partnerships, to link up in collaborative arrangements is perhaps the oldest, strongest, and most fundamental force in nature. There is no solitary, free-living creature, every form of life is dependent on other forms."[11]

Reflection

As humans, many of our relationships with other species are strictly utilitarian, but not always. I was reminded of this when, on a recent hike, I stopped by a stream to rest and have lunch. As I sat eating, a tiny yellow spider dropped down in front of me on a silken thread. I enticed her onto a pencil and then had a close-up look with my hand lens. She was magnificent, delicate, almost translucent—with a beautiful red spot on her abdomen. As I observed, she played out a silk thread from her spinneret, dropped down a few inches and then reeled herself back up to the pencil tip. Using the hand lens again, I was able to see fine bristles on her legs and the complex articulation of her mandibles.

Eventually, I released her from this pencil prison and watched her scurry onto my jacket. I returned to my lunch. A few minutes later, I felt something behind my ear and I instinctively rubbed it away. Then, I remembered the delicate yellow spider. The creature, which a few minutes earlier had enthralled me, was now smeared across my fingertips. Accidentally killing a spider normally wouldn't have mattered to me, but this time, because I had closely observed this particular spider and had marveled at her life force, I felt a bond, a relationship; and I was dismayed by her death.

Questions for Reflection

- In what ways do you experience yourself as part of life's web?
- In what ways might your connections with life's web extend beyond mere utilitarianism?

Practice: Seeking Understanding by Studying Life's Web

Mike, a student, knocked on my door not long ago with a mystery. There was a walnut tree in his backyard, which was bearing a heavy crop of walnuts. Climbing up into this tree, Mike was surprised to find walnuts sitting in the crooks of many of the tree's branches; some crooks had several walnuts. Mike thought it unlikely that walnuts were simply falling into the branch crooks and wanted to know what I thought. Knowing that questions often teach more than answers, I suggested that he spend some time observing his walnut tree at different times of the day, paying special attention to animal visitors. The next day Mike was back with a big smile. At dusk he had seen a squirrel caching walnuts in one of the largest branch crooks.

Mike found a "web string," a connection between a squirrel and a tree. Such strings are everywhere, but to see them takes patience. In Mike's case, it took sitting quietly for an hour to simply observe the tree and its visitors. He found that he could solve his mystery if he took the time.

It helps to get in the habit of solving nature's riddles while we are still young. This point was reinforced when two young girls stopped me as I was jogging through a wooded park near my home. They wanted to know if the staccato sound they heard was a rattlesnake. I embraced their question and offered to investigate with them.

"Where is the sound coming from?" I queried.

"It is up there above us," the smaller of the two responded.

So, I enjoined them to explore the branches of the white pine above us. Soon they located the source of the "rattle." It wasn't a rattlesnake; it was a red squirrel.

"Why do you suppose he is making that racket?" I asked.

"He is mad about something," came the response.

So, we talked about how we feel when we get mad and what helps us to cool down. This led to the idea that maybe the red squirrel was perturbed because we were in his space. We moved away, and the red squirrel stopped chattering. It was a simple mystery but an important one for the two girls to solve. They learned that squirrels have home places or territories. Just as important, they got to "strum" on one of life's web strings.

It is not always easy to decipher the connections in life's web. First, it requires a commitment of time; and, second, it necessitates that we turn our attention to creatures and processes outside ourselves. But like any difficult practice, the rewards can make it worth the effort. Start, like Mike did, with something that grabs your attention; then, sit with the mystery. For example, if a certain spider attracts your attention, spend two hours watching the spider build her web. Then, return the next day and the day after that to deepen and expand your observations of the spider's activities.

Don't make the common mistake of finding out (from a book or an expert) what the spider is called and then thinking that you know the spider when, really, all you know is a name. To "know" the spider requires that you come to understand how the spider is present in the world. It is in returning to observe the spider day after day that you will come to understand her. For example, on that first visit you might wonder how long the spider's web will last, what kinds of insects her web will capture, and whether she will recycle her web. When you return the next day, you might have the beginnings of an answer to some of your questions. As you log more time with the spider, you may begin to wonder about her predators, nocturnal habits, sex life, response to cold, sources of water, and on and on. If you are diligent, it is entirely possible that you will make a discovery new to science. Indeed, some of what we know about the ecology of plants and animals comes from investigations conducted by "citizen scientists."

As we deepen our practice of natural history study, it often happens that we enlarge our capacity to experience the world from the perspective of other organisms. I had my first taste of this when I was in graduate school studying plants. At the time I observed that most of the seedlings produced from plant seeds die, and I wondered why this was so. Hearing my question, a professor challenged me to imagine that I was a freshly emerged buttercup seedling, just a quarter-inch across, trying to gain a foothold in an abandoned farm field. As I descended into that microworld, I felt the hot sun reflected off the boulder-sized sand grains, desiccating my first tiny leaves. I longed for relief and visualized a heavy rain falling on the field; but when the rain came, it didn't bring relief. It brought more traumas. With my new seedling "eyes," the individual rain drops came crashing down, slamming into the soil, splashing "my" tender leaves with clay particles and creating minicraters, which exposed "my" fragile roots. In addition to offering me a new, empathetic perspective on plant seedlings, this simple visualization alerted me to new research opportunities.

German biologist Jacob von Uexkull created the word *umwelt* to signify the world around a living being as that creature experiences it. *Umwelt* has no English translation, so I suggest simply appropriating *umwelt* to signify the unique way that each organism experiences its environment.

Endeavoring to see the world from the perspective of other organisms exercises our imagination and expands our ecological consciousness. For example, knowing that insects can see in the ultraviolet range, we can exercise our brains by imagining how a flower-filled meadow would appear to a bumblebee. Or, realizing that a rattlesnake has two pits on its head that are very sensitive to small differences in temperature, we can imagine that objects that look the same to us (e.g., the floor below our feet) would be very nuanced (i.e., experienced as a mosaic of different temperatures) to a rattlesnake.

One gateway to experiencing another organism's *umwelt* comes in understanding how that organism locates itself in space. For example, we humans can tell the difference between up and down, front and back, right and left because of a balance organ—the semicircular canals located in the middle of our head. These canals contain liquid and are lined with hairs. When the liquid is sloshed about, those hairs, which are normally dry, become wet, thereby signaling to the brain that we are off balance. Now, what if you were an ant? No semicircular canals for ants! How might you know where your body was in space? Judith and Herbert Kohl in their delightful book, *The View from the Oak*, provide some clues:

> Ants have tiny hairs at all their joints. When the ants' bodies are straight and their legs are down in a resting position, these little hairs are not being rubbed or pressed on by any of the joints. But as soon as the ant lifts a leg, the leg hairs are touched and the ant knows something about the relationship between that leg and the ground. Should he turn his abdomen sideways, he knows he is headed in a new direction. When he lowers an antenna to feel a pebble in his path, the hairs at the antenna joint tell him the height of the pebble in relation to himself. It is touch not sight that has primary importance in the ant's umwelt.[12]

If you spend the time to develop an *umwelt* practice, you will gradually come to understand that each organism perceives the world with its own specific set of sensory tools. And when you take a stroll through a patch of forest on a Sunday afternoon, you will understand that you are experiencing only a sliver of the fullness of the forest. The doe, hidden from sight, will be experiencing this same forest in a very different way from you; so it is for the wood thrush, the earthworm, the moth, and the soil bacterium—each experiencing a distinct face of the universe. And you will know that it is a mistake to put the doe and thrush in your world, imagining that yours is the only world. The forest is not one world; it is a rainbow of worlds. Getting to know the lifeways of other creatures is a wonderful initiation to the web of life, as well as an invitation to expand ecological consciousness.

FOUNDATION 3.3: HUMAN SURVIVAL DEPENDS ON LIFE'S WEB

Your body is a swirling, pulsating ecosystem filled with other organisms. Just consider:

> An adult human's body is home to about 100 trillion microbial organisms, ten times more alien creatures than the number of cells of the body itself! They are everywhere: eyes, ears, teeth, gums, between the toes, in the groin. They harbor by the millions in the prairies of the skin, the woodlands of the scalp, the rain forests of the armpits. A spelunking tour of the body would take us through the respiratory tract, the oral cavity, the gastrointestinal tract, and the outer part of the urinary tract, each with its swarming population of microorganisms.[13]

Overall, several hundred distinct species of bacteria and fungi consider the human body as their home (see box).

Many species of bacteria take up residence in the human mouth. Some are adapted to the environment at the tip of the tongue; some prefer the back of the tongue; others, the cheek pouches; yet others, teeth plaque. Another raft of bacteria make their home in the human gut. Some of these microbes help in the digestion of food and in the manufacture of vitamins K and B complex; others help tune the human immune system and keep harmful microbes in check.

None of our "houseguests" is better known than the bean-shaped bacterium, *Escherichia coli*, which takes up residence in the gut. Estimations suggest that, if laid out end to end, there are enough *E. coli* in each of us to stretch from Boston to San Francisco. Although diminutive, *E. coli* is a fully functional creature: it moves using whiplike appendages; it feeds; it reproduces; and it recognizes and communicates with its own kind.[14]

Bacteria also inhabit the outside surfaces of our bodies—thousands per square inch on our legs, arms, trunk, and face. They feed on the mix of salt, water, nitrogen compounds, and oils that continually seep from the human body.

Certain mite species also depend on humans for nutrition. There is one species—just discovered a few decades ago—that uses its eight tiny legs to

---

MADURA FOOT

The idea that our bodies are "habitats" for other creatures is something that I accepted in an intellectual sort of way; but I didn't fully grasp it in a bodily way until I discovered "pulgas" living inside me.

It all began with a slight itch on the heel of my left foot. I was living in the village of San Carlos on the banks of the Rio Negro in southern Venezuela. Over a period of a week, as the itch intensified, a white spot, about a quarter-inch across with a dark center, developed on my heel. Soon, I realized that it wasn't just one itch but several and that both of my feet were involved. I decided to consult *Merck's Manual,* the "bible" for diagnosing tropical ailments and diseases. As I sat hunched over a table reading by candlelight, I ruled out one thing after another, but then I came to "madura foot." "Madura" means "ripe" or "rotten" in Spanish. Madura foot is a fungal disease that transforms the human foot into a mass of oozing sinuses and rotting flesh. It seemed plausible, based on the book's description, that I was already in the early stages of "madura foot." As I was fretting over my fate, my field assistant, Getulio, stopped by. He took one look at my feet and said, "pulgas." Then, he put a match to a needle and dug open one of the white spots. Soon I could see a cluster of very tiny eggs. Getulio explained that these eggs were deposited in my foot by a common tropical flea that burrows into human feet. The eggs hatch; the larvae feed; then they exit the foot as adult fleas. Although I found this somewhat disconcerting, I was relieved to be "habitat" for a flea rather than for a fungus that could turn my feet to mush.

---

fasten itself to the base of eyelashes. This species does especially well when it has access to nutritious eyeliner. Another recently discovered mite takes up residence in carpets and beds where it feeds on the tiny skin flakes that are continually released from the human body.

Humans Depend on Other Species

Just as some species depend on us, we depend on other species in a dizzying assortment of ways. Biologist Gretchen Daily suggests a "thought experiment" (inspired by John Holdren) to fully appreciate the scale of our dependence on Earth's species.[15]

Imagine that you have been put in charge of an expedition that will create an independent, self-sustaining civilization on the moon. To make the challenge a bit less daunting, assume, first, that the moon has a friendly atmosphere just like our own with lots of oxygen; second, that the moon's surface is composed of pulverized rock (the basis for soil); and further, that it has water. Your specific job is to decide which organisms you will need to take from planet Earth to ensure that your expedition can thrive on the moon for thousands of years into the future. The technical commander assures you that your spaceship is very large so that you will have ample room for whatever you need to bring.

Food is the first thing that comes to your mind. So you set about gathering seeds of grains, vegetables, fruits, and spices: for example, corn, wheat, oats, rice, and barley; carrots, onions, kale, potatoes, and celery; bananas, oranges, apples, grapes, kiwis, and mangoes; and thyme, rosemary, black pepper, ginger, and garlic. Then, of course, you would need coffee and tea, as well as hops and yeast (for beer). And, oh yes, you would need sugar cane and beets for sugar, and cacao for chocolate. And finally, you mustn't forget an array of mushroom types.

In the process of making your list and gathering the necessary plants and seeds, you come to realize, perhaps as never before, that humans rely on hundreds of different species of plants. And not just plants! You remember that you will need to take chickens, goats, cattle, sheep, and turkeys, as well as tanks filled with trout, salmon, shrimp, flounder, clams, and oysters.

At last, you think that your job is done, but then you remember that you will need to make clothes; and so you add cotton and hemp plants to your list and perhaps silk-producing caterpillars and various plant species that can provide dyes. You will also need to build things from wood as well as make paper, so onto your list go dozens of different tree species. "There, that should do it," you say to yourself. But then an alarming thought occurs to you: What are you going to feed all your horses, chickens, cattle, and sheep? So you add alfalfa, forage grasses, nutritious herbs, and hardy shrubs to your list.

At this point, you are pretty sure that you have everything covered, but to be certain you decide to imagine yourself actually on the moon. There you are with all your seeds; but as you go to plant them, you realize that the soil is sterile—barren of the bacteria and fungi that make up the decomposer link, which is essential for nutrient cycling (see chapter 2, foundation 2.3, p. 52).

As you continue with your visualization, imagining your crops and fruit trees growing, you realize with a jolt that you forgot the insects, birds, and bats that will be necessary to pollinate your food plants. Indeed, more than two-thirds of our crops require pollination by animals. So you set about compiling a list of all the pollinator species you will need. Once finished, you suddenly wonder what your various pollinators will survive on during times when their preferred crops are not supplying pollen and nectar. This worry sends you in search of additional plant species that can supply the various pollinators with food throughout the entire lunar year.

Finally, one night, just a month before the scheduled departure date, you awaken from a nightmare. In your dream, insect pests are destroying all the crop plants. You realize that the domestic animals and plants that you will be taking to the moon will undoubtedly harbor some microbial and insect pests. Since synthetic pesticides will not be effective in the long run, you will need to include all of the natural enemies of all imaginable crop pests.

Now, as you are beginning to really think for the long term, you consider the possibility that environmental conditions (e.g., climate) will change on the moon, just as they have changed on Earth, over time. Hence, it won't work to take just a single variety of each species of plant and animal. Instead, you will need to have as many strains as possible—that is, genetic variation—to ensure that the raw material for evolution is available.

Our poor understanding of the myriad ways in which we depend on Earth's ecosystems was illustrated by the recent failure of the Biosphere 2 mission.[16] Biosphere 2 was designed to function as a completely self-contained three-acre artificial "biosphere" within Biosphere 1, the Earth. It was constructed under a transparent, airtight bubble in the Arizona desert. The $200 million–plus unit featured zones of agricultural land, rain forest, desert, and savanna, along with wetlands and a tiny "ocean" with coral reefs. Biologists were hired as consultants to design the various subsystems of Biosphere 2 and to determine the array of plants, animals, and microbes to be "seeded" into the unit.

After everything was set, eight people were sealed into the unit. The team intended to stay for two years, but the experiment had to be terminated early in part because the atmosphere in the bubble deteriorated; the oxygen concentrations dropped to 14 percent—a level typical of elevations of 17,500 feet—and nitrous-oxide concentrations rose to levels that can impair brain function. Meanwhile nineteen of twenty-five vertebrate species of Biosphere 2 went extinct along with all pollinators, thereby dooming most of the plant species to eventual extinction. In addition, algal mats polluted the water; and ants, cockroaches, and katydids experienced population explosions.

The Biosphererians learned a basic lesson the hard way: Humankind is not only dependent on thousands of species in direct ways for food and fiber, but also on tens of thousands, perhaps millions, of other species for basic life-support services—that is, the services that maintain healthy ecosystems, such as nutrient cycling. Whether we know it or not, we are deeply insinuated into the living body of Earth.

Reflection

The thought that millions of organisms live both on me and in me is mind boggling. When I first heard this years ago, I was repulsed—spirochetes in my saliva by the tens of millions, mites on my eyelashes. How disconcerting! Now, I see it differently. I endeavor to visualize the myriad invisible creatures that live with me. They are everywhere, filling up and defining my body. It is no longer possible to see myself as a separate entity; I am a walking community of beings. What's more, thousands of web strings emanate from my body to all the Earth beings that I depend on for sustenance and well-being.

Questions for Reflection

- How does the realization that your body is a swirling, pulsating ecosystem, providing habitat for millions of creatures, affect your understanding of who you are?
- And how does this understanding affect your relationship with the other-than-human world?

Practice: Seeking Understanding by Entering Life's Web

Each October I invite my students to literally enter the web of life. We gather in an oak–hickory forest in the early afternoon. I tell them that the year is 1000 B.C.E. and that we are members of a band of Woodland Native Americans. There are forty of us, but the very young and very old of our clan (seven people) remain in camp caring for one another and tending the fire. In addition, eight of our members dedicate their time to hunting. We can count on these hunters to supply about 20 percent of our food needs. It's up to the rest of us (i.e., the twenty-five students in the class) to supply the remaining 80 percent of our clan's needs by foraging for edible nuts in the forest: acorns, hickory nuts, and walnuts. I remind them that it is now the gathering season and that soon the snows will be upon us. In the next two months we will need to gather the nuts that will sustain us through the winter. I give the students equations so that they can calculate the total amount of energy in Calories our band of forty will require to survive the coming winter. The task for the afternoon is to figure out if we will be able to gather the necessary Calories from the forest.

Students form small groups and begin to search for nuts. Often these foraging parties are gender-specific. Some groups move systematically through the forest, back and forth, searching the ground; others look around for big oaks and hickories to concentrate their foraging there; some groups fashion branches into brooms and then sweep the leaves away from the forest floor to expose nuts; some divide tasks—one person scouts, one sweeps, and one collects. In these different ways, the foraging parties experiment with simple technologies that might optimize their foraging efforts.

There are usually large differences in foraging effectiveness among the foraging parties. Groups composed of women often gather more nuts than male groups. Some groups are very quiet and earnest; some make up songs and chants; others simply chat back and forth, exchanging information. The "chanters and chatters" often gather more nuts.

After two hours, I call the groups back together and provide them with scales so that they can weigh their harvest. Again, I give them equations so that they can calculate the amount of Calories harvested per hour spent foraging; they then can compare this to their earlier calculation on their clan's caloric requirements for the upcoming winter. The upshot of all these calculations is a verdict—survival or death for their clan.

Many students are deeply moved by this experience. They talk about how the activity led them to observe squirrels more closely (e.g., the presence of squirrels in a certain tree is often a sign that there are nuts on the ground underneath). Some admit to being surprised by how peaceful they felt while foraging for nuts. Others confess that they found it hard to make the transition from the earthy, intuitive, right-brain gathering process to the rational, left-brain analytic process that involved scales and calculators.

For me this exercise has become a kind of practice. I first did it with my daughter when she was twelve. We made flour from the acorns that we gathered. I continue to forage and eat acorns each fall, noting the sweetness of the white oak acorns and the bitterness of the red's. Although I live in a modern world, a part of me is at home foraging for acorns, which is not so surprising. Hunting and gathering, as environmental philosopher Paul Shepard points out, have been a fundamental part of human evolutionary history. The selection pressures that honed our physiologies and behaviors were those favoring a being who was an alert hunter and an effective gatherer—one who could supply food for himself and his kin.[17]

Coupled with the practice of gathering wild foods, we might also choose to reenter life's web by practicing our stalking and tracking skills. "The human mind came into existence tracking," notes Shepard.[18] You don't need to be an expert to begin this practice; you just need to activate your senses and pay attention. Imagine, for example, that you are having a picnic lunch in a meadow. It is a beautiful autumn afternoon. As you sit finishing your sandwich, a large beetle comes along. You watch the beetle for a while and then wonder if this little fellow leaves tracks. You get down on your belly to have a look at the soft, sandy soil where the beetle just passed. Sure enough there are tracks! They look like the imprint from a very fine zipper.

Suddenly, for the careful observer, the soil surface becomes a book embedded with stories. As you nose along, smelling and examining the ground, you pick up a new set of tracks—those of a small field mouse—and you decide to follow them. The closely spaced mouse tracks meander through the grass, and after a few feet you see some tiny mouse droppings. The droppings are soft and moist, apparently fresh. Where will this story end? As your excitement grows, you can't help thinking about how ridiculous this is—you, a sophisticated adult, crawling along the ground after a mouse and thrilled with your discovery of mouse poop! Soon you come to a spot where the mouse tracks disappear altogether. You look for a mouse hole or nest but there is nothing. You closely examine the spot where the tracks end and notice that the final hind imprint is particularly deep. Maybe the mouse jumped, you speculate. Indeed, ten inches ahead you see tracks—a jumble of them. What happened? You are perplexed, but then you see some familiar zipper-like tracks entering from the right. "Beetle" you connect. "My mouse must have pounced on a beetle." The tracks suggest that the beetle and mouse had quite a scuffle. The mouse tracks continue on, but the beetle tracks come to an end. You had your picnic lunch here, and apparently the mouse did too.[19]

Stalking doesn't necessarily require that we crawl around on our hands and knees. Simply training your binoculars on a robin as it forages in your backyard is a form of stalking. Likewise, stopping to carefully examine some animal droppings is a form of stalking. The droppings of mammal species are distinct, and with a little practice you can learn to distinguish among them. By carefully studying animal scat you can "track backward." For example, if on a summer afternoon you encounter the seeds of wild grapes embedded in fresh bear droppings, you will know that earlier in the day that bear must have been feeding on wild grape vines, perhaps close by.

Fishing can also be a practice that allows us to employ stalking skills and enter into life's web. I experienced this recently when my friend Anders invited me to go fly-fishing with him on Pennsylvania's Little Juniata River. As a boy, I had dropped my hook, worm, and bobber into many ponds, but fly-fishing, I soon learned, was something altogether different.

When Anders looked at the Little Juniata, he was making connections and creating a story. For him, the river wasn't just water over rock, nor was it what I saw—distinct parts made up of swifts, pools, eddies, and riffles. Anders' river was alive. In his mind's eye, he seemed to see stonefly and caddis fly nymphs—wedged on the underbellies of river rocks—filtering wisps of food from the river. On the river's surface, he knew where the skittish water striders would be and noticed the mayflies rising from a pool into the misty morning air. Joining intuition with experience, Anders sensed where the brown trout were. It was as if he visualized himself as a big brown and then went fishing for himself. Engaging in this sort of mindful fishing can be a very deep practice.

CONCLUSION

When you really think about it, everything is connected—it's all connected.

—Everglades woman, 1970

Everything really is "connected" as that Everglades woman observed back in 1970. The monarch, the blue jay, the field mouse—all creatures large and small—are part of the one true worldwide web of life. We are in the web, and the web is in us: this is what our best science is revealing to us. This understanding also converges with wisdom from the Taoist, Hindu, Buddhist, Christian, Judaic, and Islamic traditions. This revelation that we are not separate from—but part of—life's web has the capacity to awaken our spirits. It bears repeating that awakenings do not occur simply by reading a chapter in a book. They are cultivated slowly through practices, such as those presented in this chapter, which allow us to fully experience our embeddedness in life's web.

NOTES

1. David Suzuki, *The Sacred Balance* (Amherst, N.Y.: Prometheus Books, 1998), 126.

2. The information on monarch butterfly life history comes mainly from Eric S. Grace, *The World of the Monarch Butterfly* (San Francisco: Sierra Club Books, 1997).

3. Robert M. Pyle, "Las Monarcas: Butterflies on Thin Ice," *Orion*, Spring 2001, 21.

4. Lincoln P. Brower, "Canary in the Cornfield: The Monarch and the Bt Corn Controversy," *Orion*, Spring 2001, 32–41.

5. Jeremy W. Hayward, *Letters to Vanessa* (Boston: Shambhala, 1997), 23.

6. Hayward, *Letters to Vanessa*, 74.

7. The information on goldenrod natural history comes mainly from Arthur E. Weis and Warren G. Abrahamson, "Just Lookin' for a Home," *Natural History* 107 (1998): 60–63.

8. I. Bossema, "Jays and Oaks: An Eco-ethological Study of a Symbiosis," *Behavior* 70 (1979): 1–117; C. Johnson and C. S. Adkisson, "Airlifting the Oaks," *Natural History*, October 1986, 41–47.

9. Lynn Margulis, *Symbiotic Planet* (New York: Basic Books, 1998), 122.

10. Margulis, *Symbiotic Planet*, 10.

11. Lewis Thomas quote is from Hayward, *Letters to Vanessa*, 102–103.

12. Judith Kohl and Herbert Kohl, *The View from the Oak* (New York: New York Press, 1997), 19.

13. Chet Raymo, *Natural Prayers* (St. Paul, Minn.: Hungry Mind Press, 1999), 127.

14. Raymo, *Natural Prayers*.

15. Daily's work is described in Andrew Beattie and Paul Ehrlich, *Wild Solutions* (New Haven, Conn.: Yale University Press, 2001), 41–48.

16. Beattie and Ehrlich, *Wild Solutions*, 41–48.

17. Paul Shepard, *The Others: How Animals Made Us Human* (Washington, D.C.: Island Press, 1996).

18. Shepard, *The Others*, 25.

19. The idea for this story came from Daniel Quinn, *The Story of B* (New York: Bantam Books, 1996).

4

NURTURING RELATIONSHIP: INTIMACY WITH LIFE

Wake up the Muggles!

Not long ago I saw "Wake up the Muggles!" scribed on a sticker. It was a reference to the "muggles" in the Harry Potter books. Muggles are people who go through life sleepwalking, often missing opportunities to experience mystery, awe, and intimacy. The first three chapters of this book have been an invitation to stand in awe before the universe, the solar system, and the web of life on planet Earth: from "discovery" (chapter 1) to "awareness" (chapter 2) to "understanding" (chapter 3), and, finally, in this chapter, to "relationship" with life.

Life's essence is relationship. We can enter into relationship with life by cultivating intimacy with our own bodies (foundation 4.1), with other species (foundation 4.2), and with our home places (foundation 4.3). The challenge for us, as the practices suggest, is to live life within the context of relationship.

FOUNDATION 4.1: INTIMACY WITH OUR OWN BODIES

It is difficult to know who we are if we don't know where we are. The first place "we are" is in our bodies. Hence, cultivating intimacy with life begins with nurturing intimacy with our bodies.

The Origins of Our Bodies

We have been taught that we evolved from earlier hominids. But our hominid form is a compilation of past forms. Consider:

> We have a head, two arms, two legs, a trunk. This form is common among all mammals. With the whales, the arms changed to fins, and the legs atrophied and became very, very tiny. But it is still the same form.
>   Go back to the reptiles and the amphibians. We are connected to all of them by this form. We don't realize how important this form is to us. It determines the kinds of tools we use, the sports we play, the furniture, buildings and automobiles we design. This form is an invention of the earth very early on—about 400 million yeas ago. We are connected with every creature since that time.[1]

If truth be told, our origins extend all the way back to the first appearance of life on Earth, almost four billion years ago. Our earliest ancestors were not hominoids or mammals or reptiles; they were bacteria. This might sound crazy, but think about it. How could it be otherwise! Life arose, evolved, and complexified until eventually the hominid form arose; but it all began with bacteria. We evolved from bacteria; we stand on their shoulders. This concept is a distressing thought to some people. After all, we have been taught to associate bacteria with germs, vermin, and pathogens (see box).

THE MAGNIFICENT BACTERIA

The basic division between life on Earth is not between plants and animals as many suspect. Rather it is between bacteria (organisms that have no nucleus in their cells) and all the other forms of life (i.e., nucleated life). Bacteria have a remarkable repertoire of capabilities; they can "swim like animals, photosynthesize like plants and cause decay like fungi."[2] It is bacteria that have given us all of life's important metabolic templates—photosynthesis, oxygen breathing, fermentation, fixation of nitrogen from the air—and they have demonstrated that it is possible to derive energy from plant fiber, animal waste, and inorganic molecules.

Bacteria have been maintaining a vast global network—exchanging genetic material, adapting, and innovating—for billions of years. Indeed, a legitimate answer to the question "What is life?" is "Bacteria!" As Lynn Margulis and Dorion Sagan remind us, "Any organism if not itself a bacterium, is then a descendent—one way or another—of a bacterium, or, more likely, mergers of several kinds of bacteria. Bacteria initially populated the planet and have never relinquished their hold."[3]

Bacteria are, in a sense, the primary operating system for all life, including the life that animates our bodies. When we look at a computer, we see the computer's body (the screen and the keyboard), often failing to see the computer's essence, its operating system. Similarly, when we look at ourselves, we often fail to see and appreciate our bacterial essence. Our cells still retain clear evidence of our bacterial origins. For example, the cells in all animals and plants contain tiny membrane-wrapped structures called "mitochondria." Mitochondria are the cells' "factories," where the food entering the cell is transformed to energy for metabolic processes. Your mitochondria give your finger muscles the energy to turn this page.

These mitochondria that populate my cells and your cells are remarkably similar to bacteria. They have their own DNA, which is very similar to the DNA of free-living bacteria. When they reproduce, they pinch and divide in two, just like regular bacteria; they also exchange genes as bacteria do. All of this and more has led scientists to conclude that between two and three billion years ago the ancestors of those little mitochondria in our cells were free-living bacteria. Just as mitochondria are believed to be bacterial in origin, so it is that chloroplasts, the locus of photosynthesis in plant cells, appear to have originated from bacteria.[4]

It is now generally (though not universally) accepted that the cells of modern plants and animals with their characteristic parts—nucleus, mitochondria, chloroplasts—came into being as a result of a series of bacterial "mergers." According to theory, one of the first mergers occurred when primitive anaerobic bacteria (unable to breathe oxygen) merged with aerobic bacterial forms (bacteria able to use oxygen). This union was akin to a "hostile takeover" at first, but the merger was ultimately successful because the two metabolisms were complementary. The mitochondria in our cells are the legacy of this merger. In other words, a veritable history of life on Earth is embedded in the cells of our bodies. Our fear of bacteria, to the extent that we experience this fear, is in a way a fear of ourselves at an earlier stage of evolution.[5]

Our Bodies Exist in a Medium of Air and Water

Air is our most fundamental environment. You and I are immersed in air as surely as the fish of the sea are immersed in water. Take a fish from the water, and it flops and dies; likewise, remove us from our native medium, the air, and we collapse. If you need reminding, you merely need to hold your breath and watch what happens. In a short time the blood vessels in your head begin to bulge, and your heart pounds as your body aches for oxygen.

Breathing is a mind-boggling activity. To see why, take a moment to bring your breathing into consciousness. Pay attention to the air as it enters your nostrils and then moves along the roof of your nasal passage, where it is humidified and warmed to body temperature. See if you detect any odor as the air arrives at your olfactory gland, the small patch of tissue rich in nerve endings that is hooked up to your brain. The air you have breathed in is chock-full of molecules, bringing you detailed information about your surroundings—sometimes stimulating your appetite, arousing you, warning you, and perhaps even triggering deep emotions and distant memories. The air goes into your throat, where it enters the trachea, which soon branches and sends "pipes" to each of your lungs. Within your lungs, the notion of boundaries, inside and outside, blurs as the planet's atmosphere enters your bloodstream on a massive scale. Once each minute, twenty-five billion red blood cells, each cell with 350 million hemoglobin molecules, flash through your lungs. Each of these hemoglobin molecules has spots to receive four molecules, which means that the average adult, with five quarts of blood, has 35×10^{18} (that's 35 followed by 18 zeros) spots for oxygen absorption and/or carbon dioxide release. So it is that with each breath your hemoglobin releases carbon dioxide, a by-product of cellular respiration, back to the atmosphere while simultaneously snatching up hundreds of billions of atoms of oxygen. And with each contraction of your heart, these oxygenated hemoglobin molecules are carried to the hundreds of billions of cells in your body.[6] "This is literally the pulse of life in us, the rhythm of the primordial sea internalized, the ebb and flow of matter and energy in our bodies."[7]

The atoms in our bodies only stay with us for a few years—a decade at most—before leaving and being replaced by new atoms from outside. With each inhalation, we take in atoms that were once worm, leaf, lake, and rock; and with each exhalation we give back part of ourselves to the earth system. We literally breathe out carbon atoms that were part of our heart, liver, kidneys, and muscles. Once free in the air, these atoms often find their way into the lungs of other creatures.

Just as air is the primary outside matrix for life, water is the primary inside matrix for the life. Biologist David Suzuki points out: "Water is the raw material of creation, the source of life. . . . Life originated in the oceans and the salty taste of our blood reminds us of our marine evolutionary birth."[8] Water is the medium for all human body chemistry; it is only in the presence of water that the cells in our bodies are able to absorb nutrients and secrete wastes.

I don't think I fully grasped my watery essence until I was lying in bed one evening during a spring rainstorm. As is often the case, I experienced an abiding sense of peace listening to the rain settle onto the land. Why was it, I wondered, that I am comforted by rain? And then, suddenly, I knew. It is because rain—water—is what we humans are! It is the medium that brought us into being. In the most intimate of human acts, spermatozoa from our father were set free to literally swim up a canal in our mother toward their target, an egg. The sperm joined with the egg, and we began to grow. Early on in this process, we briefly sprouted gills in a recapitulation of our aquatic origins. And for those first nine months we lived in a primeval salty sea of amniotic fluid.[9] And now, as adults, our bodies are still 60 to 70 percent salt water. Yes, we carry around some forty quarts of water in our trillions of cells:

> The water molecules that perfuse every part of our bodies have come from all the oceans of the world, evaporated from prairie grasslands and the canopies of all the world's great rain forests. Like air, water physically links us to Earth and to all other forms of life.[10]

The Wonders of the Human Body

Bringing our attention to routine daily acts, like breathing or drinking water, prepares us to appreciate the remarkable capabilities of the human body. I marvel when I learn that saliva (which I once regarded as mere spit) contains an enzyme called "alpha-amylase," which breaks down complex carbohydrates in my food; in addition, saliva also carries compounds such as immunoglobulin A, which defend my body against pathogen attack.[11] I react with amazement when I learn that the simple act of forming an image of my friend's face with my eyes involves tens of millions of neurons engaged in complicated feedback loops involving several distinct parts of my brain. I stand awestruck when I remember that my liver performs more than thirty thousand enzymatic reactions per second. Is it any wonder that the esteemed scientist Lewis Thomas observed that he would prefer to be given the controls of a 747 airplane, knowing nothing about how to fly, than to be put in charge of the functioning of his own liver.[12]

Taken as a whole, our bodies are a universe unto themselves, consisting of hundreds of billions of cells with extraordinary powers of self-regulation. We are the daily beneficiaries of this self-regulation. For example, consider what happens when you play a physically demanding sport. First, your heart beats faster to deliver more blood and oxygen to your muscles so that you can continue to play. Then, as your body heats up, you begin to sweat, which is how your body avoids overheating. Then, when you finally finish playing, you feel thirsty, which is your body's way of telling you to replace the fluids you have just lost. All of these responses are interconnected and self-regulating. Whether you are digesting your food, exercising, fighting infection, or snoozing, millions of self-regulating processes are occurring each second throughout your body.[13] And if you think the body's intricate, interconnecting mechanisms are phenomenal, then you will find that the capacities of the human mind are equally complex and astounding (see box).

THE MYSTERIOUS POWERS OF THE HUMAN MIND

Western scientists are now verifying what it seems that sages, shamans, and enlightened beings have known for a long time—that the body–mind complex has immense natural intelligence and power. One of the most remarkable illustrations of this connection comes from the Princeton Engineering Anomalies Research Laboratory, where two scientists, Robert Jahn and Brenda Dunne, have been studying the capacity of human beings to affect physical processes at a distance. In one carefully controlled series of experiments, these scientists used a simple machine to generate a random set of zeros and ones—like heads and tails. In a series of a thousand numbers, the machine randomly generates approximately five hundred ones and five hundred zeros; and the probability that the totals might differ by any amount other than an exact fifty–fifty (e.g., 490 ones vs. 510 zeros) can be calculated with precision. When left to run unattended, the machine prints out a completely random list of zeros and ones, just as one would expect. However, when the researchers invited human beings to sit in front of the machine and use their thought and intention to bias the machine's output toward more zeros or more ones, they found that in many cases the machine's output was no longer random. In other words, the intention of the human mind was able to affect the output of the machine.

> The Princeton work, as well as that of other laboratories, confirms that there is a very real effect here. The experiments have been reproduced by 68 different investigators in a total of 597 experimental studies. It has been estimated that when the results of all these studies are included, the odds against chance being the explanation for [the anomalous results] are 1 in 10^{35} (i.e., 1 followed by 35 zeros).[14]

How might it be that human beings can affect physical processes at a distance? The Princeton researchers speculate that the mind may have a particle function localized in the body, as well as a wavelike aspect; that is, it extends out from the body, spreading over time and space and interacting with other waves—for example, with other minds.

In sum, the human body is truly an embodiment of the cosmos: The most abundant atom in your body, hydrogen, is the most abundant element in the universe; the most abundant molecule in your body, water, is the most abundant molecule on Earth. The most fundamental unit in your body, the cell, is a summation of the entire evolutionary history of life on Earth.

Reflection

Let's take a moment to review what's been said about our human origins:

Chapter 1—We have come forth out of a dynamic and generative universe; elements in our bodies originated in the breakdown of early stars; we are, quite literally, stardust.

Chapter 2—We inhabit a living planet; our bodies participate in the earth's breathing, metabolism, and nutrient cycling.

Chapter 3—We are not separate and above life; instead, we are utterly dependent on Earth's web of life.

Chapter 4—Our lineage stretches back almost four billion years; we are an expression of the living body of Earth.

When I allow myself to fully absorb this new science-based knowledge, I discover that my biological essence is utterly different from what my culture has taught me. Contemporary culture and public schooling taught me not only to see myself as an independent, free-standing entity but also to see the world as composed of separate parts. This is a cosmology of difference, a worldview based on separateness.

Questions for Reflection

- What happens, now, as our cultural cosmology of *separateness* is challenged by the new revelations from the sciences—discoveries that reveal that life is grounded in *connection* (not separation)?
- How does this change the way that you look upon your own body? The bodies of other beings? The living body of Earth?

Practice: Relating to Our Bodies

Often, we don't fully accept our bodies. We think we are too fat, too hairy, too spastic, too ugly, or too old. Rather than judging our bodies, we can simply accept them, warts and all, as miraculous manifestations of Earth's generative powers.

To cultivate a sense of presence in our bodies requires practice. Strange as it sounds, we have to practice being in our bodies. In a sense, we have to "re-member" ourselves. A simple body-awakening practice is called the "body scan."[15] It takes about a half hour, and it involves bringing attention, sequentially, to each part of the body.

Start by spreading a blanket and pillow on the floor, lying down, and making yourself comfortable. Close your eyes, and let your palms face up (or down, whichever way feels best); then allow your feet to fall away as they would naturally. In other words, allow yourself simply to relax and sink into the floor. Next, bring your attention to your breath. Note the air coming in through your nostrils, filling your lungs; and then, as your diaphragm drops, feel your belly rising. Let go and relax on the out-breath, allowing your body to become heavy, sinking into the floor.

Continue to breathe in and out for a few minutes. With each inhalation, life-giving oxygen is entering your body and flowing to all of your cells to keep you alive. Simply experience your breath. Don't judge it: there is no right way; there is only your way, right now.

Consider that your breathing happens on its own; you are not breathing so much as you are being breathed. To grasp this concept more fully: with each exhalation, avoid anticipating your next breath; simply allow yourself to be surprised by each new breath as it arises. Indeed, you can't really stop breathing. If you hold your breath until you fall unconscious, your body will begin breathing again. The universe both breathes you and breathes through you.

When you are ready, bring your attention to the toes. Note any sensations in your toes—coldness, numbness, itching, warmth, tingling. Sense each toe individually. Feel your breath spreading from your lungs, down your legs to your toes; and, then, on the out-breath, feel it leaving the toes and returning to the lungs. Move your attention to the soles of your feet. The nerve endings here connect to every part of your body. Allow your awareness to sink into your feet, feeling their sensitivity and intelligence. Bring your attention to your calves, your knees, and your thighs. As you focus your attention on each of these places, feel each place individually; breathe into and out of each place. Don't think. Just feel whatever is occurring without judging it or identifying with it.

Next, bring your attention to the entire area of your pelvis—buttocks, anal sphincter, genital area, bones of the pelvic girdle—and feel this area, accepting your feelings without judgment; breathe into and out of this region; and, finally, let go and sink deeper into the floor. Continue to feel, focus, and breathe into the lower back, spine—feel the subtle differences from vertebra to vertebra—abdomen, chest, and shoulder regions of your body. As you proceed, place attention on your insides as well as your outsides. What do your insides feel like? Solid? Watery? Are they buzzing? Numb? Allow yourself to be aware of all the sensations—tingling, pain, tightness, softness—at each place, breathing with each sensation and then letting go. From the shoulders, go to your hands, pausing there to feel life buzzing in your fingertips. Continue on to your palms, wrists, forearms, elbows, and biceps. Then, move to the neck and throat, taking time to note subtle differences— coolness, warmth, dryness—in each place. Then, go sequentially on to the jaw, mouth, ears, nose, eyes, forehead, back of the head, noting the tensions that you carry in your face. Bring your focus to the brain as well, sensing its weight and density. Take your time, moving attention to your lips, the inside of your mouth, each tooth, breathing into and out of each place, feeling the sensations—heat, pressure, numbness—and then letting go.

Finally, move your attention to the top of your head and imagine that you have a hole there about the size of a silver dollar (e.g., like the blowhole of a whale) and that you have a unique connection between this hole and your nose. As you breathe in through this hole, allow the air to pass to your nose and then down to your lungs. Once you can visualize this image, let the connection to your nose dissolve away so that when you breathe in through the top of your head, the air flows freely down through your body. On the outbreath allow the air to continue to flow down your body and out through your toes. Then, the next time you breathe in allow the breath to enter through your toes and flow upward through your entire body, escaping through the top of your head. Do this back and forth, allowing breath and energy to flow through your body, experiencing your entire body's breathing.

At the end of a "body scan," one often feels more awake and whole. Physician Jon Kabat-Zinn observes that "each time you scan your body in this way, you can think of it or visualize it as a purification or detoxification process, a process that is promoting healing by restoring a feeling of wholeness and integrity to your body."[16]

As I have experimented with this practice, I have found that it is possible to do miniscans, or what I think of as body "check-ins" at any time during the day. After a particularly difficult encounter, I might note a tightness in my lower back; then I may take a few moments to relax and bring attention to the body sensations in this region, breathing in and out, accepting, and letting go.

Physical touch also has great power to induce comfort and bodily wellbeing. Many of us forget that we can befriend our bodies by simply touching them with caring attention. A beginning practice in this vein is to behold one's hands with awe and tenderness. Look at your hands now as if you are seeing them for the first time. Then, caress your left hand with your right hand exploring and discovering, perhaps for the first time, its wondrous nature. Take your time with this exercise. When you are ready, switch your attention to your right hand. Be open to new revelations as you explore the wonders of this hand that has been so much a part of you and the world.

All of our body parts—neck, arms, abdomen, head, legs, feet—thrive on touch and attention. The same is true for our internal organs, and there are certain practices to extend awareness and care to these organs as well. For example, you might accompany your breath by saying, "Breathing in, I am aware of my (kidneys, heart, ovaries, stomach, etc.); breathing out, I smile to my (kidneys, heart, etc.)."[17] By breathing in this way, we extend love to ourselves—not in a narcissistic, self-indulgent way, but in a compassionate, self-caring way. If we can't love and care for our own bodies, it is unlikely that we will be able to extend these capacities out into the world.

On those occasions when we are in pain, we have a special opportunity to listen to and learn from our bodies. The patterned response to pain in modern culture is to run to the medicine cabinet. I can speak from personal experience. For many years, whenever I got a headache, I would pop aspirin. I was, in effect, treating symptoms, not the underlying problems that caused my headaches. As Kabat-Zinn points out:

> This practice of immediately going for a drug to relieve a symptom reflects a widespread attitude that symptoms are inconvenient, useless threats to our ability to live life the way we want to live it and that they should be suppressed or eliminated whenever possible. The problem with this attitude is that what we call symptoms are often the body's way of telling us that something is out of balance. They are feedback about disregulation. If we ignore these messages, or, worse, suppress them, it may only lead to more severe symptoms and more serious problems later on. What is more, the person doing this is not learning how to listen to and trust his or her body.[18]

When I was popping those aspirin, it never occurred to me to see my headaches as messengers with lessons to teach me. Now, when I get a headache, I ask myself, "What is this headache telling me about my body and mind in this moment?" I "check in" with my overall emotional state: Am I depressed, angry, fearful, tense? I also reflect on what I have been doing over the previous hours or what I am preparing to do that might be causing distress.

By paying closer attention to my body, I am now usually able to detect the very first signs of a headache. Rather than drowning the impending headache with drugs, I now try to do something different. I stop and acknowledge that my head is beginning to ache. I try, with varying degrees of success, not to resist the pain. Instead, I explore it, locate it, and feel its nuances. I note that its character changes from moment to moment; and, on a good day, I breathe, relaxing around pain. I know that worrying about and resisting the headache will simply bring it on more powerfully by constricting the blood vessels in my head, which will in turn impede blood flow and trigger the pounding pressure of a full-blown headache. I still sometimes resort to aspirin; but whatever happens, I try to see my headache as a process that has come forth (with a lesson to teach) and that will soon pass away. In sum, cultivating practices that bring skillful attention to our bodies expands ecological consciousness.

FOUNDATION 4.2: INTIMACY WITH THE EARTH'S BIOTA

Every now and then most of us have the wonderful experience of seeing something "ordinary" with fresh eyes. A minute ago, you saw just a bush on the corner, but now you see a bush on fire with life. Or, perhaps, you saw an ant crawling over your toe, which normally you would have brushed away; but now that ant suddenly holds your full attention. It is not that the bush or the ant changed; rather, you changed the way you saw them. In effect, you saw them more fully.

We Are Similar to Trees in Many Ways

When it comes to thinking about other species, we are good at noting the differences between us and them; but we are less inclined, it seems, to note similarities and relationships. For example, if I were to ask you to think about the ways that trees and humans are related, your mind would probably first race to all the differences: We can walk, and they can't; we have a brain, and they don't; we can talk, and they can't; and so forth. Only with some prodding might you recall that humans have shared an intimate relationship with trees throughout our evolutionary history. Our primate ancestors lived in trees; our more recent ancestors built whole civilizations from trees; and today the forests of the world continue to regulate water flow, supply forest products, and ameliorate climate.

But there is much more to our relationship with trees, and to plants in general, than mere utilitarianism. Indeed, we share a great deal in common with plants. Consider the likenesses:

> *Elemental composition.* No part of gross human anatomy looks anything like a plant leaf; yet, when we compare the ratios of nitrogen, carbon,

phosphorus, and other vital elements in leaves to the ratios in the human body, the two are remarkably similar.[19] So it is that the greens on our plate offer us something quite close to what our bodies need.

Biochemistry/genetics. Tree tissues, like our tissues, are composed of carbohydrates, fats, and proteins. Plants use the same amino acid building blocks to construct proteins that we use. They employ very similar, sometimes identical, biochemical pathways for the synthesis and breakdown of compounds in their cells. And like humans, their cells contain a nucleus with DNA packaged in chromosomes. Also, just like us, they have an array of hormones that act as messengers, turning cellular processes on and off. [20]

Sex. It is not just humans and other animals that have sex; trees have sex, too. Flowers are the genitalia of trees and plants in general. The minuscule pollen grain is analogous to the entire male reproductive tract. When pollen arrives to a flower (transported on the wind or carried by an insect), it extends a penislike tube down into the flower's ovary; then sperm descends this tube and fertilizes one of the "eggs" in the flower's ovary.

Population expansion. Successful mating in humans and trees leads to the production of offspring (seeds, in the case of trees). Although trees are stationary during their adult lives, their seeds can be carried considerable distances by wind, water, or animals. So it is that trees, like humans, are effective at invading new territories.

Defenses. Trees, like humans, have defenses to protect themselves. Without some sort of defense system, most plants would be eaten to death. A solution is to produce special chemicals to deter herbivores. Some of these plant compounds are familiar to us as drugs (caffeine, nicotine), poisons (cyanide), or insecticides (rotenone). Many of them simply taste bad; some interfere with digestion; others cause reproductive abnormalities, paralysis, or blindness. All send the message: "Stay away!"

Communication. The investigation of plant defenses has revealed another remarkable similarity between trees and humans. Namely, that plants are able to communicate with each other (see box).[21]

Not only can plants communicate with one another, some species can communicate with insects. When these plants are being defoliated by caterpillars, they release a compound that is detected by parasitic wasps. The wasps follow the scent trail back to the plant, find the caterpillar feeding on the plant's leaves, and inject their eggs into it. The wasp larvae then feed on and kill the caterpillars and in the process save the plant from complete defoliation.

As we come to recognize the many commonalties between ourselves and trees, we slowly build the scaffolding for a relationship with the plant world. Suddenly a tree is much more than an isolated, brown and green object sticking out of the ground. The air that moves in and out of the tree, the water

"TALKING" TREES

In the early 1980s, David Rhoades, an ecologist at Princeton, was studying tent caterpillars when he made an important observation: Tent caterpillars grew very slowly when they were fed leaves from intact willow trees that were growing in close proximity to heavily defoliated willows. Based on this, as well as other observations, Rhoades speculated that the defoliated willows were emitting some signal that induced the nearby undamaged willows to produce substances in their leaves that would deter insect defoliators. It was as if the damaged trees were saying: "Hey neighbor, I am getting eaten. If you want to survive, you better make some toxins quick." Another ecologist, Jack Schultz, decided to test Rhoades' idea with a greenhouse experiment using three sets of sugar maple seedlings and two experimental chambers. Schultz placed the first two sets of maple seedlings in one chamber. One set was subjected to mild defoliation. Specifically, two of the twenty leaves of each maple seedling were torn in half; the second set of seedlings was left untouched, but it was in the same air space as the defoliated seedlings. The third group of seedlings was also left untouched and placed in the second chamber with its own air supply.

After cutting the leaves in the first set of seedlings, Schultz went to the greenhouse every few hours and removed several seedlings from each of the three "treatment" groups to monitor changes in leaf chemistry. As expected, there were no changes in the leaf chemistry of the "control" seedlings placed in the separate chamber over the seventy-two-hour experimental period. The story was different for the seedlings with the torn leaves. Chemical tests on the undamaged leaves of these seedlings revealed increases in tannins and phenols—compounds that render leaves less palatable to insect defoliators. Most interesting of all, Schultz also detected significant increases in concentrations of phenolic compounds in the set of seedlings in the first chamber that was not harmed at all but which shared air space with the damaged set of seedlings.

Overall, this experiment revealed that sugar maple trees are somehow able to communicate, presumably using airborne chemical signals. The capacity to communicate in this way is thought to be beneficial to trees because it means that they don't have to expend a lot of energy making defense compounds when the danger of defoliation is low.

coursing through its vessels, the sunlight penetrating its leaf chloroplasts, the insects fertilizing its flowers, and the mycorhizal fungi shunting nutrients to its roots are all integral parts of the tree. The tree is not separate, but part of a larger whole.

The Cloud in the Sheet of Paper

As we learn more about our common bonds with trees, and plants in general, we may come to regard paper, a derivative of trees, with new eyes. The Vietnamese monk Thich Nhat Hanh writes:

If you are a poet, you will see clearly that there is a cloud floating in this sheet of paper. Without a cloud, there will be no rain; without rain, the trees cannot grow; and without trees, we cannot make paper. . . . If we look into this sheet of paper even more deeply, we can see the sunshine in it. Without sunshine, the forest cannot grow. . . . And if we continue to look, we can see the logger who cut the tree and brought it to the mill to be transformed into paper. And we see wheat. We know that the logger cannot exist without his daily bread, and therefore the wheat that became his bread is also in this sheet of paper. The logger's father and mother are in it too. When we look in this way, we see that without all of these things, this sheet of paper cannot exist.[22]

After reading this, I picked up a piece of paper and looked at it with a hand lens. I was then able to see the wood fibers that make up the paper. Plant anatomists call these fibers "tracheids." Tracheids are large, as cells go, often a quarter-inch or more in length. At maturity the tracheid's cytoplasm and nucleus degenerate, leaving a hollow tube. These tubes, lined end to end, serve as the passageways for water and nutrients within tree trunks. So, the millions of fibers (tracheids) making up this page were once a tree's water pipes—canals that channeled life-giving water, from the soil where the tree's roots absorbed it, all the way to the tree's leafy canopy.

In any given moment a tree might be silently engaged in the construction of tens of thousands of tracheid cells. Making a tracheid is not a simple task insofar as a single tracheid contains two trillion glucose molecules. The construction process takes approximately thirty days, which means that approximately 770,000 glucose molecules are produced per second as a tree assembles a single tracheid cell (2×10^{12} glucose molecules/2,592,000 seconds in 30 days = 771,600 glucose molecules/second). There is more: Six carbon dioxide molecules are required to make one glucose molecule. Hence, 4.6 million carbon dioxide molecules must be taken up from the atmosphere each second by a tree in the building of just one growing tracheid cell (770,000 glucose units × 6 carbon atoms = 4.6 million carbon dioxide molecules/second). Biologist Bernd Heinrich offers a context for understanding the scale of this carbon dioxide uptake:

> The carbon dioxide is taken up from the air, masses of which sweep across continents in a matter of days. Since these air masses mix together as in a giant blender, it stands to reason that those 4.6 million carbon dioxide molecules taken up by just one tracheid cell in just one second could have come from a decaying log in the Amazon, a car on a Los Angeles freeway, a coal-burning power plant in Utah, a hornbill in Indonesia, and a baboon in Tanzania. If we could put a pinhead-size red dot on a map of the world to indicate the source of each of those 4.6 million molecules [taken up second-by-second over the last month] by one growing tracheid cell, then the whole map from pole to pole would be colored solid red.[23]

Seen in this way, the page that these words are printed on contains the thumbprint of the planet.

As we gain familiarity with trees, the notion that we humans are somehow a "higher," or superior, form of life becomes less and less tenable. Certainly, if we define "higher" in terms of cognitive intelligence, there is no contest. We outshine trees by a mile. But trees outshine us in many respects—most significant, by their capacity to animate the world by making complex carbohydrates from water, air, and solar photons. No capacity of ours comes anywhere close to this in terms of its significance for the unfolding of life on planet Earth.

Reflection

Over the years, I have come to see that how I perceive the world and my place in it determines, to a significant degree, my actions. For example, if a tree is simply a resource—a commodity to be harvested—then I will treat it shabbily. If, on the other hand, I perceive the tree as a fellow being with its own unique life force and integrity, I will be more inclined to treat it with the same respect that I expect to receive from others.

These days I am drawn to trees and feel an inclination to lean against them, touch them, and even embrace them. I realize that this practice isn't "weirdness," but rather an awakening to my "wiredness" within life's web. As I consider those two words, "weird" and "wired," I note that it is only the shifting of the location of the letter "e" that creates new meaning. Through a shifting and slight deepening of consciousness, I have come to see tree hugging as an expression of connection—a manifestation of "wiredness," and not as something "weird." In this, I take solace in the words of English professor Scott Russell Sanders:

> I confess that I do hug trees in my backyard and any place else where I happen to meet impressive ones. I hum beside creeks, hoot back at owls, lick rocks, smell flowers, rub my hands over the grain in wood. I'm well aware that such behavior makes me seem weird in the eyes of people who've become disconnected from the earth. But in the long evolutionary perspective, they're the anomaly. Our bodies were made for this glorious planet, tuned to its every sound and shape.[24]

Questions for Reflection

- What would happen if you allowed yourself to see and focus on the similarities, instead of the differences, between yourself and the trees?
- How might it be for you if you allowed yourself to hug a tree in the dead of night?

Practice: Relating to the Earth's Biota

We "receive" the world through our senses. Our bodies are continually bombarded with electromagnetic waves (light), vibrations (sound), gases (odors), and infrared radiation (heat). Our sense organs and neural processing centers translate this "information" into colors, sounds, objects, and sensations. Oftentimes, though, our heads interfere with our ability to directly receive Earth's signals. For example, you may look at a sunflower and think, "There is a sunflower," all the while failing to truly see and experience the reality of the sunflower. Instead, you might only be "seeing" a memory of "sunflower" and not the real thing before you. But suppose you really take the sunflower in. In this case, you see the honeybees foraging on the flower head, the aphids affixed to the stem, the caterpillars munching on the leaves, and the cluster of shiny insect eggs at the base of the stem. And you bring your other senses into play: gently moving your fingers back and forth over the fine hairs that adorn the stem; placing a petal on your tongue to experience its taste; listening to the rasping as the wind rattles the plant's leaves; breathing in the subtle fragrances of the flower head, aware that molecules from the sunflower are merging with your olfactory cells—that is, you are literally taking the sunflower into your body.

Paying attention to what attracts us and then exploring those attractions cultivates intimacy. Here is a simple practice: The next time you are in a natural setting, look around; pay attention to the things that you find pleasing. Then, when you are ready, choose something that attracts you. Walk over to it. Ask permission to spend time with this "being" that you have chosen. This exercise may seem strange, but think about how you would feel if someone came barging into your house, raided your refrigerator, and generally swaggered around as if he owned the place. Contrast this to how you would feel if the person politely knocked and asked if he might visit with you. Asking permission, then, is a way of extending to beings in the natural world the same respect that we expect others to extend to us.

Most people who engage in this practice sense a clear response when they ask permission: either a "yes," a feeling of safety and spaciousness; or a "no," a feeling of constriction or imbalance. If you sense a "no," simply move on and choose another being; perhaps explore possible reasons for the "no" at a later time.

Once you have received a sense of permission, explore what it is that attracts you to this "being." For example, if it is an active ant mound, sit by it and reflect on the connections, or "web" strings, between you and the ant colony. This is a time to look both outward and inward. When you are ready to end your encounter, take a moment to simply express gratitude to the being you have spent time with.[25]

When I first did this practice, I was immediately attracted to a leaf on a nearby tree branch. I noted that this particular leaf was a bit tattered, but still vital. "Hmm, that's kind of how I see myself," I thought. Next, I decided to do something I had not done since childhood. I explored the leaf with each of my senses. The leaf's touch, smell, and taste—I was careful just to lick the leaf and not chow down!—were all different from what I imagined they would be. Then, I reflected on how the leaf serves as channel for life by receiving and transforming photons from the Sun, water from the earth, and carbon dioxide from the air. It occurred to me that, in writing this book, I have been engaged in a similar process—using the scientific literature, the writings of social thinkers, and the body of my personal experience as raw materials as I endeavor to serve as a "channel for life" (see box).

As our intimacy with the other-than-human world deepens, we open ourselves to inspiration. Wild things can serve as our teachers:

[We] learn about flexibility from trees, about the path of least resistance from water, about the twinship of life and death from volcanoes, about receiving help from birds riding the thermals like surfers. [We] learn humility from fossils and from the distances that stretch between the stars. [We] learn something about call-and-response from thunder, which answers only to lightning, and from cobbles at the ocean's edge, which won't clatter until rushed at by a wave.[27]

NEVER TOO LATE

It is never too late to cultivate intimacy with the plants and animals that live around us. Take the case of Steve Talbott. At age fifty-one, Talbott wanted to do more than identify the birds in his backyard. He wanted "to make contact with them," so he sat by a patch of shrubs in his backyard offering seeds to the birds, acting as a sort of "human feeding station." In a few days he had chickadees, juncos, titmice, and nuthatches feeding from his hands. Talbott is quick to point out that he had no special skills and that his approach was utterly mechanical. But he was, nonetheless, deeply moved by his experience: "All this has been an epiphany for me. . . . I suggest that a bird in hand—and a pine cone, and a rock, and a crawdad, and a snowflake—are the counterbalance we need if our alienation from nature is not to become more than the world can bear."[26]

In sum, to become intimate with the earth's biota, you may find it helpful to cultivate a spirit of innocence, openness, and imagination—what Buddhists call "beginner's mind." Rather than seeing the bush, the ant, the bluebird, and the red spruce through the veil of your fixed thoughts and opinions, endeavor to see them with an uncluttered mind and fresh eyes. Then—who knows!—there may even come a day when your experience of the earth is literally the earth experiencing itself.

FOUNDATION 4.3: INTIMACY WITH OUR HOME PLACES

When someone asks where it is that we live, most of us respond by giving our "indoor" codes—our street address, phone number, fax number, e-mail address, and so forth. What we seldom describe, in any meaningful way, are the physical and ecological characteristics of the places where we live: the lay of the land, the geological features, the types of trees, the varieties of insects, the special smells and seasonal sounds of our home places. In his book *The Hidden Heart of the Cosmos*, Brian Swimme provides a "local universe test" to gauge our connection to our places:

> It's easy to do. You simply invite someone to visit you who lives at least twenty miles away and who has never visited you before. You can give verbal instructions on how to get to your abode over the telephone, but the one rule is this: In your directions you may refer to anything but human artifice. You may refer to hills, oak trees, the constellations of the night sky, the lakes or ocean shores

or caves, the positions of the planets or any ponds, trails, or prairies, the Sun and Moon, cliffs, plateaus, waterfalls, hillocks, estuaries, bluffs, woodlands, inlets, forests, creeks, swamps, bayous, groves, and so on. Whenever your friend gets stuck, she is free to phone you for more directions, but the rule for her is that she must describe her location without referring to any human artifice.[28]

Swimme's "local universe test" coaxes us to cultivate greater intimacy with our home places. Ironically, the so-called homeless in American society would probably perform better on Swimme's "local universe test" than the affluent. Naturalist Jesse Wolf Hardin explains why:

Ask [the homeless] where they live, and instead of rattling off a numbered address they're more likely to respond with a litany of landmarks: next to the river past the ruins of the old brick factory, under the old fir tree behind St. Martin's, a five-minute walk from the tracks. They know what hour each bakery throws out its unsold bread, what days the supermarkets rotate their milk, how often the police patrol a particular street, what air ducts supply heat in the winter, and where the overhangs provide the most relief during the hottest months. With no lawn of their own, they become familiar with the flowers and layout of everyone else's, and can often recite names of the local children they come into contact with. They can, and often do, find their way home to their camp in the dark. They usually have a deeply realized sense of where they "belong," and exhibit a profound intimacy with their local environs. Often frequenting the same neighborhood for decades on end, some can recount the succession of families moving in and out of any particular house, identifying each by the way they took care of their yards, the style of their cars, the attitude of their dog, or a habit such as what time a renter would always come out to pick up his newspaper. Judging by their relationship to place, they may be without houses, but they are hardly "homeless."[29]

For many of us, the outdoor places that we most often come in contact with are the planted lawns gracing parks and surrounding our homes, schools, churches, and town centers. This community green space is an important entry point for cultivating intimacy with our home places.

The Industrial Lawn

The ideal lawn in America, as defined by the lawn-care industry, is uniform (weed-free), deep green, and lush; that is, the ideal lawn is the result of pesticides, frequent watering, and frequent fertilizer applications. It is not wild, but controlled; not complex, but simple. Frank Bormann, professor emeritus at Yale University, refers to this ideal lawn as the "industrial lawn."[30]

The industrial lawn is very different from a natural ecosystem. For example, in a wild grassland (figure 4.1), the only inputs are natural—solar energy, rainfall, rock-derived minerals—and nutrients cycle efficiently with very little leakage out of the system. Now, consider what happens when a diverse natural ecosystem, like this wild grassland, is transformed into a uniform, tame industrial lawn (figure 4.1). In this case, artificial inputs—fossil fuels, irrigation water, pesticides—are added on top of the natural inputs. These extra inputs are necessary to force the ecosystem into the new "in-

NATURALLY OCCURRING GRASSLAND VS. INDUSTRIAL LAWN

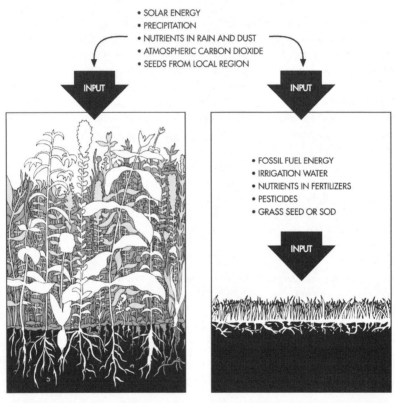

- SOLAR ENERGY
- PRECIPITATION
- NUTRIENTS IN RAIN AND DUST
- ATMOSPHERIC CARBON DIOXIDE
- SEEDS FROM LOCAL REGION

INPUT   INPUT

- FOSSIL FUEL ENERGY
- IRRIGATION WATER
- NUTRIENTS IN FERTILIZERS
- PESTICIDES
- GRASS SEED OR SOD

INPUT

- PROVIDES HOME FOR MANY SPECIES OF PLANTS, ANIMALS, AND MICROBES
- REDUCES GLOBAL CLIMATE CHANGE THROUGH NET REMOVAL OF CARBON DIOXIDE FROM THE ATMOSPHERE
- RETAINS NUTRIENTS; TIGHT NUTRIENT CYCLES
- FREE OF SYNTHETIC CHEMICALS

- LOCAL PLANT SPECIES DISPLACED BY TURF GRASS AND TURF-ADAPTED ANIMALS AND MICROBES
- CONTRIBUTES TO CLIMATE CHANGE BECAUSE OF FOSSIL FUEL DEPENDENCE
- NUTRIENTS LESS EFFECTIVELY RETAINED; LEAKY
- RELIES ON PESTICIDES THAT CAN CONTAMINATE FOOD WEB AND AFFECT HUMAN HEALTH

Figure 4.1.   A naturally occurring grassland versus an "industrial lawn"[32]

dustrial lawn" configuration. But artificial inputs inevitably lead to artificial outputs—leakage of nutrients and pesticides from the lawn ecosystem— and these outputs have ecological consequences, such as loss of biodiversity and contamination of water and soil.

The concept of the uniform, weed-free, lush lawn was created by the lawn-care industry; and it has turned out to be an illusive goal. For example, in a study of eight hundred lawn owners in Virginia, 80 percent were not satisfied with their lawn's present condition. This dissatisfaction creates the stage for an infinitely expanding array of lawn-care products and accessories. Nowadays, the average homeowner spends three to four hundred dollars per year on lawn care—half for maintenance and half for equipment purchase.[31]

The Freedom Lawn

Homeowners have an alternative to the industrial lawn. Bormann calls it the "freedom lawn," a low-energy, low-maintenance lawn composed of a diverse mix of stress-tolerant grass and herbaceous species.[33]

At the risk of self-indulgence, I will use the story of my old yard to illustrate the difference between the norm (industrial lawn) and the alternative (freedom lawn). The story begins in the early 1990s when Beto, a friend from Latin America, visited my home and asked to know the name of the beautiful yellow flower (dandelion) in bloom all over my front yard. He was right; dandelions in bloom are beautiful.

Just for fun, I decided to see what else was blooming in the yard. There were violets and chickweed in the shade under the pear tree; a large patch of veronica and gill-over-the-ground in a slight depression under the clothesline; some daisy fleabane and plantain on the front slope going down to the sidewalk; patches of Canada thistle, lamb's-quarter, and amaranthus in the back corner by the compost pile; and a variety of clovers and grasses everywhere else. Yes, it was a kind of freedom lawn.

The "weeds" infesting my lawn were wild beauties, each with stories to tell. Many were exotics from Europe. Some, such as dandelion and white clover, were surely brought intentionally to serve as salad greens or as forages for introduced livestock. Other "weeds," such as amaranthus and lamb's-quarter, were natives that had served as food for Native Americans, long before the arrival of European settlers.

The careful observation of any of these plant species reveals an impressive array of adaptations that help ensure growth and reproduction in the presence of humans. The dandelion is a case in point. After forming a rosette and basal flower bud, it pushes its stalk six to twelve inches up into the air and opens its yellow inflorescence in a matter of just a few days. After pollination, the flower closes and the stalk bends down, nearly prostrate (i.e., out of the way of any whacking mowers). As the seeds mature, the stalk straightens again, and the seeds are lifted away by the wind. Observing these things closely can bring us into closer relationship with the plants growing around our homes.

With each passing year, a freedom lawn—powered by solar energy, irrigated by natural rainfall, and fertilized by the natural cycling of nutrients—becomes wilder. Of course, every now and then a bare patch may develop, but it won't be bare for long. Seeds lying dormant in the soil will quickly germinate, and new plants will take over the patch (see box on page 113).

I mowed my yard from time to time, but I opted for a push mower over a power mower. I still remember the day that I made the switch. It was mid-July. The push mower whirred softly through the grass and herbs. It stopped when I stopped. Neighbors walking by paused to talk. One even asked if she might borrow my mower. In the absence of power-mower exhaust fumes, I was able to smell the fragrances of cut clover, seasoned on occasion with the aroma of mint. As I moved along, I could sense the different plant textures, some soft, some fibrous, that the mower blade was cutting through. I could also feel the little ups and downs, the microtopography, of the land.

When I finished mowing, I'll admit I was more tired than I would have been had I used a power mower; and the lawn was not cropped close and uniform. Rather, it had a sort of roguish aspect. Some dandelion flower stalks and ryegrass seed stalks stood up here and there, too slippery or tough to be sliced by the mower's blade. Meanwhile, some previously unnoticed ground-hugging flowers had become visible.

When we give the land at our doorsteps the freedom to develop its own character, we, in a sense, give ourselves freedom. Yes, the land at our doorsteps provides the starting point for developing an affection for and intimacy with the earth.

Reflection

From time to time, I forage for wild foods in my neighborhood. In the process, I have learned that the dandelions that grace many yards in America offer a smorgasbord of edibles. The roots can be roasted and ground as a coffee substitute, and the young leaves make a welcome addition to spring salads, along with wild chickweed and lamb's-quarter. Furthermore, dande-

---

THE SPREAD OF THE FREEDOM LAWN

I am not alone in my experimentation with the freedom lawn. Many people throughout the United States are showing how citizens can reclaim their connection to the land at their doorsteps.[34]

- Laurie Otto of Bayside, Wisconsin, decided to allow a corner of her yard revert to its wild state so that neighborhood children would have a place to pitch tents and hang their hammocks.
- The Stewarts of Potomac, Maryland, simply decided to stop cutting their lawn and let it turn into a meadow.
- The Blums of Athens, Georgia, have gradually transformed the land around their suburban home into a woodland bird sanctuary.

Whole communities are also beginning to see the wisdom of freedom lawns. For example, Millford, Connecticut, now gives an annual award for the best freedom lawn. At the outset, they only had ten entries. These days they have more than fifty.[35]

---

lion blossoms can be used to make wine. A while back I set about to pick a gallon of the blossoms (not a very difficult task where I live). Afterward, I poured boiling water over the blossoms and allowed them to soak for a few days. Then, I separated the liquid from the blossoms and added sugar, citrus, and yeast. When winter rolled around, this sweet, smooth front-yard wine was ready for sipping.

Another favorite wild food of mine is violets; their blossoms make a beautiful addition to salads, and their leaves are nutritious as well. As I was gathering violet blossoms one evening, an Asian couple walked by. They asked what I was doing; and as I explained, I noticed that the woman was holding some onion grass that she had apparently foraged in the neighborhood. She was pleased to tell me that she added finely chopped onion grass to her eggs. Before continuing on, she kindly offered me some of the onion grass, and I offered that she pluck some violet flowers from my yard as a garnish for her eggs.

Questions for Reflection?

- What democratic ideals, if any, might be violated by the "industrial lawn"?
- If the "industrial lawn" is a manifestation of culture, what does the "freedom lawn" portend?

Practice: Relating to Our Home Places

The place we call "home" is also home for many other beings. In fact, humans represent well less than 1 percent of what scientists term Earth's "biomass"—the weight of living things on Earth. One way to begin practicing intimacy with our home places is to "learn the language" of the other-than-human beings that we live side by side with. For example, by carefully attending to the sounds and songs of the birds in your neighborhood, you can gradually learn to recognize the various bird species that live in your home place. With further practice, you can learn to discern the time of day and season of the year based on which species of birds are singing.

Because many of us live essentially "indoor" lives, it is helpful to have physical reminders that tether us to the natural world of wind and sun, bone and beetle, rain and rock. I was first awakened to the power that natural objects can hold when Trish Shanley, an ecologist colleague, gave me a beautiful palm seed. It was dark, smooth, and hard, the size of a bird's egg. Trish suggested that I keep it in my pocket, adding, "When you are having a difficult moment in life, touch the seed and ground yourself." She referred to the seed as a "touch stone." Part of me, my head part, thought that this was pretty silly; but another part, my heart, was moved by what she said. I believe that all our hearts are yearning for touchstones—aching to feel and be connected with that which has meaning for us.

The natural objects that I have collected over the years mark moments when my relationship with the earth has deepened. I don't search out these objects; rather, they seem to call to me. For example, one day I was sitting in the woods near my home, and my eye was drawn to a small, fist-sized rock. Something about the rock attracted me, so I picked it up. Why am I picking up this dead, inert material? I wondered. What is this subtle attraction? Then, as I handled the rock, I recalled being asked by my older sister many years earlier whether I thought rocks were alive. Strange question, I had thought at the time. I knew that my sister wasn't naïve. What was she getting at? Her question, like a thorn lodged deeply in the flesh, had stayed with me.

Now, with rock in hand, perhaps the time had come to remove that thorn. According to my biology texts, if the rock didn't reproduce or respond to stimuli, it was dead. But maybe that definition was restricting my thinking. Examining the rock closely, I saw lichens, an interplay of fungi and algae, growing on its surface. Lichens release organic acids that slowly dissolve rock; the minerals released in this process become part of the lichen: that is, rock becomes lichen. As I munched on a berry from a wintergreen plant growing near where my rock had lain, it occurred to me that some of the minerals in the berry were probably, not long ago, part of the rock that I was holding in my hand. Rock becomes berry, and berry becomes me.

Sitting with this rock, I was reminded that life isn't an either–or proposition. It knows no clear boundaries. Like it or not, we are all caught up in a symphony of relationships and mergings. Because I wanted to hold onto this understanding, I placed the small rock in my backpack and carried it home. In addition to physical markers, temporal markers help to anchor us to home (see box).

As we spend time in a particular place on Earth and come to call it home, we imbue that place with meaning. It might even become a kind of "sacred" place for us. When such places are desecrated, we might actually feel like we have been personally violated. When I ask my students to describe the wild nooks of their childhood—the spreading oak, the little patch of woods,

---

A DAY TO COME HOME

In a world grown highly secular, the Sabbath, or Shabbat, has unfortunately become, for many, the day to do all the things that couldn't be squeezed into the work week—washing clothes, shopping, paying bills, and so on. The faster we move, the more we distance ourselves from our soul's deep yearnings. But there is an antidote. As Rabbi Michael Lerner points out, "You don't have to think of yourself as religious or a believer in God to get the benefits of the Bible's most brilliant spiritual practice"—Shabbat, or the Sabbath.[36] The practice consists of taking a full day each week to celebrate the wonder of being, which means removing yourself completely from worldly concerns. Lerner offers a list of things to avoid:

- Don't use or even touch money.
- Don't work or even think about work.
- Don't cook, clean, sew, iron, or do housework.
- Don't write, and don't use the computer, e-mail, telephone, or any other electronic device.
- Don't fix things up or tear things down. Leave the world the way it is.
- Don't organize things, straighten things up, or take care of errands. Put your "to do" list away for a day.

What should you do? Dedicate the day to joy, celebration, and the expression of gratitude. "Focus on pleasure. Good food, good sex, singing, dancing, walking, playing, joking or laughing, looking at the magnificence of creation, studying spiritual texts, communing with one's inner voice, or whatever else really generates pleasure—all this is on the Shabbat agenda."[37]

the hidden pond—many confide that these childhood sanctuaries have been snuffed out, replaced by highways, malls, parking lots, airports, and subdivisions.

Many of these tragedies can be avoided if people's attachments to the special places in their communities can move beyond personal sentiment to public expression. This realization is giving rise to a new practice: People are beginning to speak openly about the places that are sacred to them while these places still exist. For example, not long ago, people from communities all around Lake Superior, the world's biggest freshwater lake, participated in a walk that ultimately circumnavigated the entire lake. This was a ceremonial walk to bear witness to the sacredness of the Lake Superior ecosystem.[38] And it was in a similar spirit that a young woman named Julia Butterfly Hill one day in 1997 climbed up high into an old California redwood that was in danger of being cut and stayed for two years (738 days, to be exact) saying, in effect, that the life of this tree was every bit as important as her life.

Scott Russell Sanders, whose childhood home was inundated when a river was impounded to create a reservoir, believes that our home grounds can be saved if we recognize them, speak about them, and celebrate them. We need, says Sanders, "new maps, living maps, stories and poems, photos and paintings, essays and songs. We need to know where we are, so that we may dwell in our place with a full heart."[39]

Start with maps that reveal the warp and woof of your home territory—for example, United States Geological Survey maps. Gather with friends around these maps and locate where your water comes from, the streams that flow through your bioregion, the principal soil types of your home place, the hilltops that offer cherished vistas, the fragile marshes and wetlands, the best fishing spots, the places where bald eagles have been sighted, the zones where rare orchids occur, the locales known for the biggest blueberries, the spots where woodcocks perform their spring mating rituals. Sit together and share stories of these places; rescue the lore of the land.

This is a practice of "re-membering" and making public what for too long has been private. Activist and bioregional planner Doug Aberley offers a vision of where this practice of "re-membering" can lead:

> Imagine this. In the town hall of your community a large atlas that describes "home" in a great variety of ways is prominently displayed. It has several hundred pages that depict layers of biophysical and cultural knowledge: climate, soils, flora and fauna, historic places, wind patterns, how much food was harvested by place and year, plus a summary of a host of related community experience. It is a well-worn tome, referred to continuously by local citizens. In the margins are penciled notes, adding new information to that which is already shown. Every year or so,

your community updates the atlas, growing another layer to the collective understanding of the potentials and limits of place. On the evening that each new edition of the atlas is unveiled, elders are invited to "speak" each map, adding stories to further animate the wisdom that the flat pages tell. There are songs, dances, ribald stories, all relating to the occupation of a well-loved territory. It is entertainment and celebration on one level; on another, it is an absolutely critical validation of larger community potential and purpose.[40]

Cultivating intimacy with our home places is a lifelong practice. It is never too late to begin.

CONCLUSION

How can we be so poor as to define ourselves as an ego tied in a sack of skin?
. . . We are the relationships we share, we are that process of relating, we are, whether we like it or not, permeable—physically, emotionally, spiritually, experientially—to our surroundings. I am the bluebirds and nuthatches that nest here each spring, and they, too, are me. Not metaphorically, but in all physical truth. I am no more than the bond between us. I am only so beautiful as the character of my relationships, only so rich as I enrich those around me, only so alive as I enliven those I greet.

—Derrick Jensen[41]

Derrick Jensen is right: Life's essence is relationship; the living earth pulsates with relationship. The notion that the earth is merely a big rock in empty space—and that life is nothing more than complex mechanics involving electrical impulses and things called "atoms"—has been a logical outgrowth of the Industrial Revolution. But it is now clear that this mindset strips life of its richness and revelatory capacity.

We will experience life's fullness to the degree that we are willing to risk intimacy. And not just intimacy with other people—intimacy with our bodies, with our home places, and with the other-than-human world. By tuning into the intelligence of our bodies, the power of our imaginations, and the wisdom of our hearts, we awaken ecological consciousness.

NOTES

1. Brian Swimme, *Canticle of the Cosmos Study Guide* (New York: Tides Foundation, 1990). Contact information: Tides Foundation, 40 Exchange Plaza, Suite 1111, New York, NY, 10005; www.tidesfoundation.org.

2. Lynn Margulis and Dorion Sagan, *What Is Life?* (Berkeley: University of California Press, 1995), 91.

3. Margulis and Sagan, *What Is Life?* 90.

4. Margulis and Sagan, *What Is Life?*

5. Margulis and Sagan, *What Is Life?*

6. David Suzuki, *The Sacred Balance* (Amherst, N.Y.: Prometheus Books, 1998).

7. Jon Kabat-Zinn, *Full Catastrophe Living* (New York: Dell, 1990), 48.

8. Suzuki, *The Sacred Balance*, 53–54.

9. Suzuki, *The Sacred Balance*.

10. Suzuki, *The Sacred Balance*, 60.

11. Suzuki, *The Sacred Balance*.

12. Kabat-Zinn, *Full Catastrophe Living*.

13. Kabat-Zinn, *Full Catastrophe Living*.

14. Jeremy W. Hayward, *Letters to Vanessa* (Boston: Shambhala, 1997), 143–44.

15. Kabat-Zinn, *Full Catastrophe Living*; Wes Nisker, *Buddha's Nature* (New York: Bantam Books, 1998).

16. Kabat-Zinn, *Full Catastrophe Living*, 88.

17. Thich Nhat Hanh, "Life Is a Miracle" in *Engaged Buddhist Reader*, edited by Arnold Kotler (Berkeley: Parallax Press, 1996), 24.

18. Kabat-Zinn, *Full Catastrophe Living*, 277–78.

19. Tyler Volk, *Gaia's Body* (New York: Copernicus/Springer-Verlag, 1998).

20. Bernd Heinrich, *The Trees in My Forest* (New York: HarperCollins, 1997).

21. Jack C. Schultz, "Tree Tactics," *Natural History* 92 (May 1983): 12, 14, 16, 18, 20–25.

22. Thich Nhat Hanh, *Peace Is Every Step* (New York: Bantam Books, 1991), 95.

23. Heinrich, *The Trees in My Forest*, 71–72.

24. J. Lee, "Honoring the Given World: An Interview with Scott Russell Sanders," in *Stonecrop* (Denver: River Lee Book Company, 1997), 29.

25. This practice was inspired by Project NatureConnect: www.ecopsych.com

26. Stephen Talbott, "Why Is the Moon Getting Farther Away?" *Orion*, Spring 1998, 44.

27. Gregg Levoy, *Callings: Finding and Following an Authentic Life* (New York: Three Rivers Press, 1998), 297.

28. Brian Swimme, *The Hidden Heart of the Cosmos* (Maryknoll, N.Y.: Orbis Books, 1996), 56.

29. Jesse Wolf Hardin, *Coming Home: ReBecoming Native, Recovering Sense of Place* (Reserve, N.Mex.: The Earthen Spirituality Project), 7, ch. 5. Contact information: The Earthen Spirituality Project, PO Box 516, Reserve, NM 87830.

30. F. Herbert Bormann, Diana Balmori, and Gordon T. Geballe, *Redesigning the American Lawn* (New Haven, Conn.: Yale University Press, 1993).

31. Bormann, Balmori, and Geballe, *Redesigning the American Lawn*.

32. This figure was inspired by Bormann, Balmori, and Geballe, *Redesigning the American Lawn*, 90–91.

33. Bormann, Balmori, and Geballe, *Redesigning the American Lawn*.

34. Bormann, Balmori, and Geballe, *Redesigning the American Lawn*; Steve Lerner, "Suburban Wilderness," *The Amicus Journal* 16, no. 1 (1994): 14–17.

35. Frank Bormann, personal communication.

36. Michael Lerner, *Spirit Matters* (Charlottesville, Va.: Hampton Roads, 2000), 299.

37. Lerner, *Spirit Matters*, 300–301.

38. Bob Olsgard, "A Walk to Remember," *Orion Afield*, Winter 2000–2001, 23–26.

39. Scott Russell Saunders quoted in Evelyn Adams, "Learning to Be at Home on Fidalgo," *Orion Afield*, Autumn 2001, 10–13.

40. Doug Aberley, "Eye Memory: The Inspiration of Aboriginal Mapping," in *Boundaries of Home*, edited by Doug Aberley (Gabriola Island, British Columbia: New Society, 1993), 16.

41. Derrick Jensen, *A Language Older Than Words* (New York: Context Books, 2000), 126–127.

Part I: Summary

EARTH, OUR HOME

There are hundreds of ways to kneel and kiss the earth.

—Rumi[1]

Part I is complete. Four chapters—"Discovering Home," "Cultivating Awareness," "Seeking Understanding," and "Nurturing Relationship"—built on twelve foundations that offer new ways of seeing our connections to the cosmos, the solar system, planet Earth, the web of life, and ourselves.

The message that runs through all the foundations in part I is: We are not separate from Earth; we are in relationship with it! To grasp this concept, think of life as one grand ocean with waves in constant movement on the surface. We are all part of life's ocean. Each of us is a unique wave, invited to dance in union with the ocean depths.[2]

Care for the earth takes on new meaning once we acknowledge that we are part of the living body of Earth. It is no longer something we do just to be nice; we do it for self-preservation. When we care for the earth we are, quite literally, caring for ourselves.

Seeing ourselves as one with Earth, instead of separate from it, requires that we relinquish old ideas and beliefs. In these chapters we have seen, for example:

- Creation was not a once-upon-a-time phenomena; it is a continually unfolding mystery.

- The Earth is not an object, a hunk of natural resources; rather, it is a happening—a living, self-regulating, evolving community of life.
- Humans do not stand above and apart from life; rather, we are the earth experiencing itself; we bring critical consciousness to the story of the universe.

When our old ways of understanding the world no longer work, we need not go blind. Instead, we can learn to see with new eyes. It is never easy. For a time we may lose our balance; but, if we persevere, we will eventually find our way to a place of expanded consciousness.

NOTES

1. From "Open Secret," translated by John Mayne and Coleman Barks.
2. Thomas Stella, *The God Instinct* (Notre Dame, Ind.: Sorin Books, 2001).

Part II: Introduction

ASSESSING THE HEALTH OF EARTH

Unless we change the direction in which we are headed, we might wind
up where we are going.

—Chinese Proverb

Welcome to part II, the bridge between awe (part I) and action (part III).
It is here that we candidly examine the condition of Earth.

A statement issued by 102 Nobel laureates in science and sixteen hundred other distinguished scientists from seventy countries warns us of the
deepening ecological crisis caused by human activities:

> We the undersigned, senior members of the world's scientific community,
> hereby warn all humanity of what lies ahead. A great change in the steward-
> ship of the earth and the life on it is required, if vast human misery is to be
> avoided and our global home on this planet is not to be irretrievably muti-
> lated.[1]

Although the general public might be startled by such a strongly worded
statement, researchers, both within and outside the United States, have
been monitoring the pulse of planet Earth for several decades; and their
findings leave little doubt that the earth is in a state of decline. Atmospheric
chemists report steady rises in greenhouse gases; soil scientists tell us that
soils are eroding in many places more rapidly than they are forming; human

physiologists report increases in foreign, sometimes disease-causing chemicals in our bodies; ecologists register the impoverishment of ecosystems and the extinction of species; sociologists observe the breakdown of families and the deterioration of communities; and philosophers and religious leaders discuss the erosion of moral principles and the alienation of humans from the natural world. In short, these researchers know that our Nobel laureates do not exaggerate when they tell us that "vast human misery awaits us" if a "great change in our stewardship of the earth and the life on it" is not forthcoming.

The concept of "health" knits together the three chapters of part II. We begin in chapter 5 by examining the earth as a medical doctor might—by listening to our planet's vital signs. In chapter 6 we assume the role of a diagnostician and seek to discern the causes of Earth's declining health. Finally, in chapter 7, we act as forecasters and consider the long-term ecological consequences of continuing on our present path.

The practices associated with this section's nine foundations focus on listening with compassion (chapter 5), discerning with wisdom (chapter 6), and speaking with courage (chapter 7). Taken together, the practices offer a way of forthrightly facing the ecological crisis that now envelops Earth; and by participating in them, we may find ourselves emerging, not fearful and immobilized, but awake and impassioned.

NOTES

1. "World Scientists' Warning to Humanity" (Boston: Union of Concerned Scientists, 1992), www.ucsusa.org.

5

LISTENING: GAUGING THE HEALTH OF EARTH

We see quite clearly that what happens to the nonhuman happens to the human. What happens to the outer world happens to the inner world. If the outer world is diminished in its grandeur than the emotional, imaginative, intellectual, and spiritual life of the human is diminished or extinguished.

—Thomas Berry[1]

I live close to Spring Creek, a spring-fed trout stream located in the Ridge and Valley region of central Pennsylvania. One afternoon, shortly after beginning my job at Penn State, I decided to take a hike from my office over to Spring Creek. I planned my walk so that I would get to see three of the creek's springs as I hiked. My first stop was Thompson Spring. It wasn't as easy to get to Thompson as I anticipated. I had to cross a road clogged with traffic, climb over a fence, and then navigate my way around a wastewater treatment facility. There at the back of the treatment plant was Thompson Spring, set in a little grove of trees. I stood spellbound, watching as hundreds of gallons of water bubbled from the earth into a lovely pool about ten feet across. But just thirty feet away, the crystalline water cascading out of this pool was overwhelmed by the voluminous effluent from the wastewater treatment plant.

Less than a half-mile from Thompson Spring, to the east, I came upon Thorton Spring; Thorton bubbles up at the base of a forested hillside. It would have been a fine picnic spot if it hadn't been for a strange smell. Thorton Spring, I soon learned, is located just below a chemical plant. For many years this facility stored its chemical waste in underground storage drums. The drums developed leaks and two highly toxic chemicals, mirex and kepone, seaped into Thorton Spring and Spring Creek.

Looping back toward my office, I encountered Bathgate Spring, located on a small farm just below Penn State's gigantesque football stadium. This spring flows cold and clear for a short distance, but then it joins the effluent from a water retention pond constructed to gather runoff from the football stadium's enormous parking lots.

Why am I telling you this? As an environmental scientist, I know that the health of our land and water directly influence our own health, which is why my colleagues and I pay special attention to the "vital signs" of planet Earth: first, by monitoring the health of important life-forms like forest trees (foundation 5.1) and songbirds (foundation 5.2); and, second, by tracking changes in the well-being of vital earth substances like soil and water (foundation 5.3).

Like a medical doctor with a stethoscope to a patient's chest, we have our ears to planet Earth, listening attentively for signs of well-being as well as danger. But one need not be a card-carrying environmental scientist to do so. Anybody can cultivate this capacity to listen. Listening is essential to developing ecological consciousness. This chapter's practices focus on the art of listening to the earth's biota, to one another, and, perhaps most important, to our own inner wisdom.

FOUNDATION 5.1: GAUGING THE HEALTH OF THE TREES

In the early 1970s, when I was working at a boy's home in Virginia, I would take the gang out on an adventure from time to time. One weekend I decided to take them to Mount Mitchell in North Carolina. I had read that Mount Mitchell was the highest peak east of the Rockies, standing at 6,684 feet. The boys were into things that were the biggest, strongest, and toughest. So this was an acceptable destination.

One road went right to the top of the mountain, which seemed like cheating to me—kind of like getting in the ring with the world champion boxer and having him forfeit. As I was considering suggesting that we park our van and run to the summit, one of the boys called out, "Hey, why are all these trees dead?" There was silence as we all looked, really looked, at the forest we had been nonchalantly cruising past. At the top of Mount Mitchell, standing dead trees, mostly spruce, were everywhere. I was mystified. There were no signs of fire (the summer had not been particularly dry), nor were there any obvious disease symptoms on the surviving trees. No one I spoke with on the summit seemed to know the cause of this scourge.

A decade later I came across an article by Hubert Vogelmann, an ecologist at the University of Vermont, which contained the following passage:

> Gray skeletons of trees, their branches devoid of needles, are everywhere in the forest. Trees young and old are dead, and most of those still alive bear brown needles and have unhealthy looking crowns. . . . As more and more trees die and are blown down, the survivors have less protection from the wind, and even they are toppled over. The forest looks as if it has been struck by a hurricane.[2]

The tree die-off that Vogelmann described at Camel's Hump in Vermont's Green Mountains sounded just like what I had observed on my earlier visit to Mount Mitchell. In fact, I later learned that conifers were dying in high-elevation forests in the Appalachians of West Virginia; the Adirondacks of New York; the Laurentians of Quebec; and the Green Mountains of New Hampshire.[3]

From the 1970s through the 1990s, hundreds of scientists were working to try to figure out the cause (or causes) of this high-elevation conifer decline. As this research progressed, it become clear that the problem extended beyond isolated conifer diebacks on mountain tops. For example, hardwood trees at lower elevations were also growing poorly in many areas of New England.

Some of the best evidence for what the scientists began referring to as "forest decline" came from tree-ring studies. Rather than cutting a tree down to "read" its rings, researchers use a long drill with a hollow center to remove a straw-like wood cylinder (core), extending from the tree's exterior to its center. By counting the number of annual rings in a tree core and the distances between individual rings, it is possible to determine a tree's age and to pinpoint the periods during the tree's life when it was growing well or poorly. For example, imagine a core showing narrow rings for many years and then a gradual change to fatter rings, followed by a gradual shift back to narrow rings (figure 5.1). This is the standard pattern for most forest trees. The tree grows very slowly early in life because it is in the shade, but then its annual rings fatten when it reaches the canopy and has access to light. Finally, toward the end of its life, as the tree begins its senescence, its growth rate again declines.

However, a core removed from a tree in a region of forest decline is likely to differ in some ways from the standard pattern just described. The early

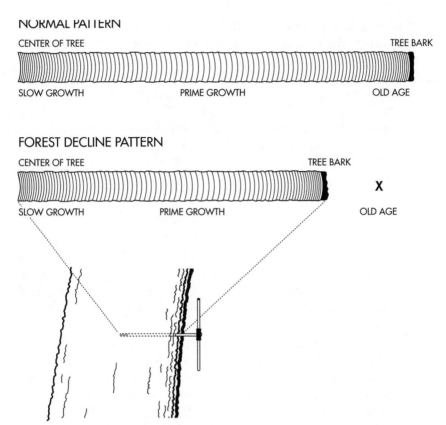

Figure 5.1. Tree-coring device and tree cores representing a standard pattern of tree growth and the forest-decline pattern

rings of a forest-decline tree would be close together (just like in the standard core), revealing that the tree started out in the shade. Then, the rings would fatten, signaling the time when the tree began to gain access to canopy light. But then, soon after the tree entered its prime, its growth rate would begin to decline, as evidenced by narrowing rings; and this decline would result in a premature death. Scientists have found that this anomalous tree-growth pattern is common for many hardwoods at lower elevations in the mountains of southern New England—for example, sugar maple, yellow birch, beech. Through painstaking studies, ecologists managed to rule out the more obvious potential causes for this retarded tree-growth pattern, such as insect attack, fire, fungal disease, and drought.[4]

One less obvious cause for forest decline and tree dieback lingered in the background—air pollution. At first, air pollution seemed like an unlikely culprit. After all, the New England mountains are far from the air-polluting impacts of industry, but airborne pollutants can travel long distances. Consider:

Inside the narrow band that extends from the ground to just 25 feet above the ground, five billion tons of air pass over the United States each day. For . . . millions of years, this air has brought to trees the carbon dioxide, hydrogen, and water they need to live. Since the beginning of the Industrial Revolution, the air has also brought increasing amounts of noxious pollutants. The burning of oil and coal by electrical power plants and manufacturing facilities, the combustion of gasoline by motor vehicles, incineration, the smelting and refining of . . . ores cumulatively pours tens of millions of tons of particles and gases into the atmosphere every year. All of the particles and gases that go up ultimately come back down either as fallout or in precipitation. . . . Since conifer's needles exchange carbon dioxide, oxygen, water vapor, and other gases with the air, there are many microscopic pores into which [pollutants] can pass.[5]

Researchers favoring the air-pollution hypothesis pointed out that U.S. emissions of sulfur dioxide and nitrogen oxides increased dramatically during the twentieth century and that the bulk of these emissions originate in fossil-fuel-powered energy plants located in the mid-Atlantic states (figure 5.2).

As sulfur-containing coal is burned in these mid-America power plants, sulfur is released to the atmosphere; it reacts with oxygen to form sulfur oxides, which then combine with water vapor to form dilute sulfuric acid (H_2SO_4). A similar process renders nitrogen emissions acidic. The prevailing winds in the central United States then funnel this contaminated air eastward toward New England. When the polluted air hits the mountains of New England, it is forced upward where it cools and condenses, often forming acid precipitation or fog.

The initial research on this air-pollution hypothesis showed a correlation between acid precipitation and forest dieback, but correlation isn't proof. So ecologists set about devising experiments to gain a fuller understanding of the possible modes of action by which acid precipitation might affect the health of forest trees. For example, some scientists reasoned that acid rain might leach essential nutrients like potassium from foliage. They then tested this hypothesis in a greenhouse experiment that involved spraying "normal" rain on one group of tree seedlings and simulated acid rain on a second group. They discovered that potassium leaching from foliage increases somewhat in the presence of acid rain and that this might play a small part in the decline.[6]

Other scientists, directing their attention to the soil, proposed that acid rain could acidify the soil and in the process cause nutrient leakage, which would lead to soil impoverishment and forest decline. The results of this work provided evidence that soil nutrient leaching, especially in the case of calcium and magnesium, was indeed occurring.[7] Researchers also verified

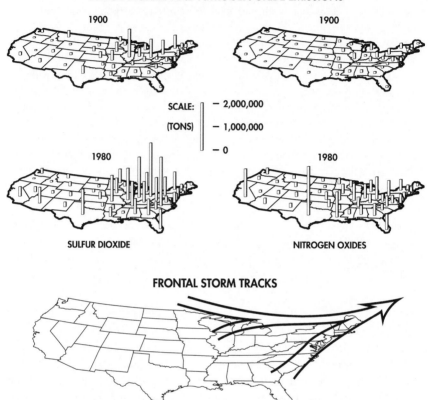

Figure 5.2.   (Top) Emissions of sulfur dioxide and nitrogen oxides in the United States during the twentieth century. (Bottom) Prevailing wind patterns in the eastern United States.[8]

that increases in soil acidity cause soil aluminum, which under normal soil conditions is not soluble, to move into solution.[9] This observation is a concern because aluminum, even in low concentrations, interferes with the ability of trees to take up calcium and magnesium, which are essential for tree survival. When scientists subjected red spruce seedlings to acid water containing small amounts of aluminum, the root tips turned brown, and lateral roots failed to develop. At the end of the experiment, the lab plants looked, overall, like the ravaged mountain-top red spruce on Mount Mitchell and Camel's Hump (see box).[10]

   The exact processes leading to forest decline are still being investigated, but the broad consensus among scientists is that the principal cause is air pollution. Acid precipitation resulting from air pollution apparently has

A HYPOTHESIS TAKES FORM

An environmental mystery like forest decline is seldom solved by one person or even by the scientists working within one discipline. For example, the development of the hypothesis that acid rain (linked to air pollution) could cause forest decline was the result of observations and research by professionals in at least four different fields:

Environmental scientists: They helped document the increases of nitrogen and sulfur emissions to the atmosphere over the past century, as well as the sources of these emissions.

Meteorologists: They played a role in showing how compounds like sulfur and nitrogen oxides can be transported long distances.

Atmospheric chemists: They helped elucidate how nitrogen and sulfur oxide emissions can combine with oxygen in the atmosphere to form acid precipitation.

Ecologists: They teased apart the mechanisms whereby acid precipitation can adversely affect forest trees.

multiple effects on forested ecosystems. It can, among other things, harm plant tissue outright; cause leaching of nutrients from tree foliage; harm mycorrhizal fungi that trees depend on; and slow down the decomposition of dead organic matter, thereby slowing the circulation of nutrients back to living trees.[11] But the most important overall effect of acid precipitation is the acidification of forest soils, which leads to aluminum mobilization and nutrient impoverishment.

Trees in many forests throughout the world, not just those in New England, are being harmed by air pollution. In sum, the trees are reminding us of an important ecological truth: Everything is connected. If we contaminate the air, we will produce "downstream" affects. Insofar as humankind ultimately depends on the plant world—plants are the foundation of the food web—the decline of forests is sending us a "message"; that is, if we care to listen.

Reflection

Bill McKibben, in his book *The End of Nature*, writes:

The character of rain has changed, the joy of watching it soak the waiting earth has been diminished. . . . Instead of a world where rain has an independent and mysterious existence, the rain has become a subset of human activity: a phenomenon like smog or commerce or the noise from the skidder towing logs. The rain [bears] a brand; it [is] a steer, not a deer.[12]

Several years back, shortly after reading McKibben's book, I was invited to be a judge for a middle-school science fair. Perhaps because of this book, I was drawn immediately to Janet's project on acid rain. Janet had two African violet plants on her table. She explained that she had sprayed one with simulated acid rain (a teaspoon of vinegar in a gallon of water) and the other with nonacidified water. She pointed out the brown speckles on the leaves of the violet that had received the acid-rain concoction. I acknowledged that there appeared to be a "treatment effect," and I asked what would happen to this plant now. She responded gravely that it was going to die. I asked if there might be anything that she could do to save it—for example, rinse the leaves and roots with pure water and repot it—but she responded resolutely that this plant was going to die. The melancholy in her voice and eyes was contagious.

It seemed that there was nothing more to say, but I lingered at her table, caught in the lethargy induced by despair. Finally, I crouched down to Janet's level, engaging her eyeball to eyeball, and I asked her if she thought that citizens could fix this acid rain problem—could they stop polluting the earth? She looked at me without blinking, like a parent preparing to give her child some bad news, and she said, "No, no, it's too late; nothing can be done." I looked beyond Janet to her parents standing off to the side, then I returned my gaze to this twelve-year-old with bangs and freckles; and with every bit of resolve and shred of hope that I could muster, I told her that I saw it differently. In fact, I was darn sure that this problem of acid rain could be beat! Then, I asked her if she and her family would be willing to spend a few extra dollars each month to purchase their energy from a "green" power company—that is, one committed to supplying nonpolluting, renewable energy from sources like wind—instead of continuing to power their home with the fossil fuels that contribute to acid rain. If it meant saving the violets and trees, would she and her parents be willing to consider this? She nodded. And did she think that her friends would consider this for the sake of the trees? A second nod. Finally, deep down, did she believe that caring for one another, and the life all around us, is what is most important to all of us? A third nod.

Questions for Reflection

- How is sitting with a dying tree different from sitting with a dying person? How is it the same?
- What are the qualities of fear, anxiety, and resolve that exist before and after a serious environmental problem has been exposed?

Practice: Listening to the Biota

The trees have a story to tell. In some places, forests are experiencing decline and dieback; in other places forests are being destroyed outright. When we hear about death and destruction, it is a natural human response to want to turn away because it seems too painful to bear. However, facing "bad" news—whether it be a personal tragedy or an environmental catastrophe—is a prerequisite for healing.

We can begin to free ourselves from the bondage of fear and despair by simply paying attention (i.e., listening) to the feelings and emotions that arise in us when we hear about different aspects of the ecological crisis now gripping Earth. What arises in you, for example, when you overhear a conversation about dying trees, shrinking aquifers, or human conflict? What happens in your body, your jaw, neck, shoulders, eyes? What happens to your breathing? What about in your mind? Is there judgment, resistance, anger, fear? Do you want to run away? Do you feel helpless? As we gain the capacity to listen to ourselves—to watch feelings arise and experience different body states, even painful ones—we become more conscious.

The Vietnamese monk Thich Nhat Hanh has said that to heal the world the first thing we need to do is hear, within ourselves, the sounds of the earth's crying. Because each of us has come forth from the earth, we *are* Earth. How could it be otherwise! Hence, our tears are quite literally Earth's tears; when we cry for Earth, we are Earth crying. It is important to grasp this concept.

To hear the sounds of the earth crying, we need to listen. When a friend is suffering from depression, we sit with her and listen. We don't rush in with advice, nor do we tell her to cheer up (if we are wise, that is). Instead, we are simply "there" for her. Likewise, if we have a friend who is dying, we sit with him, empathic, ready to listen, yet comfortable with silences. By simply being present, we serve as a compassionate witness. We don't turn our back to life; we remain awake to each moment as it unfolds.

Now, at the beginning of the twenty-first century, when suffering is everywhere—in the wars that divide peoples, in the greedy exploitation that defiles ecosystems, and in the psychic depression that shrivels hearts—we have, arguably, nothing more important to do than to practice listening and bearing witness.

The most neglected recipient of our attentive listening is the wild earth. Listening to the other-than-human beings that make up the earth's community helps us to forthrightly face the environmental tragedies, both large and small, that are now a part of daily existence. For example, if the tree that you climbed in and played under as a child is being cut down, then you are bearing witness to an environmental tragedy. It is not an exaggeration to say that a part of *you* is being cut down because a relationship is being extinguished; part of what has given your life meaning is being severed. Of course, you won't become aware of this loss unless you allow yourself to experience it—to hear it.

In her book *The Attentive Heart*, environmental studies professor Stephanie Kaza writes about her response when she learned that the elm tree in the backyard of her childhood home was to be cut down. She wanted to somehow mark the elm's passing, but she was unsure what to do. Our culture offers no instructions on how to hold a wake for a beloved tree.

> I stood outside in the gray dawn light and gazed at the tree, rain falling gently on my face and shoulders. What could I possible do? I felt obliged to mark this tree's passage from life to death; something inside was crying for the tree. . . . I thought a ceremony might at least deflect my grief. With three sounds from a small bell, I began. I walked slowly around the tree nine times, breathing deeply. . . . Last rites, these were last rites. I lit a small candle and offered four sticks of incense to the four directions, placing them at the base of the tree. The tree was the centerpiece of its own altar, the altar of its death. . . . I chanted a dedication under the dripping rain, a request for forgiveness. . . . I asked for compassion for those who are uncertain about how to care for tree beings and for those who suffer the consequences of loss of tree friends. . . . Three last bells and the short ceremony was over. It was a quiet act of intention that did little to reverse the fate of the tree. But at least the elm did not die alone.[13]

Several years ago I learned that an interstate highway was going to be placed right down the middle of the beautiful valley where I live in central Pennsylvania. The decision had already been made. The route was flagged; soon the explosives and the heavy machinery would arrive. "Nothing could be done," I was told. Although the road appeared to be "inevitable," I felt called to somehow witness and mourn the loss of the hay meadows and forests and farms that this road would destroy. Like Kaza, I wasn't sure how to do this. I finally decided to simply walk the land that would become interstate and to listen. In so doing I thought that I would be able to experience this land, thank it, and say good-bye to it. The idea of a good-bye walk resonated with others in the community. So it was that a group of fifty citizens gathered on a cold Sunday in February to walk the land that would soon become I-99. We walked in silence, stopping here and there along the way to listen to the land and to write our thoughts. Our walk ended late in the afternoon on the banks of Spring Creek. The silence continued, enriched from time to time by expressions of grief, tenderness, and sorrow:

Walking along we notice birds, silence, and recently cut stumps—the 200-year-old-oak and its acorns getting ready to sprout. We bear witness.

—Alice Crawley

It is time to extend our funeral practices to include the dying of real places. As we learn to listen deeply to the "sounds of the earth's crying," we expand our capacity to feel empathy for the other-than-human world, and our doing so expands our ecological consciousness.

FOUNDATION 5.2: LISTENING TO WHAT THE BIRDS TELL US

It is not just the trees that have a story to tell. Birds can also serve as indicators of planetary health. The idea of using birds to read the health of the environment is not a new one. Coal miners used to take canaries with them into the mines. If the canary stopped singing or, worse still, croaked, the miners cleared out quickly, knowing that carbon monoxide had reached dangerous levels.

The Health of Migrant Bird Populations

Today, U.S. scientists are using birds to gauge the health of the earth. Not just any bird species will do. Ideal candidates for study are migratory bird species—that is, species that spend spring and summer in North America (where they breed) and then fly south when the weather turns cold. Examples of migratory bird species include red-eyed vireos, wood thrushes, scarlet tanagers, ovenbirds, and many species of warblers, hawks, eagles, and shorebirds. Collectively, migratory bird species sample an immense range of environments throughout the Americas. Scientists reason that if such wide-ranging bird species are healthy, then it is likely that the many environments these species depend on are also in good condition (see box).

Just as a medical doctor might perform several different tests to assess the health of a patient, scientists have employed several approaches to assess the status of migrant bird populations. One simple approach is to census birds in the same parcel of land year after year. For example, the breeding birds of Rock Springs Park in Washington, D.C., have been censused each spring for the last fifty years. The censusing is done in the spring by people who walk through this large woodland park, day after day, identifying all the birds that are nesting. Although this may sound tedious, it can be exciting to be out in the early morning, listening to birds sing and observing their courting and nest-building behaviors. However, in the case of Rock Springs, the enjoyment has been tempered by concern because the researchers have discovered that the number of breeding migratory bird species in the park has dropped by one-third in recent decades.[14]

Of course, it could be that the bird species that have disappeared from this park are now thriving in other areas. It would be necessary to have a thousand Rock Springs–type studies spread across the United States before firm conclusions could be drawn about the status of migratory birds. In fact, something very similar to this has been done as part of the Fish and Wildlife Service's Breeding Bird Survey. Initiated in 1965, this survey enlists citizen volunteers to conduct bird counts along more than fifteen hundred prescribed transects spread throughout the United States and Canada (each transect is twenty-five miles long). The observers stop at half-mile intervals along these transects and note all the birds they see or hear during a three-minute period. In this survey, some seventy-five thousand geographic locations are being sampled throughout North America each year (50 stops per transect × 1,500 transects = 75,000). The information compiled in this ongoing study corroborates the Rock Springs study: Three-quarters of North America's forest-dwelling migrants have experienced population declines in recent years.[15]

WHY MIGRATE?

Why do many familiar birds migrate to the South each winter? The quick answer is: to escape the cold . . . and there is some truth in this. But if these birds go south for warmth, why do they bother to come back to the North? Why not just stay where it's warm? The answer is food. In particular, it takes a lot of excess food to produce offspring each year. These migrants return North to partake in a banquet. With the onset of spring, insects hatch out by the gazillions in the North. These insect hatches are timed to correspond with the unfurling of new leaves, the food source for most insect larvae. Imagine for a moment that you are a bird and that you arrive to a forest with literally thousands of insect larvae on every tree! What bliss! Now, imagine that you are one of those trees with newly emerged caterpillars beginning to munch on your newly unfurled leaves. What terror! How grateful you would "feel" to the migrant birds who come north and, almost as partners, pluck off many of the caterpillars that are eating your body. Of course, trees don't experience "terror" or "gratitude," nor do birds (as far as we know) experience "bliss"; but in human terms, the seeming reciprocity in the interactions among plants and animals sometimes evokes these anthropocentric interpretations.

Sidney Gauthreaux Jr. of Clemson University has provided a third piece of corroborating evidence for declines in migratory bird populations. Gauthreaux, aware that migratory bird flights are detected on radar screens, reasoned that if migrant bird numbers are truly declining, this diminishment would show up in radar records. He proceeded to collect and analyze radar data from U.S. meteorological stations scattered along the Gulf Coast in the southern United States and, indeed, found that only half as many migratory waves were passing the Louisiana coast in the late 1980s as had passed in the mid-1960s.[16]

Causes for Declines of Migrant Birds

Documenting a decline in the populations of many migrant bird species has been only one part of the research challenge; explaining the causes of this decline has been the other part. One obvious place to look for causes has been the tropics (i.e., the wintering ground for many North American migrants). We have all grown up hearing about tropical deforestation, and surely a connection exists between the disappearance of migratory birds and the disappearance of rain forests in Central and South America.[17]

But tropical deforestation and associated habitat loss is only part of the story. The birds also face threats in the North. For example, field biologists have found that many migrant species these days have a difficult time reproducing. The birds return to the North in the spring; they are successful in finding mates; the females successfully lay eggs; but, in the end, the breeding pairs often fail to fledge young. This was puzzling to many field biologists.

Ecologist Dave Wilcove hypothesized that animals such as raccoons, opossum, and feral cats were eating the eggs and nestlings of migratory birds. These animals have proliferated in the open, suburban types of environments that have been rapidly expanding throughout eastern North America in recent years. Meanwhile, the continuous tracts of forests that once covered much of the East have been sliced and diced until all that remains in many areas are isolated patches of forest in a sea of "developed" land. Wilcove tested his nest-predation hypothesis by placing quail eggs in artificial nests and scattering these nests in forest patches ranging in size from a few acres to a few thousand. Wilcove used the Smoky Mountain National Park (five hundred thousand acres) as a "control" site—that is, to represent the natural condition before the onset of forest fragmentation that now characterizes many parts of eastern North America. After a time Wilcove went back to check all the nests that he had set out. He found that only one nest in fifty was discovered and raided by nest predators in the Smokies; but in the smaller forest patches, located in rural and suburban environments, predation was much higher, rising to 100 percent in some of the smallest forest fragments.[18] Wilcove's experiment doesn't prove that nest predation is *the* cause of migratory bird declines; it simply offers some experimental evidence linking forest fragmentation to nest predation and nesting failure (see box).

Another factor contributing to nesting failure of migratory birds, specifically songbirds, is nest parasitism by brown-headed cowbirds (figure 5.3). Historically, cowbirds followed American buffalo herds, eating the insects kicked up from the grass as the buffalo moved throughout the Great Plains. When it came time to reproduce, cowbirds simply dropped their eggs into the nests of other bird species that happened to be nesting in the vicinity; then they continued to follow the buffalo. Because cowbirds don't have to spend energy raising young, they are free to put all their reproductive effort into egg production. Thus, a cowbird female might distribute as many as fifty eggs among the nests of other bird species during a single breeding season. The cowbird, in effect, tries to trick another mother into raising her children. The alien cowbird chicks often emerge from their eggs before the migrant songbird chicks; and because of this and their relatively large size, these chicks tend to take over the nests of the songbirds. Cowbird chicks may be the only nestlings who survive.[20]

LANDSCAPE FRAGMENTATION MAY HURT MORE THAN BIRDS

At a time when humankind has been led to believe that it has won the war against infectious diseases, it is sobering to learn that since 1980 the World Health Organization has identified more than thirty new infectious diseases. It now appears that the proliferation of new diseases is due, to a significant degree, to human disruption of the natural environment. Epidemiologists, studying the outbreak and spread of new ailments such as Lyme disease and mad cow disease, often link these outbreaks to human-induced phenomena like landscape fragmentation, industrial farming, and global climate change. Take Lyme disease as a case in point. In the early 1970s, it was virtually unheard of in the United States, but today it is no longer unusual. The insect that transmits Lyme disease to humans is a small tick that normally lives on deer and mice. Deer and mice thrive in fragmented, suburban-type landscapes, and in recent years their populations have been exploding; and, apparently, so have the tick's. It now appears plausible (although by no means certain) that the spread of suburbia, with its attendant landscape fragmentation, could be a significant factor favoring the expansion of tick populations (Lyme disease) by encouraging the growth of mice and deer populations.[19]

Figure 5.3. A female brown-headed cowbird (inset) lays her eggs (two speckled eggs on top) in the nests of other bird species

In the past, when forests occurred as large uniform expanses, few cowbirds were found in eastern North America because very little cowbird habitat existed. Now, with all the "development" activity in the East, the landscape has changed, and cowbirds abound. Although these cowbirds prefer open habitat, they will venture into small forest fragments in search of nests to parasitize (e.g., into those only several hundred yards in diameter).

In sum, birds tell a story similar to that told by the trees. Many migratory bird populations are in decline. Are raccoons to blame? How about cowbirds? Is it the people in Central and South America who are cutting down their forests? How about multinational corporations that cut down forests to plant bananas or coffee? Or should we point the finger at the land speculators and developers in the North who are slicing apart forests with roads, malls, and suburban housing developments? In fact, these are all only proximal causes of bird decline. The challenge, as we proceed in part II, is to discern, not the proximal causes, but the ultimate causes of the tree and bird declines and the myriad other manifestations of planetary deterioration.

Reflection

It occurs to me that an important connection exists between that morning cup of coffee that so many of us enjoy and the well-being of migratory songbirds. But the connection is easy to miss.

Coffee, one of the world's most traded agricultural commodities, is grown in the tropics. The land area devoted to coffee cultivation is equivalent to a one-mile-wide strip extending all the way around the equator. Those of us who consume one cup of coffee each day require roughly ten coffee bushes to satisfy our annual coffee consumption.[21]

It turns out that there are two general approaches to coffee cultivation: traditional and industrial. In the traditional approach, coffee bushes are cultivated under a forest canopy on small family farms; chemicals are used sparingly, if at all. This method contrasts with the industrial approach, in which coffee is grown in large, company-owned clearings in "full sun" and tended by hired laborers.

Crop yields are higher on industrial coffee plantations, but so are the ecological impacts—for example, deforestation, water contamination by agrochemicals, and biodiversity loss. Because traditional coffee farms maintain a forest canopy, birds, insects, mosses, and other life forms thrive. For example, in traditional coffee farms in Central America, researchers tallied as many as 150 bird species, many of which were migrants from North America. The nearby industrial coffee plantations, lacking a forest canopy and receiving regular dousings of pesticides, had fewer than ten bird species.[22]

These considerations are becoming important to an increasing number of coffee drinkers who, it seems, want their coffee and their forest, too. Now, companies like Equal Exchange are offering consumers "songbird friendly" coffee—that is, coffee that is produced in the shade of forest trees by small-scale farmers, without the use of toxic pesticides.[23]

Questions for Reflection

- In North America we consume far more than coffee. Each item that we purchase has an ecological story embedded within it. What, for example, might the ecological underbelly of a can of soda, a pair of jeans, or the Sunday newspaper look like?
- What connections might you make between your personal consumption habits and the ecological health of Earth?

Practices: Listening to One Another

The whole epic story of evolution, from bacteria to the complex life forms of today, can be seen in one sense as the story of the evolution of consciousness. Listening enlivens consciousness; it brings us to life.

Deborah Lubar's commitment to listening led her to knock on doors in her suburban neighborhood outside of Boston. She introduced herself and explained that she wanted to hear people's concerns about the world. Then she posed questions like, "What do you think is the greatest problem facing the world today?" and "What do you think will make our country safe and strong?" As Lubar listened, she discovered, to her surprise, that many people had not had the opportunity to talk about their concerns and feelings about the world (see box on page 142).[24]

Listening without judgment creates enormous space for understanding and reconciliation. It is an act of healing. Indeed, it has been said that if you hate someone or condemn someone, you just haven't heard their story yet. Since the 1980s U.S. citizen delegations have been traveling to places like Nicaragua, Guatemala, Israel, and Palestine simply to listen to people's stories. Organizations such as Witness for Peace and the Compassionate Listening Project nurture and coordinate these efforts. According to Leah Green, who is involved with listening in the Middle East:

> The fundamental premise of compassionate listening is that every party to a conflict is suffering, that every act of violence comes from an unhealed wound. . . . What we're doing is creating an environment conducive to peace-building through deep, empathetic listening. It is no simple thing. We work to see through any masks of fear or hostility to the sacredness of each individual. . . . When we listen with the intention of building empathy and understanding, we also quickly build trust, and possibilities emerge.[26]

A GOOD QUESTION SETS THE STAGE FOR LISTENING

We all know what it feels like to be asked a good question, one that takes us into the depths of our being. Suddenly we experience a shift, and we may say things that we have never said before—things that we didn't even know we held within us.

The questions that lead to insight are always spacious; the person asking the question is not pushing an agenda. For example, imagine that you are telling your friend about a stuck place in your life, and she responds, "Have you considered . . . ?" In this case, your friend is more interested in coaxing you toward what *she* thinks is best, rather than in listening to you. But imagine how you might feel if your friend asks, instead: "How would you like it to be?" or "What is it that you would like to move toward?" These types of questions are "spacious" because they allow ideas and energy to come from you.[25]

Listening compassionately is especially challenging when our own buttons are being pushed; but with practice it is possible, and the results can be transformative. For example, imagine the following situation: You are sitting in a coffee shop when the guy at the next table throws down his newspaper and exclaims, "Enviro-tree-huggers are frigging nuts. They are trying to stop everyone from making a living." This guy has been reading about an environmentalist group that is seeking a court injunction to stop a clear-cutting operation. You look over and catch his eye, signaling a willingness to listen. This is not going to be easy because your brother works for the very environmental group that this guy has been bashing. You know arguing won't work, so you decide to try something new. Instead of getting hung up on the man's words (which only push your buttons), you place your attention on the feelings under his name-calling and for the needs that these feelings point toward. So it is that you are able to nod empathetically and say, "You're pissed off about the 'enviros'?"

He shoots back, "They are a bunch of spoiled brats, and they don't know jack."

"They really piss you off," you repeat.

"Damn right, I'm pissed. My brother and uncle are both loggers, and their jobs are on the line if this stuff keeps happening."

You pause; and then once again, focusing on what you hear as the man's feelings, you say, "You're worried about how the stuff these enviros are doing could affect your people."

"Yeah, ya-know it's not easy being a logger these days," he responds.

"You care about your brother and uncle," you offer.

"Well, yeah. Ya-know, they're family."

You can feel a softening in the man. One of the biggest challenges in listening to people in pain is to help them distinguish their judgments from their feelings. As you listened compassionately to this man, you were able to go beyond his initial complaint about the "friggin' enviro-tree-huggers" (which was actually a judgment born of the pain he was feeling) to his underlying feelings. It turned out that he was feeling a combination of anger, worry, and care. When you were able to reflect his underlying feelings back to him, he felt heard.

When we listen in this way, we can detect the unmet needs that are always lurking below feelings and judgments. Marshall Rosenberg, in his book *Nonviolent Communication*, suggests a practice to become more attuned to the connections among judgments, feelings, and needs:

> List the judgments that float most frequently in your head by using the cue, "I don't like people who are . . ." Collect all such negative judgments in your head and then ask yourself, "When I make that judgment of a person, what am I needing and not getting?" In this way, you train yourself to frame your thinking in terms of unmet needs rather than in terms of judgments of other people.[27]

For example, if you are in a conversation with someone who keeps talking *at* you a-mile-a-minute, you may judge him as self-centered and boring. But what if you followed Rosenberg's suggestion, and rather than sitting in judgment, you simply asked, "What am I needing and not getting in this conversation?" In this case, you might discover your need for a conversation that is alive and participatory, which would leave you with two options: excuse yourself from the present conversation, or express your need for a conversation that has heart and meaning. Either way you are respecting your own need for interactions that are life-giving.

In addition to judging others, we are continually judging ourselves. Hence, Rosenberg suggests that we make a list of the ways that we judge ourselves and then translate these judgments into unmet needs. Finally, we should also consider how we are frequently judged by others, and we should understand that their judgments are simply expressions of their unmet needs. In sum, the practice of translating judgments into unmet needs is a way of developing "need consciousness" and can dramatically hone one's listening skills.

Once we understand that our judgments are invariably grounded in unmet needs, we may find it easier to think about ways of resolving differences. For example, continuing your imaginary conversation with the man in the coffee shop, you could move from feelings to needs by saying,

"Sounds like it is important for you to know that your brother and uncle aren't going to be out on the street without a job."

To this the man responds: "Yea, they both have young kids. Without a job they'd be screwed."

As you continue to talk, it becomes clear that this man doesn't harbor any ill will toward environmentalists per se. In fact, he confesses that he likes to fish and that all the runoff and erosion associated with forest clear-cutting is messing up one of his favorite fishing streams. Ultimately, this man's need is for the safety and well-being of his family members; it isn't for logging. In this he shares immense common ground with the so-called enviros, including your brother, that he had been bashing.

In sum, helping others explore the feelings and needs lying below their expressions of judgment is a form of deep listening that leads to deeper understanding and fuller consciousness. As we listen, we heal one another, which leads to the healing of the earth.

FOUNDATION 5.3. LISTENING TO WHAT THE SOIL AND WATER TELL US

When most people hear the word "capital," they think of money, cash flow, investments, and so forth. But there is another form of capital that is far more fundamental than money—namely, "natural capital." Natural capital refers to the stocks of such things as fertile soils, ocean fishes, productive forests, and clean water. In other words, natural capital calls our attention to the resources that our quality of life ultimately depends on. If natural capital stocks are steadily deteriorating, the performance of the stock market, no matter how "bullish," means very little over the long haul. Therefore, as we continue our assessment of the earth's health, it is appropriate to examine the current status of Earth's natural capital stocks.

Soil and Water "Capital"

Topsoil is the form of natural capital that humans rely on for growing almost all of their food. Farmers know that new soil "capital" can accumulate if organic debris like manure and compost are added to the soil; soil formation also occurs, albeit slowly, via the weathering of the rock already in the soil. Soil capital can also be depleted, as when topsoil is eroded away by the combined forces of wind, water, and gravity. Ideally, new additions of soil capital will either balance or exceed soil losses. However, when more soil is eroded away each year than is formed, soil stocks decline. Think of it as money in your bank account. If you are always taking money out and seldom putting new money in, your money stocks will be eroded away. The United States Department of Agriculture has been monitoring the status of U.S. soils for many years. Their data reveal that, on average, 5.6 tons of soil are lost per acre each year from U.S. croplands. Meanwhile, new soil is formed at an average rate of a half-ton per acre per year, which means an average net loss of about five tons of soil per acre per year from U.S. cropland. One inch of soil spread over an acre (an "acre-inch") weighs approximately 150 tons. Hence, on average, one inch of topsoil is lost from U.S. cropland every thirty years.[28] It is important to underscore the word "average." On some farms, the loss of soil is greater than this average; on others, the loss is less; and on some farms, there is a net *gain* of soil because of conscientious soil conservation practices.

Of course, if soil stocks were huge, it would not much matter if they were overdrawn a bit each year. Unfortunately, though, Earth's soil stocks are not huge. For example, the topsoil on many U.S. farms is only six to twelve inches deep. Hence, given the average erosion rates, six inches of topsoil could be lost within a span of only 180 years.

In many parts of the world, average soil erosion rates are even higher than in the United States. In addition, the world also faces a great pressure on water capital. Scientists estimate that close to half of the earth's renewable water resources are already being used by humans. Indeed, water might prove to be the resource that ultimately puts a cap on the earth's capacity to support humans.[29]

In the United States the depletion of water capital is evident most dramatically in the case of the Ogallala Aquifer, an immense underground water reserve located beneath the Great Plains. This water was deposited slowly over hundreds of thousands of years during the Ice Ages. More than fourteen million acres of farm land in America's "breadbasket" are irrigated with water from the Ogallala Aquifer. New water enters this aquifer each year in the form of rainfall, which slowly percolates down through the soil; but irrigation water is being "mined" from the Ogallala much more rapidly than new water is being deposited. In fact, the average thickness of the Ogallala Aquifer was sixty-six feet in 1930, but at present the average thickness is less than ten feet.[30]

Exploring a Paradox

It seems paradoxical that the cash economy of a nation like the United States can thrive while its natural capital stocks are being depleted. A concrete

example illustrates how this is possible. Imagine that you own one hundred
acres of forest and that this forest requires one hundred years to grow back to
maturity after it is cut down. Fortunately, your ancestors exercised restraint by
only cutting one acre each year. Having been bequeathed these one hundred
acres of forest, you realize that each acre is in a distinct age class—from one
to one hundred years. But now imagine that, instead of cutting one acre each
year as your ancestors did, you decide to cut ten acres each year. You can do
this for a while, and your "personal economy" will surely flourish; but in a
decade your system will begin to come tumbling down because your greed for
wood will have exceeded the land's carrying capacity—that is, your demand
will have exceeded what the land is able to sustainably produce.

Figure 5.4 tells the story. First, note the point of intersection between
the consumption curve and the carrying-capacity curve. This point defines
the limit of sustainable consumption. For our forest example, this limit is
an annual harvest of 1 percent (or one acre) because the forest requires
one hundred years to grow back. As we see in this example, it is possible
to "overshoot" the carrying capacity of the forest (or planet) for a time and
never know it (i.e., there is no sudden collapse).

A key point made manifest in this figure is that carrying capacity actually declines when humans overexploit a natural resource like timber.
This fact is registered by the gradual downward tendency in the carrying

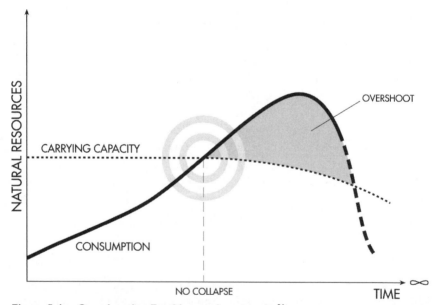

Figure 5.4.   Overshooting Earth's carrying capacity[31]

capacity line over time. To understand why this happens, let's return to our forest example. Imagine that you continue to cut ten acres of your one-hundred-acre tract year after year. After forty or fifty years of this relentless exploitation, your forest will be so degraded—so eroded and trampled—that very little wood fiber will accumulate in the short ten-year intervals between successive cuts; that is, the forest will no longer have the "capacity" to produce new wood at the same rate as it did under the earlier sustainable harvest regime.

On a planetary scale, humans now exceed the earth's carrying capacity. We are drawing down natural stocks of water in Earth's aquifers, soil in Earth's agricultural lands, timber resources in Earth's forests, and fish in Earth's oceans. We are also exceeding the "capacity" of Earth's atmosphere to absorb our wastes as evidenced by ozone thinning, acid precipitation, and climate change. Slowly but surely these overdrawn natural capital accounts will come due:

> Declaring that current levels of consumption can be maintained, on the basis that we are already consuming at this rate, [calls to mind] the warped logic of a bad joke. It is like the stuntman who, jumping from the top of a 50-story building, declares to onlookers as he passes the fifth floor that his stunt is a perfect success because, so far, he has not been hurt.[32]

In spite of this somber analysis, humans (especially those in Western culture) have a strong tendency to believe that our technological prowess will somehow lead us to find substitutes for the "natural capital" assets that we are depleting. Some economists insist that natural and manufactured capital are interchangeable. They support this claim by pointing out that human ingenuity has already resulted in the discovery of effective substitutes for some important metals. This is certainly true, and it is likely that technological breakthroughs will continue to help resolve some ecological shortfalls; but there are no substitutes for Earth's myriad life-forms nor for many of Earth's ecosystem services. For example, it is difficult to imagine how humans would ever be able to invent a machine that can do everything a green plant can do: sequester carbon from the atmosphere, incorporate nitrogen gas into its tissues, manufacture complex carbohydrates using solar rays, purify water, self repair, and reproduce more of itself. Likewise, it is tough to fathom how humankind, acting alone, would ever be able to replace such fundamental ecosystem services as the creation of topsoil, the decomposition of organic materials, the recycling of nutrients, and the pollination of field and tree crops.[33]

Reflection

I find it confusing to decide if conditions are improving or deteriorating on planet Earth. Depending on what I pay attention to, I can be convinced that things are steadily improving or that they are steadily deteriorating. For example, if I limit my focus to the United States, I see that some things are clearly improving—for example, average life expectancy rose from forty-seven years in 1900 to almost eighty years in 2000—while, at the same time, other things are getting worse: personal bankruptcies are at record highs; hunger and homelessness are rising; and U.S. prisons are bulging.[34]

In recent years, with the benefit of insights passed on by systems-thinker Donella Meadows, I have come to understand that the reason that analysts reach very different conclusions regarding the condition of the earth is that they are measuring different things. In effect, they are measuring different types of "capital." Some measure capital in a monetary sense. When dollars accumulate in bank accounts or stock portfolios, these analysts conclude that things are improving. Other researchers, recognizing that "money capital" is really a surrogate for physical capital, focus on physical assets—the things that money buys, builds, and maintains (e.g., roads, buildings, machines, houses). They know that when bridges fall into disrepair, when roofs of public libraries begin to leak, and when factory machinery becomes antiquated, conditions are getting worse—irrespective of what money accounts might say.

But assessments that are limited to considerations of money and physical capital tend to be misleading because they omit three fundamental forms of capital in their assessments. First, they miss human capital—the skills and health embodied in people. If a society's money is used to build machines at the expense of human health and education, physical capital may rise; but human capital will decline. And unlike physical capital, human capital has a huge multiplier effect. Use money to buy a machine, and the outcome will simply be the output of the machine; but use money to invest in the education of a child, and the outcome could be something as grand as a revolutionary new approach to manufacturing or an utterly original contribution to the arts that elevates the human spirit.

Second, traditional assessments of well-being fail to consider social capital—the capacities for creative problem solving and mutual caretaking that come when people trust one another and share talents and resources. Communities rich in social capital are able to transcend, to a significant degree, the limitations imposed by limited access to financial and physical capital (for examples, go to chapter 9, foundation 9.2, p. 276).

Finally—but, really, first—comes natural capital. Yes, all of the four preceding forms of capital depend on the wealth of the earth. Without fertile soils; clean water; productive forests, estuaries, and lakes; and the replenishing cycles of the earth's ecosystems, we would have no other forms of capital.

Now, as I employ this five-part taxonomy of capital, I understand why it is so easy to become confused about whether things are getting better or worse on planet Earth. Meadows, in one of her last "Global Citizen" columns before her death, summed up the challenge when she observed that what the myriad research reports on the state of the earth reveal is

> a world fixated on building up money capital, and in most places physical capital, and for a minority of the population, human capital, at the expense everywhere of social and natural capital. Those who count money and the welfare of the privileged see good news. Those who count nature and societal health and the welfare of the entire population see bad news. If there is to be any future for us, we have got to learn to admit all the news. . . .[35]

Questions for Reflection

- What do you know to be true about the condition of the earth?
- What is the status of social and natural capital in your household? Community? State?

Practice: Listening to Our Own Beings

There is a Zen story about a man on a horse. The horse is galloping down the road, and it appears that the man on the horse is going somewhere very important. Another man, standing alongside the road, shouts out, "Where are you going?" And the man on the horse replies, "I don't know. Ask the horse." This, increasingly, is the story of modern life. Rushing has become a habit; speed has become a way of life (see box).

I have heard it said that how we spend our time is how we spend our lives. If we spend all our mental energy rushing around worrying about money, our shortcomings, past mistakes, future hopes, and so forth, then these things become our life meditation—that is, our awareness and consciousness become imprisoned in fear, worry, self-abasement, and longing.

This self-imprisonment need not be. We can learn to place our awareness and life energy in the present moment, which really is the only moment there is. As alluded to in chapter 2, many awareness practices focus on the breath, which makes sense: breathing is what animates all life; paying attention to the breath is a way of listening to our own beings.

NATURE'S PACE

Most processes in the natural world—the decomposition of leaves, the growth of trees, the pollination of flowers—are slow by human standards. Even things that appear fast, such as falling rain, move slowly in the embrace of nature. The rainwater dripping down to the forest floor slowly soaks into the soil and gradually percolates down to the water table, recharging the soil aquifer. But what happens when humans replace forests with parking lots? In this case, the rainwater is not able to recharge the aquifer. Instead, it rushes across blacktop into culverts and then to streams. Hence, when rainwater moves too quickly, it is lost from the ecosystem, and soil aquifers fail to recharge.

Environmental educator David Orr notes that there is a lesson in this for our times. When things move too quickly in modern life, the essential well-springs of community life are not replenished. Take money aquifers, for example. When businesses are locally owned, money moves slowly, passing from person to person; it circulates within the community and thus has a multiplier effect. But when local businesses are replaced by the likes of Wal-Mart and Sam's Club, money quickly exits the local economy. Like water over blacktop, outside businesses cause money to speed away; they fail to recharge local wellsprings of prosperity.

The same is true for knowledge aquifers. It takes time to read, contemplate, and exchange ideas. When communication and knowledge seeking take the form of sound bytes, clichés, skim reading, and Internet surfing, there is not enough time for information to mature into understanding and wisdom. Speed works against the emergence of understanding, and ultimately it depletes the knowledge and wisdom aquifers, which in turn has an impact on a community's moral aquifers. Moral convictions and integrity are cultivated in an atmosphere of reflection and community dialogue. When these elements are present, citizens remain clear on their deeply held values—the things they would be willing to stand up for. In their absence, citizens often become disempowered; they are unable to act with moral agency.

In sum, much of what is important in life—thinking deeply, creating works of art, making difficult moral decisions, practicing democracy, raising children, caring for each other—can only be done slowly. Slowing down is a revolutionary act; it allows us to live our lives with more attention and awareness.[36]

The most basic breathing practice requires that one simply sit upright but comfortably in a chair, with head in alignment with spine. Once you are settled into an alert, relaxed position, bring your attention to your breath. Feel your breath coming into your body, your diaphragm lowering, your chest rising; then feel it leaving your body, your diaphragm rising, your chest settling. Don't try to control your breath; simply witness it as it moves in and out. There is no need to verbalize or conceptualize. Simply feel the breath as it passes your nostrils, as it expands and contracts your chest, as it resonates in your belly. "Focusing on the breath at your belly can be calming. . . . Your belly is literally the 'center of gravity' of your body, far below the head and the turmoil of your thinking mind."[37]

Watching the breath in this way, you will undoubtedly find that your attention wanders, which is not surprising because we humans are so habituated to thinking. Indeed, much of our "waking" life is spent in our heads, thinking about the past or the future. When your thoughts wander, simply note that your thoughts have wandered without making any judgments. When you return to the breath, you are automatically placed back in the present.

> Observe the breath closely. Really study it. . . . Don't observe just the bare outline of the breath. There is more to see here than just an in-breath and an out-breath. Every breath has a beginning, middle, and end. Every inhalation goes through a process of birth, growth, and death and every exhalation does the same. The depth and speed of your breathing changes according to your emotional state, the thought that flows through your mind, and the sounds you hear. Study these phenomena.[38]

By stopping to spend some time each day in inner stillness—say, a half-hour—you, in effect, step outside of time. Then, when you return to your activities, you can do so with a more spacious perspective; you will be able to flow with time, rather than fight against it or feel driven by it.[39]

Practices that cultivate awareness can also be incorporated into the flow of daily life. For example, the simple act of walking from place to place can become an opportunity to practice awareness. To experience this as a practice, locate yourself at one end of an open space, either indoors or outside. Stand for a moment, bringing attention to your posture and your breath. Hold your arms in any way that is comfortable—at your side, in front, or in back. When you are ready, walk across the space paying attention to the physical sensations of movement. Soften your gaze; walk with your head up and your neck relaxed at the slowest pace that is comfortable for you; allow your face to settle into a light smile.

> Put all of your attention on the sensations coming from the feet and legs. Try to register as much information as possible about each foot as it moves. Dive into the pure sensation of walking, and notice every subtle nuance of the movement. Feel each individual muscle as it moves. Experience every tiny change in tactile sensation as the feet press against the floor and then lift

again. Notice the way these apparently smooth motions are composed of a complex series of tiny jerks. Try to miss nothing. . . . Don't think, just feel. Just register the sensations as they flow.[40]

You may find it helpful to break your movements down into discrete components, coordinated with your breathing. For example: first, the heel of the right foot lifts (in-breath); then the right foot rests on the toes (out-breath); next the right foot swings through the air (in-breath); finally the right foot comes down to the floor (out-breath). The idea isn't to look pretty or to be graceful; it is to simply be alert and aware. At first, you may have some difficulty with balance because you are using your leg muscles in un-accustomed ways. Be patient with yourself. With practice, it is possible to walk in this mindful way throughout life—deeply at ease and grounded in the earth. Even when you are walking briskly you can learn to do so with awareness.[41]

Of course, in the flux of daily life, it is easy to lose connection with ourselves. Hence, it helps to have "markers" to bring us back into a state of awareness. Some people use door handles as markers. Each time they come to a door, they feel the temperature and texture of the handle and mindfully note that they are leaving one place and moving toward something new. Bathroom activities, like toothbrushing, are especially good markers because they engage our senses. Each time you reach for your toothbrush, see it with fresh eyes; behold the water flowing from the faucet; delight in the physical movements of your hand and wrist as they manipulate your toothbrush; and revel in the tactile sensations as the brush titillates your mouth and gums. Any activity or object that we encounter on a regular basis can serve as a marker—the mailbox, the computer, the tea bag, the church bells, the car ignition, the stoplight. Markers invite us back into the present moment.

As we learn to pay attention and listen to ourselves, we become more "centered." In other words, we have the experience of being at the "center" of our bodies and our lives—living in a place of integration at the hub of our power and awareness. To appreciate what this feels like, imagine yourself as a wagon wheel, with the spokes of the wheel representing your various facets. When your attention and energy are scattered, you are living out on the ends of a few of those spokes, whirling about and becoming more dizzy, disoriented, and out of control with each spin of the wheel. On the other hand, when you operate from your center, you reside at the wheel's hub; and all of your spokes (facets and powers) are available to you. In this centered state, you are balanced and more fully conscious.[42]

CONCLUSION

No other quality is so urgently needed today [as listening]: Millions cry out to be heard, to be listened to, to be allowed some say in a harsh and brutal world. Planet Earth itself cries out amid its pain of pollution, exploitation and desecration. . . .

—Diarmuid O'Murchu[43]

Environmental scientists, like vigilant doctors, have been listening to the pulse of planet Earth for more than fifty years, and the messages they have been receiving—from the trees and birds, the soil and water—are cause for concern. Man, the agent, has been "acting" on the earth: burning fossil fuels, which cause acid rain and, in some places, forest decline; cutting down forests and smearing roads, malls, and subdivisions across the land, which, among other things, affects the well-being of migrant bird populations; and engaging in aggressive farming practices, which deplete Earth's natural capital stocks of soil and water. But humans are more than simply "agents"; we bring reflective consciousness to Earth.

When a dynamic balance is struck between doing and being—action and presence—human consciousness deepens and expands. However, the hyperaccelerated pace of modern life has caused humankind to temporarily lose this balance. We can recover our balance by cultivating listening practices—paying attention to the biota, to one another, and to our own beings. Listening has great untapped power; it enlarges us, and in so doing it allows us to bring spaciousness and generosity into the world.

NOTES

1. Thomas Berry, *The Great Work* (New York: Bell Tower, 1999), 200.
2. Herbert Vogelmann, "Catastrophe on Camels Hump," *Natural History* 91 (November 1982): 8.
3. Robert A. Mello, *Last Stand of the Red Spruce* (Washington, D.C.: Island Press, 1987).
4. Mello, *Last Stand of the Red Spruce.*
5. Mello, *Last Stand of the Red Spruce*, 18–19.
6. R. M. Klein and H. W. Vogelmann, "Current Status of Research in Acid Rain," in *Technical Symposium on Acid Rain: Transport and Transformation Phenomena* (Burlington: University of Vermont, 1984) as described in Mello, *Last Stand of the Red Spruce.*

7. Frank H. Bormann and Gene E. Likens, "Catastrophic Disturbance and the Steady State in Northern Hardwood Forests," *American Scientist* 67 (1979): 660–669.

8. Source for sulfur dioxide and nitrogen oxides data in this graph is G. K. Gschwandter, K. Gschwandter, K. Eldridge, C. Mann, and D. Mobley, "Historic Emissions of Sulfur and Nitrogen Oxides in the United States from 1900 to 1980," *Journal of the Air Pollution Control Association* 36, no. 2 (1986).

9. C. S. Cronan and C. L. Schofield, "Aluminum Leaching Response to Acid Precipitation: Effects on High Elevation Watersheds in the Northeast," *Science* 204 (1979): 304–306.

10. Mello, *Last Stand of the Red Spruce.*

11. Mello, *Last Stand of the Red Spruce*

12. Bill McKibben, *The End of Nature* (New York: Anchor Books, 1989), 61, 89.

13. Stephanie Kaza, *The Attentive Heart* (Boston: Shambhala, 1996), 95–96.

14. John Terborgh, "Why American Songbirds Are Vanishing," *Scientific American*, May 1992, 98–104.

15. Terborgh, "Why American Songbirds Are Vanishing."

16. Terborgh, "Why American Songbirds Are Vanishing."

17. Terborgh, "Why American Songbirds Are Vanishing."

18. Terborgh, "Why American Songbirds Are Vanishing."

19. Mark J. Walters, "What Bugs Us," *Orion*, January/February 2003, 14–15.

20. Terborgh, "Why American Songbirds Are Vanishing"; Scott K. Robinson, "Nest Gains, Nest Loses," *Natural History* 105 (July 1996): 40–47.

21. Eldon Kenworthy and Eric Schaeffer, "Coffee: The Most Teachable Commodity?" *The Journal of Environmental Education* 31, no. 4 (2000): 46–50.

22. Accessed at www.audobon.org.

23. Accessed at www.equalexchange.com.

24. Katrina Shields, *In the Tiger's Mouth* (Gabriola Island, British Columbia: New Society, 1994), 49.

25. Fran Peavey, *Strategic Questioning* (San Francisco: Crabgrass Organization, 2001). Contact information: Crabgrass Organization, 3181 Mission St., Number 30, San Francisco, CA 94110.

26. Leah Green, "Just Listen," *Yes! A Journal of Positive Futures*, Winter 2002, 21, 23.

27. Marshall Rosenberg, *Nonviolent Communication: A Language of Compassion* (Encinitas, Calif.: Puddle Dance Press, 1999), 147.

28. Soil erosion data are U.S. cropland average for nonfederal rural land for 1997 (accessed at www.nrcs.usda.gov).

29. Joel Cohen, *How Many People Can the Earth Support?* (New York: W. W. Norton, 1995).

30. Michael L. McKinney and Robert. M. Schoch, *Environmental Science: Systems and Solutions* (New York: West, 1996).

31. This figure was adapted from Mathis Wackernagel and William Rees, *Our Ecological Footprint* (Gabriola Island, British Columbia: New Society, 1996), 54.

32. Nicky Chambers, Craig Simmons, and Mathis Wackernagel, *Sharing Nature's Interest* (London: Earthscan, 2000), 48.

33. Paul Hawken, *The Ecology of Commerce* (New York: Harper Business, 1993).

34. Donella Meadows, "Things Are Getting Better or Worse Depending on What You Count," *The Global Citizen*, November 23, 2000 (accessed at www.sustainabilityinstitute.org/meadows/).

35. Meadows, "Things Are Getting Better."

36. David Orr, "Speed," *Conservation Biology* 12, no. 1 (1998): 4–7.

37. Jon Kabat-Zinn, *Full Catastrophe Living* (New York: Dell, 1990), 52–53.

38. Venerable Henepola Gunaratana, *Mindfulness in Plain English* (Boston: Wisdom, 1991), 81.

39. Kabat-Zinn, *Full Catastrophe Living.*

40. Gunaratana, *Mindfulness in Plain English*, 175–176.

41. Gunaratana, *Mindfulness in Plain English.*

42. Molly Y. Brown, *Growing Whole* (Center City, Minn.: Hazelden Educational Materials, 1993).

43. Diarmuid O'Murchu, *Our World in Transition* (New York: Crossword, 1992), 152.

6

DISCERNMENT: DECIPHERING THE CAUSES OF EARTH'S BREAKDOWN

I am only a child, yet I know if all the money spent on war was spent on ending poverty and finding environmental answers, what a wonderful place this Earth would be. . . .

—Severn Cullis-Suziki, age twelve[1]

When humans hear about a certain problem—a company going bankrupt, a divorce, forest decline—they seek an explanation. Usually, they hunt for a specific cause rather than a systemic cause. So we conclude that a cash-flow problem causes the company to go bankrupt; the affair causes the divorce; acid rain causes forest decline; and so forth.

For many years analysts glibly concluded that population growth was the primary cause of the environmental problems plaguing Earth (foundation 6.1). Then, in the 1980s and 1990s, the "experts" came to see that it is not the number of humans that exerts an impact on the earth so much as it is their individual lifestyles—that is, the amount that each person consumes (foundation 6.2). But until recently, we had no way to quantitatively assess the relative ecological impacts of different lifestyles. This, however, has changed. Using the "ecological footprint" metric, scientists are now able to determine the area of biologically productive land and sea required to meet individual needs, and thus collective needs, and thereby gauge the capacity of Earth to satisfy—or not satisfy—present-day human demands (foundation 6.3).

The theme of discernment connects this chapter's foundations with its practices. The practices offer ways of cultivating discernment by enlarging our capacity to see connections, question assumptions, and reframe problems as opportunities. Through practicing discernment, humans transform information into knowledge and knowledge into wisdom.

FOUNDATION 6.1: POPULATION EXPLOSION

In the early years of the third millennium, the human population is growing by eight thousand people each hour, almost two hundred thousand every day, and more than one million each week. Before long the human population will exceed eight billion. This so-called population explosion is a very recent phenomenon. The human population didn't reach one billion until 1850. At about that time, infant and childhood mortality rates began to decline because of dramatic improvements in sanitation and medicine, such as development of vaccines, and concomitant reductions in childhood diseases. With the resultant population surge, human numbers doubled from one billion to two billion in just eighty years, between 1850 and 1930. Then in the forty-year span from 1930 to 1970, human numbers doubled again from two to four billion; and by the end of the twentieth century, human numbers had topped six billion (figure 6.1).

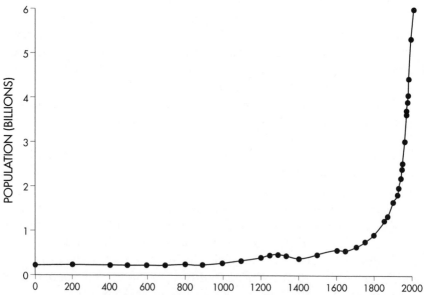

Figure 6.1. The growth in the human population on Earth over the past 2,000 years[2]

If you are anything like me, you have a hard time fathoming really big numbers like a billion. A million is within my grasp. For example, I can imagine filling a major metropolitan sports stadium (capacity of one hundred thousand people) ten times—that's a million. What about a billion? Even when I recall that a billion is one thousand million, I still have difficulty grasping just how much this is. It helps if I think of it in terms of seconds. A million seconds would occupy nine days, but it would take thirty-two years for a billion seconds to tick off.

Human population growth is as serious a concern in the United States as it is in many other countries. The United States now adds approximately four million to its population each year (half of this through immigration). Between 1970 and 2000, the U.S. population grew by approximately 60 percent, from 170 million to 280 million; and by 2025, the U.S. population is projected to reach 335 million. These are sobering numbers for a nation that has generally considered rapid population growth to be a phenomenon that occurs in other countries, not its own.

Exponential Growth

Human population growth in recent times has often been referred to as "exponential growth"—growth at a fixed rate or percent. Things growing exponentially have fixed doubling times. For example, imagine an island with a human population of two thousand and a population growth rate of 5 percent per year. If the island has 2,000 people at the beginning of the year, its population will have grown to 2,100 people by the end of the year (0.05 × 2000 people = 100 new people added to the population). After two years the population will have grown to 2,205 people (0.05 × 2100 = 105 new people). If you continue with these calculations, you will discover that in fourteen years the population will have grown to 4,000 people. In other words, given a 5 percent growth rate, the island's population will have doubled in just fourteen years!

In fact, using the so-called rule-70, you can calculate the doubling time of any population or phenomenon if you know the growth rate. The population doubling time is simply the number "70" divided by the percent growth rate. Hence, in our example 70/5% = 14, or the number of years it took the population of the island to grow (double) from two thousand to four thousand people (see box).

In 2003, the human population stood at 6.3 billion, and the population growth rate was 1.2 percent. How long might it take for the human population to double from six to twelve billion? The rule-70 gives us the answer. We can expect twelve billion humans on Earth sometime around

THE BACTERIA IN THE BOTTLE

Things that are growing exponentially can sneak up and surprise us. Consider the often-told story of the bacteria in the bottle. Imagine that a single bacteria cell divides, forming two bacteria cells every day. The two bacteria thus formed will divide, so after two days the population will be composed of four bacteria; at the end of the third day, the population will consist of eight bacteria, and so forth. Such a population has a doubling time of one day. Imagine that you know that the bottle containing the bacteria will be full—no more room for bacteria to grow—after thirty days. When do you suppose that the bacteria in the bottle might "sense" that things are getting a bit crowded? Will it be on the twenty-seventh day, when the bottle is only one-eighth full? Or on the twenty-eighth day, when the bottle is only one-fourth full? Or on the twenty-ninth day, when the bottle is still half-empty? Metaphorically speaking, human beings are the "bacteria in the bottle," and the time is nearing the thirtieth day. Just as the finite size of the bottle ultimately puts a cap on the size of the bacteria population, the finite size and resources of planet Earth may ultimately place a limit on human consumption and, by extension, on human population size.

For arguments sake, though, let us imagine that the bacteria in our example are able to locate three more bottles, each with just as many resources for growth as the first bottle. How long would this new bonanza last them, assuming prevailing technology? On the thirty-first day, their population would double again, and they would occupy all the resources in the first and second bottles. After just one more day (one more doubling), they would have usurped all the resources in the third and fourth bottles! Hence, when a population is in exponential growth mode, finding new resources (e.g., new natural gas fields, new planets) or developing new energy-saving technologies often only postpones, for a short time, the final reckoning with limits.

2060 (70/1.2% = 58 years from the year 2000), unless population growth declines significantly.

Walking in the Shoes of the World's Poor

Among those of us in "developed" countries, there is a tendency to blame the world's poor for what we have dubbed the "population problem." While it is true that poor people often have large families, seldom do we ask, "Why is this so?" But place yourself for a moment in a poor family living, say, in the highlands of Ecuador. You are fifteen, the eldest among five children. Your family has no electricity; your meals are cooked in a clay stove; you rely on wood to warm your small shelter. You have no telephone in your home, nor in your small village. Your days, and those of your siblings and parents, are spent engaged in such activities as gathering wood and water from distant sources; cultivating, weeding, and harvesting tubers, vegetables, and grains; creating and repairing terraces for your crops; processing and preparing food; herding goats and making goat cheese; weaving clothes and making handicrafts; constructing and repairing tools; and tending to and caring for one another. If you had the wherewithal to construct a time budget—an accounting of how the seven people in your family spend their time—you would realize that more children means more hands and legs and heads to help with the many time-consuming tasks necessary to run a household. Seen in this more empathetic light, the Ecuadorian family is not illiterate or reproductively irresponsible; rather, these people are doing just as you would do if placed in similar circumstances.

In a Paradox Lies an Insight

What might it take to reduce family size and thereby put an end to human population growth? For many years family planning experts believed that birth control technologies—condoms, sterilization, and birth control pills—were the way to control family size in lesser developed countries (LDCs). It is now generally accepted that these technologies, although part of the solution, fail to address what in many cases is the root cause of population growth—namely, systemic discrimination against women. Women in LDCs, as well as those born into poverty in the "developed" world, routinely receive less education, get paid less money, work longer hours, and generally exercise less control over their lives than men in the same countries or circumstances. Hence, population experts in the UN and other international agencies are in broad agreement that any lasting solution to the population problem must be geared toward the empowerment of women.

The state of Kerala in India offers a clear example of how empowerment can lead to a reduction in family size. Kerala is similar to lesser developed countries in terms of its low per capita income (three to four hundred dollars per year). However, in spite of this meager per capita income, Kerala looks a lot like the United States in terms of key quality-of-life indicators such as life expectancy, birth rate, and literacy (table 6.1). How could this be? Science writer Bill McKibben, who traveled to Kerala to try to unpack this paradox, puts the mystery this way:

> Kerala undercuts maxims about the world we consider almost intuitive: Rich people are healthier, rich people live longer, rich people have more opportunity for education, rich people have fewer children. We "know" all these

Table 6.1. Quality-of-Life Indicators for Lesser Developed Countries, Kerala, and the United States[4]

Quality-of-Life Indicators	Lesser Developed Countries	Kerala State	United States
Per capita income ($/year)	400	350	25,000
Life expectancy (years)	58	70	72
Birth rate (N°/1000/year)	40	18	16
Literacy (%)	50	96	96

things to be true—and yet here [in Kerala] is a counter case, a demographic Himalaya suddenly rising on our mental atlas. It's as if someone demonstrated in a lab that flame didn't necessarily need oxygen, or that water could freeze at 60 degrees. It demands a new chemistry to explain it, a whole new science.[3]

Kerala's new "chemistry" has ingredients that are rarely seen in LDCs—effective government, redistribution of wealth, equal rights for women, and full access to education and health care. The educational opportunities afforded to all of Kerala's citizens have been especially important for women because educated women are better able to take charge of their lives. As a reflection of this, the average woman in Kerala doesn't marry until twenty-two years of age versus eighteen years for the rest of India. Kerala's reforms, especially the emancipation of women and comprehensive health care for children, seem to lead quite naturally to smaller families.[5]

Just as poverty, lack of education, and discrimination are insidiously linked to large family size in many LDCs, so it is in the United States. For example, U.S. college graduates have 2.0 or fewer children, on average, while high school graduates have 2.7 kids, and those who fail to finish high school average 3.2 children.[6]

Fortunately, in recent years women in many parts of the world have been having fewer children as they become more assertive in their families and gain more control over their reproductive lives. According to current optimistic United Nations projections, the human population might top out somewhere around eight or nine billion people. This projection assumes—perhaps correctly, perhaps incorrectly—that women in the developing world will soon average just two children apiece.

In sum, as understanding of the dynamics surrounding human population growth has grown, simplistic assumptions and prescriptions from an earlier time are no longer tenable. It is now clear that the "population problem" is symptomatic of a larger suite of problems rooted in inequality among the sexes and, as we will soon see, inequity in the distribution and control of wealth.

Reflection

One day a student stopped by my office and said that she really didn't see why I was talking about population growth in class because the population problem was solved. In fact, she had read in the *New York Times* that the human population growth rate is declining. This student was being confronted with two streams of information that didn't seem to jive, and this made her uncomfortable. She wanted resolution. I told her that I would bring her question to the class the next day.

As students entered the classroom the following morning, I had the following "mock" newspaper headline on the blackboard:

> *Decline in Human Population Growth Rate*
> *Heralds End to Population Growth*

Once they were settled, I asked them to examine the assumption explicit in this headline—namely, that a decline in growth rate will bring an "end" to population growth. Then, to address this assumption empirically, I invited them to "crunch" some population numbers with me. First, I gave them the estimate for the earth's human population in 1990 (5,284,000,000) along with the population growth rate in 1990 (1.6 percent/year), and I asked them to calculate the increase in human numbers for that year (5,284,000,000 × 0.016 = 85,000,000 net population increase in 1990). Next, I gave them the estimate for the human population for the year 2000 (6,080,000,000 people) and the population growth rate for that year (1.3 percent), and I asked them to calculate the increase in human numbers during the year 2000 (6,080,000,000 × 0.013 = 79,000,000 net population increase). Conclusion: Even when the population growth rate was declining from 1.6 percent to 1.3 percent, as was the case between 1990 and 2000, the human population was still increasing.

As we saw earlier in the chapter, a steady growth rate of 1.2 percent will cause the human population to double in about sixty years. What would the population growth rate have to be in order for human population growth to stop? Zero! The *New York Times* headline that would truly herald an end to population growth would read: "Population Growth Rate Declines to Zero" or better yet "Population Growth Rate Turns Negative." Some countries, such as Italy, Spain, and Russia, are already experiencing genuine population declines and thus show that lowering world population size is a possibility. In fact, if all couples on Earth were suddenly to limit their family size to only one child starting today, the human population would drop to between two and three billion by the end of this century. In other words, just as the human population increased dramatically during the twentieth century, human numbers could decline significantly in a similar time—which is not to say that it will happen, but only that such things are theoretically possible.

Questions for Reflection

- Why do you suppose it is that some people prefer to use the term "Two-Thirds World" instead of the more conventional terms, "lesser developed world" or "Third World"?
- If you had unlimited access to data and technical expertise, how might you go about figuring out the number of people that Earth can support?

Practice: Searching for Connections Leads to Discernment

Discerning connections is an important practice for expanding ecological consciousness. We humans make connections all the time. For example, you might awaken in the morning and make a connection between frost on the window and the temperature outside, between the smell coming from the kitchen and pancakes, between the smile on your housemate's face and her mood.

The more critical and discerning our thought processes, the greater our capacity to make connections. Often, though, it is difficult to make connections between our everyday actions and the deteriorating condition of the earth because our myriad ecological dependencies are frequently hidden from view. For example, if your friend drives fifteen miles in a SUV to visit you and then fifteen miles back home, he will burn two gallons of gas. Your friend will likely be sensitive to the economic impact of his cross-town visit. After all, two gallons of gas costs a few dollars. On the other hand, he might be totally oblivious to the trip's ecological impact, failing to note that the burning of those two gallons of gasoline introduces almost forty pounds of the greenhouse gas carbon dioxide into the atmosphere.

The daily newspaper serves as wonderful source material for this practice of discerning connections. Day after day the headlines come at us: collapsed economies, toxic spills, new computer technologies, civil wars, corporate mergers, genetic engineering breakthroughs, acts of terrorism, record growth, and so on. These headlines catch our attention, but many of us are unable, or unwilling, to see the connections among the various news events. For example, I looked at the front page of my local newspaper one November morning and saw headlines announcing, first, that the temperature of the earth had reached a record high; and second, that holiday shopping was breaking all past records. In class that same day, I asked my students if they saw any connection between these two records—a hotter Earth and record holiday shopping. At first they seemed baffled, but then someone pointed out that all the stuff society is consuming is made, transported, and disposed of using energy (fossil fuels); and the burning of these fossil fuels releases carbon dioxide to the atmosphere, which is causing climate destabilization. Such connections are everywhere. As educator David Orr cogently observes:

They are part of a larger pattern that includes:
—shopping malls and decaying downtowns
—sprawling suburbs and the loss of farmland
—crowded freeways and climate change
—overstocked supermarkets and soil erosion
—cheap electricity and acid rain
—hazardous wastes and childhood cancers.[7]

One way to deepen this practice of seeing connections is to identify something that you find troublesome in the world. Pick anything and then ask, "Why is this happening?" It is especially stimulating to do this with a group of friends. For example, I recently explained to some friends that I was deeply troubled by the terrible suffering of the world's children—thirty thousand innocent children die each day of malnutrition and easily preventable illnesses. "Why is this happening?" I wondered aloud. Marie talked about the lack of food and medicine in specific places; however, she correctly observed that in the big picture humankind has more than enough food and more than enough medicine to care for its children. We concurred that the poor distribution of food and medicine is part of the "why."

What else is at work? Joel mentioned that corporations like Nestlé aggressively market infant formula to Third World countries as a high-status substitute for mother's breast milk; poor women sometimes fall for this ploy. If the water they add to the formula is contaminated, as is sometimes the case, children may die of dysentery. So perhaps disempowered women and corporate irresponsibility are also aspects of the "why."

Julie observed that many people in the United States seem to regard those who are far away and of a different race or culture as somehow inferior. Hence, we included prejudice as a possible part of the "why."

Tressa then observed that organizations like UNICEF and Oxfam (Oxford Committee for Famine Relief) can save the lives of children. In fact, she informed us, it would only take two hundred dollars to help a sickly two-year-old transform into a robust six-year-old—offering safe passage through childhood's most precarious years (even recognizing that a portion of this money would go to administration).[8] So, selfishness also seems to be part of the "why." As we probed deeper into the question, I wondered if part of the explanation might lie in our own deadness: Are we able to let innocent children die because we, ourselves, are dead—our eyes blind, our hearts numb, afraid to feel the pain of the children's suffering?

All of these observations led us to deeper-level "why" questions. Why don't we solve the food distribution problem? Why are mothers disempowered? Why do corporations sometimes behave irresponsibly? Why are people of means unwilling to do more to help their poor brothers and sisters in distant lands? As we went deeper and deeper into the "why," the ecological, political, economic, and psychological layers and connections began to reveal themselves.

The importance of asking "why" is illustrated in a story told by Fritz Stern, about Primo Levi, a Jewish man who was forced to endure a long journey in a cattle car on his way to a concentration camp during World War II. Levi was hungry and very thirsty. Deep into the journey, the train stopped, and spotting an icicle, Levi reached out to break it off; but before he could bring it to his lips, a hulking guard grabbed his arm and snatched the icicle away. Levi looked at the guard and asked, *"Warum?"* ("Why?"). The guard responded, *"Hier ist kein warum"* ("There is no 'why' here").

Responding to this incident, Stern writes: "This 'Hier ist kein warum' stands against everything that is human and constitutes a form of verbal annihilation." The story reminds me that asking "why" is a precious right, as well as a powerful tool in helping us discern connections and deepen understanding.[9]

FOUNDATION 6.2: CONSUMPTION EXPLOSION

The rapid growth of the human population in recent times has been accompanied by an extraordinary rise in the production and consumption of toasters, fertilizers, plastics, cars, buildings, guns, rice, along with tens of thousands of other items. This growth, in turn, has led to an increase in the by-products of production: solid waste, hazardous waste, freshwater pollutants, greenhouse gases. The links between waste and population are so strong that garbage production, by itself, is a good predictor of human population size (see box).

THE GARBAGE–POPULATION LINK

In the 1980s, the U.S. Census Bureau contracted archeologists at the University of Arizona to help them estimate population size in difficult-to-census areas like inner cities. The research team proceeded to collect garbage from sixty-three households over a five-week period. All the garbage was separated, household by household, and sorted into sixteen categories—for example, newspaper, food waste, yard debris, textiles, plastic, glass, and so on—then it was weighed. When the team analyzed their data, they found that total garbage weight was a good predictor of the number of people living in the households; but an even better predictor was plastic: "During any given period of time, every man, woman, and child in America generates about the same amount of plastic garbage, usually in the form of many, many, small items."[10] In other words, knowing the amount of plastic in a region's garbage will allow one to predict the population size of that region with surprising accuracy. There is even a formula: number of people = (0.285 × plastic), where "plastic" refers to the weight of plastic in pounds generated over a five-week garbage-collection period. "The projected total population estimate derived from this equation . . . will be accurate to within plus or minus 2.5%. This is considerably better than the Census Bureau can do in many places."[11]

Not everyone consumes resources and generates garbage at the same rate. Environmental analyst Alan Durning has divided the world's peoples into three broad economic classes: "poor," "middle income," and "consumers." In Durning's classification, the world's poor include all households that earn less than seven hundred dollars per year, per family member. They are mostly rural Africans, Indians, and South Asians. They eat grains, root crops, and legumes; they drink mostly unclean water. They live in huts, or worse; they travel by foot; and their meager possessions are constructed of materials from the local environment, such as stone, wood, and clay. The poor represent 20 percent of the world's people, but they earn just 2 percent of the world's income.[12]

The people in Durning's "middle income" class earn between seven hundred and seventy-five hundred dollars per family member, per year; they live mostly in Latin America, the Middle East, China, and East Asia. Their diet is centered on grains; they have access to electricity; they travel by bus, railway, and bicycle; and they possess a modest stock of durable goods—for example, several changes of clothes, a radio, a refrigerator, some basic furniture, and so on. Collectively, this class represents approximately 60 percent of the world's people and claims about one-third of the world's income.

Finally, the "consumer" class—the richest members of global society—includes all households whose income per family member is above seventy-five hundred dollars per year. If you are reading this book, the chances are good that you are in the consumer class. Consumers dine on meat and processed foods, and they drink soft drinks and other beverages from throwaway containers. They spend much of their time in climate-controlled buildings, and they generally travel in private automobiles. Consumers spend more time shopping, paying bills, and caring for their possessions than they spend with their children. Overall, this class represents roughly 20 percent of the human population while commanding two-thirds of the world's wealth.

The consumer lifestyle results in the consumption of prodigious amounts of energy. When humans lived by hunting and gathering, each person consumed about twenty-five hundred Calories each day to survive, essentially all of it in the form of food. This amount is roughly equivalent to the daily energy intake of the common dolphin. But these days, someone in Durning's consumer class uses about 180,000 Calories each day, most of it in the form of fossil fuels, which is equivalent to the caloric intake of a sperm whale. For consumers, it is as if our stomachs have expanded seventyfold. We have become giants.[13] Referring to our colossal consumption, the renowned ecologist Eugene Odum suggested that we change our name from *Homo sapiens* to *Homo colossus*.[14]

Overall, the rate at which humans are consuming the earth's resources is much greater than the rate of human population growth. This statement bears repeating: Even if the human population were stable, the human impact on Earth would still be growing because each human, on average, is consuming more and more materials each year. For example, since the 1950s, the worldwide per capita consumption of copper, energy, meat, steel, and timber has approximately doubled; car ownership and cement consumption, quadrupled; plastic use, quintupled; aluminum consumption, grown sevenfold; and air travel, increased thirty-three-fold.[15]

For those of us in Durning's "consumer" class, we have a tendency to avoid thinking very much about what goes into the things we buy. But like a boat crossing a bay, each product that we purchase leaves behind an ecological wake (see box).

What may seem like small purchasing decisions (e.g., buying a can of Coke) often add up to big environmental impacts because of the waste created in the production of all the stuff that consumers purchase. For example, in the mid-1990s, each U.S. citizen was consuming, on average, "120 pounds—nearly their own body weight—every day in natural resources extracted from farms, forests, rangeland, and mines."[17]

TELL ME ABOUT THAT CAN OF COKE I JUST BOUGHT

I dug some change out of my pocket and bought a Coke. My cola was 90 percent water. The rest was mostly high-sugar corn syrup. A corn milling plant
used water, enzymes, and acids with heat, grinders, and centrifuges to turn
the corn kernels into starch and then corn syrup. A bottling plant added water, citric acid, preservatives, caffeine, artificial coloring, and secret flavors to
the corn syrup.

My Coke was in an aluminum can. Massive machines—with twelve-foot
high tires and shovels, big enough to scoop up a car—mined the aluminum
ore from the earth. Near the mine, the ore was crushed, washed, dried,
mixed with caustic soda, heated, filtered, and roasted with calcium oxide. The
cleaned ore was transported to a smelter, which used massive amounts of
electricity to produce giant ingots of aluminum. The ingots were then transported to another factory and pressed into a thin rolled sheet of aluminum;
afterward, they went to yet another factory where the aluminum sheets were
formed into cans and printed with a colorful design. Ovens then baked the
can to dry the printing. Later, at the bottling plant, machines filled the can
with near-freezing soda and immediately crimped the top on. I finished my
drink seeing, for the first time, the ecological saga behind what I thought was
just another can of Coke.[16]

The spread of the consumer lifestyle around the world over the last
fifty years represents a significant departure from past patterns of human
existence. This change has been particularly dramatic in the United
States, where many homes have been transformed from production centers to consumption centers. As recently as the 1950s, U.S. homes were
often designed with pantries, storage cupboards, root cellars, laundry
rooms, sewing rooms, and workshops. In other words, homes were built
as production centers. Modern American homes, by contrast, have few,
if any, of these production features. Instead, as Durning points out, they
are designed as consumption centers—places to eat (e.g., cook food in
the microwave), sleep, bathe, and be entertained between work shifts.[18]

The glorification of material consumption has been a part of human
culture for a long time. According to anthropologist David Gilmore, who
has studied the images of manhood in traditional cultures from around
the world, boys in most cultures need to prove themselves to earn the
right to manhood. They do so through their physical prowess and by being able to gain access to or produce an abundance of goods. Specifically,
the high-status male is usually the one who produces more materials—

crops, meat, fiber—than he consumes. Gilmore asserts that, in culture after culture, the most respected men have traditionally been "those who give more than they take; they serve others. Real men are generous, even to a fault."[19]

Along the path to modernity, the focus shifted from the acquisition of resources to serve the community to the acquisition of resources to serve one's self. So, in the contemporary industrialized world, status is accorded to the male who has accumulated the most capital, cars, houses, stocks, and so forth; the "big man on the block" is the one who has the most money, power, and material goods. As author Fritjof Capra observes, "The association of manhood with the accumulation of possessions fits well with other values that are favored and rewarded in [contemporary] culture—expansion, competition, and an 'object-centered' consciousness."[20]

Overall, the key point is this: Human numbers, by themselves, don't exert an impact on Earth; the impact is exerted through the combination of population *and* consumption. A person in the consumer class takes twenty to thirty times more from the earth than one person in the poor class. Thus, the consumers apparently have a much more severe "population problem" than the world's poor people.

Reflection

As I was writing this chapter, I became interested in the psychological basis of consumerism. A book that helped clarify my thinking is *Why They Buy: American Consumers Inside and Out* by Robert Settle and Pamela Alreck. This book was written for "marketers"—professionals who specialize in getting people to want to buy specific products. The authors identify sixty-odd psychological needs that humans have, including the need to

- be visible to others,
- establish one's sexual identity,
- engage in forceful bodily activity,
- win acceptance,
- engage in the unusual,
- play,
- be amazed,
- learn new skills,
- see living things thrive,
- win over adversaries,
- be free from the threat of harm.

The challenge for the marketer is to take this list of needs and identify products that can meet each type of need. For example, if humans truly have a need "to be free from the threat of harm," then it follows that marketers and production specialist would come together to offer us locks, double bolts, alarm systems, motion sensors, personal revolvers, watchdogs, outdoor lights, mace spray, self-defense courses, and on and on.[21]

Reading Settle and Alreck's book, I came to appreciate that all the stuff that I see in catalogs and stores—the nail polish, customized clothes, pet supplies, perfumes, how-to books, exercise equipment—really does meet a need, so to speak. But this realization does not mean that it is necessarily a good idea to meet all of our needs. For example, if I ate a bar of chocolate every time I experienced my chocolate "need," I would be one poor, fat man.

Questions for Reflection

- How might you distinguish between your "needs" and your "wants?"
- Might it be possible to meet some of your needs without the purchase of additional goods or services.

Practice: Questioning Assumptions Leads to Discernment

Almost every day the news media seems to present distressing information about planetary degradation. I find it depressing to learn, for example, that the fertilizers and herbicides that we apply to our lawns can have deleterious effects on nearby streams and lakes, and that our increasing dependence on the automobile causes climate destabilization. We have developed sophisticated, largely unconscious strategies to avoid facing up to such unsavory truths. One such strategy is the consensual lie—the culturally sanctioned lie that we embrace to avoid the pain of having to face the truth. Social activist Sharif Abdullah tells a story that illustrates the pervasive nature of the consensual lie in our culture:

> When my eldest daughter was in fifth or sixth grade, she asked me a homework question: "What is the principal cash crop in Florida?"
> My answer was immediate. I didn't even look up from my newspaper. "Marijuana."
> The next day her teacher sent a note home asking for a conference. At the conference, I supplied the teacher with the statistics that conservatively estimated the marijuana crop at twice the value of citrus products. She said, "That maybe true, but it's not the answer we want. It's too controversial."

The teacher expected us to participate in the lie that our economy is not based [in part] on drugs, pornography, stealing, or other activities we label criminal or antisocial. By our refusal to consent, the teacher had to face her own acquiescence in the lie. . . . The consensual lie is practiced so constantly in our society that we begin to think that telling the truth is wrong.[22]

Courageously questioning the everyday assumptions embedded in our culture is a powerful practice for rooting out consensual lies. Certain assumptions have become so completely kneaded into our psyches that we find it is easy to mistake them for essential truths. For example, we often assume that

- the role of humans is to control and dominate the earth;
- the industrial economy can continue to grow forever;
- capitalism is the best economic system humans can aspire to;
- technology will solve all of humankind's problems;
- individuals don't have the power to change the way things are.

Once an assumption has been identified, the next step is to scrutinize it. For example, consider the following assumption: "We, as a society, are making progress." In examining this assumption, I find it helpful to ask myself "What is meant by progress?" followed by "What are the signs that we are progressing?" In doing so, many unsettling questions arise. For example:

- Are we progressing when the average citizen in the consumer class requires at least fifty times more energy—almost all of it from fossil fuels—to supply his or her "needs" than was required by U.S. citizens two hundred years ago?
- Are we progressing when citizens know almost nothing about the origins of the food that they put into their bodies: where it was grown, who grew it, what chemicals it might be contaminated with?
- Are we progressing when one-third of U.S citizens die of cancers, many of them environment-related?
- Are we progressing when we dwell in cities that have been converted to cement-scapes to accommodate a polluting machine (the car) that causes the direct death of approximately forty-five thousand Americans each year?
- Are we progressing when humankind's quest for material goods has resulted in the extinction of some five hundred thousand fellow species in the past one hundred years alone?

- Finally, what kind of progress is it when young people, upon graduation from college, enter the workforce with a high likelihood of spending much of their lives sitting in front of a computer?

The point here is that many facets of "progress" or "development" may,
in the end, not represent a movement toward something comprehensively
better. For example, is having two cars in your garage and grapes from
Chile on your table a manifestation of genuine human progress? It may be,
but it would be naïve to simply assume so without first thinking about it
long and hard.

As we raise questions surrounding our cherished assumptions, our understanding often becomes more robust and nuanced; we come to see a
bigger picture. But make no mistake: Questioning assumptions is not easy.
It can be exciting, but it is also disorienting because when you question assumptions, you are often questioning the authority of our culture; and it is
usually much easier to obey authority than to question it. As author Derrick
Jensen points out, our culture censors us when we question its sacred icons:
"Question Christianity, damned heathen. Question capitalism, pinko liberal. Question democracy, ungrateful wretch. Question science, just plain
stupid."[23]

Most of us understandably find it very difficult to question authority.
It is a tendency that was powerfully illustrated by Dr. Stanley Milgram's
research, conducted at Yale University in the early 1960s. To experience
the import of Milgram's work, imagine that you are a student at Yale. The
year is 1962, and you see a sign asking for volunteers for a psychology experiment. You need cash and decide to sign up. You are told that you will
help with research aimed at determining "the effects of punishment on
learning."

After an orientation, the work begins. You are seated facing another person who is taking a test. This person wears electrodes, and a control panel
rests on the table where you are seated. You are instructed to monitor the
test. If the test taker marks a wrong answer, you are to deliver a shock to
him; a second mistake merits another somewhat more potent shock, and so
on. You look at the switch board in front of you—thirty separate shock in-
tensities ranging from "slight shock" to "danger—severe shock!" . . . This is
a true story.

What you and the other volunteers are not told is that the test taker is an actor. No real jolt will be administered when you pull the various switches. The
actor will pretend to be in pain, but you won't know that he is pretending.

The experiment begins. A wrong answer. You pull the first switch (slight shock). Then come two correct answers and then another mistake; you pull the second switch. Then another mistake, and another. You look at the test taker, and you see that the level-five shock clearly caused him pain. Three correct answers in a row, and then the test taker makes another mistake, and another, and another, and another. You are uncomfortable and turn to the scientist in a white lab coat who is supervising the experiment. He responds that "the experiment requires that you continue." And so you continue—wrong answer, wrong answer. . . . Even when the test taker exhibits agonizing pain, you pull the switches.

You are not alone: 30 percent of the other volunteers continue on, inflicting excruciating pain on another human sitting in front of them. And two-thirds of the volunteers continue inflicting severe shocks when the test taker is not immediately in view. Here's how Dr. Milgram interpreted the results:

> It is the extreme willingness of adults to go to almost any lengths on the command of an authority that constitutes the chief finding of the study. . . . Ordinary people, simply doing their jobs, and without any particular hostility on their part, can become agents in a terrible destructive process.[24]

In sum, a fundamental lesson of our culture, absorbed to varying degrees by all of us, is to follow orders: do the job you were hired to do; don't rock the boat.

When we stop questioning fundamental cultural assumptions, we stop taking responsibility for our lives. We are no longer the captains of our destiny. Author and poet David Whyte uses a story to describe the consequences for ourselves and for others when we give up our captaincy:

> A six-month-old child is admitted to the hospital with early congestive heart failure. The doctor prescribes Rogoxin which steadies the heart rate but can be lethal above a certain level. The doctor places the decimal point in the wrong place and prescribes 0.9 mg instead of 0.09. An experienced nurse catches the error and consults with another nurse. They both say it is too high; they take it to a second doctor for a second opinion; he does the recalculation and says the first doctor was right. They give the Rogoxin at the higher dose and the child dies. Who had the captaincy? Somewhere inside themselves the nurses thought the doctor was the real captain no matter the outward circumstances and that they were powerless. They were not; they had the captaincy, but not the courage of a captain's convictions.[25]

Questioning assumptions forthrightly and challenging outside authorities—whether that authority be vested in individuals, institutions, or in culture—is a pathway to clarity, discernment, and captaincy. Practiced over a lifetime, it can contribute to personal growth and the expansion of consciousness.

FOUNDATION 6.3: UPSHOT—
OUR EXPLODING ECOLOGICAL FOOTPRINT

Finally, we arrive at the heart of the matter—the most important question: Does Earth have the capacity to accommodate humanity's burgeoning numbers and resource demands? Fortunately, we do have a metric that reveals if humankind is living within Earth's supply limits. This metric is called the "ecological footprint." A person's ecological footprint is the area of bioproductive land and sea necessary to produce all the things he or she consumes plus the space necessary to accommodate all the waste he or she generates. The footprint concept is based on two straightforward assumptions: first, it is possible to accurately estimate both the stuff people consume and the waste they produce; and, second, these consumption and waste flows can be expressed in terms of Earth-area equivalents.

In practice, calculating an ecological footprint is an accounting exercise: Consider the common hamburger that you might order at a fast-food joint (ignoring, for the time being, the burger's bun and condiments). The beef patty came from a steer. During its early life, this steer required land to graze on; and later, when it was being fattened in a feedlot, land was needed to grow the steer's feed. This steer was killed and processed at a meat-packing plant, and this required space for the facility and energy to power the plant. Next, the processed beef patty was transported to your local fast-food restaurant. Here, too, land was required for roads, parking, and so on; and energy was needed for transportation. Still more energy was needed in the restaurant for both refrigeration and cooking. The ecological footprint of this burger can be expressed by summing the various components of your hamburger's ecological history in terms of land-area. To further demystify the ecological footprint concept, you may find it helpful to actually calculate the footprint of a daily newspaper (see box).

Newspaper is just one part of a person's footprint. Each of us also needs land for food, housing, roads, energy, and so forth. In most countries, energy constitutes the lion's share of the human footprint. Researchers take two different-yet-related approaches when calculating energy footprints. One is to determine the land-area necessary to grow biofuels. For example, in Brazil huge expanses of land are planted to sugarcane, which is distilled to produce ethanol that then goes to power vehicles. Crops can also be grown for fuel in the United States; in the footprint example (see box on page 176), it was assumed that the energy to manufacture newspaper would come from biofuels. It is also possible to calculate footprints when fossil fuels are the energy source of choice. In this case, analysts determine the energy footprint by calculating the forest land-area necessary to assimilate the carbon dioxide released in the

DEMYSTIFYING THE FOOTPRINT CONCEPT

Let's calculate the amount of land necessary to supply an individual with his or her daily newspaper over the course of a year. Let's start with the amount of land needed to grow the pulp wood for the newspaper. If a daily newspaper weighs 0.66 pounds on average, then total newspaper consumption over the course of a year would amount to 241 pounds (0.66 pounds × 365 days = 241 pounds/year). Next, knowing that, in one year, one acre of natural (i.e., nonplantation) temperate forest produces enough pulp to make approximately twelve hundred pounds of paper, it follows that 0.2 acres of forest is necessary to supply the pulp for the newspaper (241 pounds of newspaper × 1 acre/1,200 pounds paper = 0.20 acres). Commonly, half the fiber in newspaper comes from recycled sources, so the actual forest area necessary to produce virgin pulp for one year's supply of newspaper would be 0.1 acre (0.20 acres × 0.5 = 0.1 acres).

Land will also be required to produce the energy to manufacture the newspaper. Given that sixty-six hundred kilocalories of energy are necessary to manufacture a pound of paper, then 1,590,600 kilocalories of energy would be required to manufacture one person's annual consumption of newspaper (6,600 kcal/pound × 241 pounds of newspaper = 1,590,600 kcal). Next, given that one acre of land can produce ten million kilocalories of energy per year, an estimated 0.16 acres would be required to produce the energy necessary to manufacture a year's supply of newspaper (1,590,600 kcal of energy to manufacture newspaper × 1 acre/10,000,000 kcal = 0.16 acres). Summing the virgin pulp acres (0.1) and the energy for paper manufacturing acres (0.16) gives one person's newspaper footprint—0.26 acres. Think of it as an area of forest roughly one hundred feet on a side, "working" day after day to supply a person with his or her daily newspaper.[26]

burning of the fossil fuels to produce a certain product (e.g., newspaper). This approach is taken because, as we will see in the next chapter, the unrestricted release of carbon dioxide into the atmosphere has begun to destabilize Earth's climate. Hence, to create a sustainable world, land must be set aside to assimilate the carbon released in the burning of fossil fuels.

Insofar as consumption levels vary from country to country, it should come as no surprise that footprint sizes also vary. The footprint for the average U.S. citizen is twenty-four acres; the average Chinese citizen, three and a half acres; the average United Kingdom citizen, twelve acres. Note that people in countries with similar standards of living can have very different ecological footprints—for example, United States (twenty-four acres) versus United Kingdom (twelve acres).

The full significance of the footprint approach is revealed when we consider it in the context of Earth's total bioproductive acreage. The total surface area of Earth is 126 billion acres. About 38 billion acres of this total is land. Of this land total, roughly 25 billion acres is productive land—that is, crop, pasture, and forest land. Most of Earth's surface is water (70 percent, or roughly 90 billion acres). About 7.3 billion acres of this area is considered productive (it's mostly associated with continental shelves); the rest is relatively barren. Summing the land and water numbers, there are 32.3 billion acres of biologically productive land and sea on Earth (table 6.2).

The relationship between population and productive land-area varies from country to country. Some countries (Brazil, Australia) have huge amounts of productive land and relatively low populations. These countries have an ecological surplus. Other countries (Singapore, Hong Kong) have large populations for their comparatively small areas of productive land. These countries are operating on an ecological deficit, which means that they have to rely on land and sea outside of their borders to supply their ecological needs. Overall, out of the 146 countries with populations over one million people, 52 have overshot (exceeded) their domestic biocapacity.[27] The United States is among those in the "red"; it has nine million square miles of "biocapacity," but its population's ecological footprint covers sixteen million square miles. In other words, ten acres of each American's twenty-four-acre footprint must be "imported" from elsewhere.[28] With all the world's countries combined, the global ecological overshoot is now

Table 6.2. Ecological Footprint Statistics

Productive Acreage on Earth (in Billions of Acres):	
—Cropland	3.7
—Pasture land	8.5
—Forest	12.8
—Productive ocean surface	7.3
—Total	**32.3**
—Adjusted total	**28.4[a]**

Summary Statistics:	
—Average productive earth surface available (acres/person)	4.5[b]
—Number of Earths necessary to support 2003 human population at U.S. standard of living	5[c]

[a] The World Commission on Environment and Development recommended that 12 percent of the productive surface of Earth be preserved in a natural state. Thus, 3.9 billion acres were subtracted from 32.3 billion acres, the productive acreage on Earth.
[b] 28,400,000,000 acres / 6,300,000,000 people = 4.5 acres/person
[c] (6,300,000,000 people × 24 acres/person) / 28,400,000,000 acres of productive surface on Earth = 5 Earths

approximately 25 percent, which means that it takes about fifteen months for the earth to regenerate the resources that humankind consumes in twelve months (figure 6.2).

Another useful feature of the footprint concept is that it offers us an opportunity to calculate the amount of bioproductive space each person on Earth would receive if everybody was to get an equal slice.[29] Although the earth has 32.3 billion acres of bioproductive space (table 6.2), we would be arrogant in the extreme (as well as ignorant) to consider that all this land should be set aside to satisfy human needs. The World Commission on Environment and Development recommended that at least 12 percent of the productive capacity of the biosphere, representing all ecosystem types, be preserved in a natural state.[30] Using this conservative 12 percent figure, the amount of bioproductive space available for humans comes to 28.4 billion acres, or 4.5 acres for each of the 6.3 billion people residing on Earth in 2003. This all begins to sound rather academic until we note that the cur-

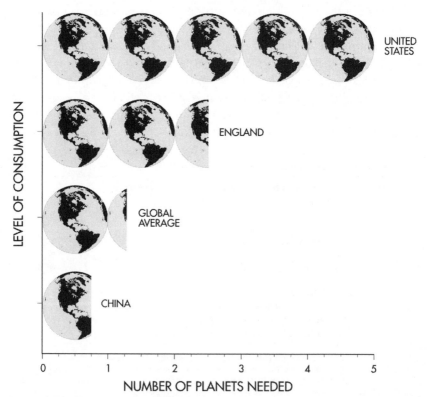

Figure 6.2. The number of "Earths" required to satisfy humankind's needs under different consumption scenarios

rent U.S. footprint, 24 acres per person, is more than five times this equitable footprint.

Another way to work with the numbers is to ask how many people the earth could support if everybody had a U.S. average footprint of 24 acres. The answer: 1.2 billion people (28.4 billion productive acres/24 acres per person = 1.2 billion people). In other words, five Earths would be necessary to support the world's current population at U.S. standards. Calculations like these reveal the fallacy of imagining that, given current technologies and consumption patterns, the earth's growing human population could ever come close to enacting U.S. lifestyles.

One final run at the numbers is warranted—this time using the optimistic assumption that the human population will top out at eight billion sometime later in this century. What might an equal share of productive space be for our children and grandchildren? The answer: 3.5 acres, 15 percent of the current U.S. footprint.

All of this "footprinting" can lead to some downright depressing conclusions, but we do have a bright side. Footprinting finally provides humankind with a tool for candidly assessing the "population problem." Indeed, footprint assessments are like bank statements, and bank statements are great tools—even when they tell us that our finances are in disarray—because they give us the information and motivation required to choose wise actions.

Reflection

Several years ago, I invited Dr. Mathis Wackernagel, one of the originators of the ecological footprint concept, to visit my university. In his public lecture, Dr. Wackernagel explained that the U.S. footprint (24 acres/capita) is five times greater than what is available on the planet per person (i.e., 4.5 acres). In accord with this, Dr. Wackernagel went on to recommend that we place a giant banner at the entrance to our university community to remind us of the most significant challenge for the twenty-first century: How can we all live well on 4.5 acres per person?

Shortly after Dr. Wackernagel's visit, I met with some friends over dinner to explore the possibilities for footprint reduction in our community. With the current footprint five times larger than the available capacity, we focused on a specific example: How might we reduce newspaper consumption fivefold? Ideas came quickly:

- Don't bother with the newspaper in the first place (70 percent of the "ink" is advertisements).

- Read the paper on the web.
- Team up with four friends. One subscribes to the paper for the year and puts it in a central place for the others.
- Stop at the library on your morning walk, and read the paper there.

Next, we considered how we might reduce (fivefold) the footprint associated with commuting back and forth to work. Again, creative solutions were forthcoming:

- Buy a house or apartment close to where you work, and then walk. This practice would result in dramatic gasoline savings, as the human body functions on the energy equivalent of two tanks of gas for an entire year.
- Bike to work. It would reduce one's transportation footprint more than tenfold.
- Carpool with four others in a fuel-efficient car. This would reduce one's commuting footprint by 80 percent.

For each of the other categories that we went on to discuss—clothes, food, entertainment—we came to see that viable solutions are available for significantly reducing footprint size. Moreover, we were intrigued to note that many of our solutions appeared to offer opportunities to interact more with others, build a sense of community, and have fun.

Questions for Reflection

- Oftentimes when people learn about the environmental problems in the world, they say that there is really nothing they can do. What do you think about this?
- Suppose that you were to simply accept that your grandchildren may very well live in a world of degraded forests; polluted oceans; water and food shortages; wildly oscillating climate; and constant social strife. Suppose, further, that this grim scenario could be avoided if knowledgeable citizens like you were willing to take steps now to significantly reduce their ecological footprint. What actions, if any, would you consider taking?

Practices: Reframing Leads to Discernment

We frame a picture to give it definition. If we chose the wrong-size frame, the picture doesn't look right. We also frame problems. Here, too, the way we frame a problem affects the way the problem looks to us. If our frame is too small—that is, if we are stuck in rigid thought patterns—we might not be able to solve the problem. Ellen Langer, in her book *Mindfulness*, provides an example of how narrow, conditioned thinking can compromise one capacities for discernment:

> Imagine that it's two o'clock in the morning. Your doorbell rings; you get up, startled, and make your way downstairs. You open the door and see a man standing before you. He wears two diamond rings and a fur coat and there's a Rolls Royce behind him. He's sorry to wake you at this ridiculous hour he tells you, but he's in the middle of a scavenger hunt. . . . He needs a piece of wood about three feet by seven feet. Can you help him? In order to make it worthwhile, he'll give you $10,000. You believe him. He's obviously rich. And so you say to yourself, how in the world can I get this piece of wood for him? You think of the lumber yard; you don't know who owns the lumber yard; in fact you're not even sure where the lumber yard is. It would be closed at two o'clock in the morning anyway. You struggle but you can't come up with anything. Reluctantly, you tell him, "Gee, I'm sorry."
>
> The next day, when passing a construction site near a friend's house, you see a piece of wood that's just about the right size, three feet by seven feet—a door. You could have just taken a door off its hinges and given it to him, for $10,000.
>
> Why on earth, you say to yourself, didn't it occur to you to do that? It didn't occur to you because yesterday your door was not a piece of wood. The seven-by-three foot piece of wood was hidden from you, stuck in the category called 'door.'[31]

The words that we use are frames of a sort. In contemporary culture, many things are cast in terms of a "battle" or a "struggle." When we reframe, we see things in a fresh light. The spirit and tenor of contemporary discourse could change if instead of talking in terms of combating disease, fighting for rights, and controlling the discussion, we spoke in terms of fostering health, cherishing our rights, and sharing ideas in the discussion.[32]

When problems are reframed, whole new arenas for action are often revealed. For example, as we saw earlier, the population problem was for many years framed in terms of poverty: Poor people were simply having too many children! The solution was birth control—plain and simple. Only recently has the population problem come to be understood in the context of a deeper set of issues, which include the oppression of women, gross inequities in the distribution and use of resources, and Earth's finite capacity to supply human needs and wants.

One way to practice reframing is to ask, "What is at the core of this issue?" The issues that are most highly polarized in our society are often the ones in greatest need of reframing. As a practice, take any issue and ask the question, "What is this issue really about?" As I write this chapter, there is a great deal of talk in the United States about the need to build more power plants to satisfy national energy needs. At the same time, there are voices that are endeavoring to reframe the issue. For example, Amory Lovins at the Rocky Mountain Institute reminds us that we don't want power plants per se, but rather what they give us—heat and electricity. Hence, building more power plants is not the only solution. Lovins has made it abundantly clear that the United States could supply all of its energy needs while dramatically reducing overall energy consumption simply by adopting state-of-the-art energy conservation strategies.[33]

It is also possible to reframe issues like energy conservation in the context of our personal lives. Imagine, if upon feeling a need to buy a second car, your family was to ask, "What's this issue really about?" On the surface, it would seem to be about the *need* for a second car, but after reframing, you might come to see it as the *need* to be able to get from place to place in a convenient fashion. Reframing would allow your family to think about other ways of meeting its transportation needs—public transportation, biking, ride sharing—and you might decide that you don't need a second car after all. Indeed, twenty-three families in Seattle recently agreed to give up using their second cars for several months as part of the city's program to encourage public transit. By the end of the study, four families had decided to sell their second car, and another six families were seriously considering selling.[34]

Finally, just as we reframe a problem by asking what's at the core of this issue, we can reframe our personal lives by asking, "What is my life all about now?" or "How is it that I am to be in the world now?" (see box).

A PERSONAL REFRAMING

For me the process of writing this book has allowed me to reframe my life. My previous professional identities—scientist, tropical ecologist, environmental researcher, sustainability specialist—no longer fully define who I experience myself to be. As I sat recently searching for a better label, the words "awakening," "healing," and "mentoring" surfaced. I was surprised because these are not the usual kinds of words used in our universities to describe what faculty do; but these words, more than any of the university job classifications I am familiar with, describe what I find myself doing. As a result of this personal reframing, I have a clearer sense of my purpose.

We make the first frame for our lives in our late teens. It is often a practical, safe frame that confines our spirit and subdues our dreams; it is constructed more by society than by our own heart's longings. If we are not vigilant, we will die in our "teen" frame. We can reframe our lives by periodically taking time to discern where our heart's deepest longings meet the world's deepest callings.

In sum, the practice of reframing offers a new, more expansive way of seeing. Instead of asking "What's the problem here?" we ask "What's possible here?" Instead of thinking of ourselves in terms of who we are, we come to understand ourselves as beings in an ongoing process of becoming.

CONCLUSION

Much argument can be made about the causes of global change, and bewildering statistics can be assembled on all sides of the debate. It's hard to know what to believe. So let me ask you this simple question: Is your childhood home in better ecological shape now than when you grew up? I'd be willing to guess that for most people the answer is No.

—David La Chapelle[35]

The ecological crisis is about growth—growth in human numbers and consumption, more growth than the planet can bear. Exercising discernment, we see that lying below the "population problem" is the more systemic ill of gender inequity. Dispassionate discernment also allows us to see that the explosive growth in the consumption of materials and in the generation of waste, particularly among the people in Durning's consumer class, has a greater impact on Earth than human population growth per se. A typical American family of four, with its estimated ninety-six-acre ecological footprint (4 people × 24 acres per person), is equivalent in terms of ecological impact to an Indian couple (2.5 acre footprint) with thirty-six children. The conclusion seems inescapable: People living in "developed" countries like the United States have a more serious "population problem" than people living in the "Two-Thirds World."

This is a painful pill. We have tried to ignore it for a long time. A part of all of us wants to make excuses and participate in consensual lies, but we do not need to dim our consciousness in this way. To be human can also mean exercising an insatiable thirst for the "why" of things. We have the capacity as humans to seek out connections, question assumptions, and reframe problems. At our best, we are all truth seekers, and by nurturing our capacity for discernment, we become whole.

NOTES

1. David Suzuki, *The Sacred Balance* (Amherst, N.Y.: Prometheus Books, 1998), 219.

2. This figure is adapted from Joel Cohen, *How Many People Can the Earth Support?* (New York: W. W. Norton, 1995), 82, 400–401.

3. Bill McKibben, "The Enigma of Kerala," *Utne Reader*, March–April 1996, 103–111.

4. McKibben, "The Enigma of Kerala."

5. McKibben, "The Enigma of Kerala"; Mark Hertsgaard, *Earth Odyssey* (New York: Broadway Books, 1998).

6. Alan T. Durning and Christopher D. Crowther, "Misplaced Blame, the Real Roots of Population Growth," (Seattle, Wash.: Northwest Environmental Watch, 1997). Contact information: Northwest Environmental Watch, 1402 Third Ave., Suite 500, Seattle, WA 98101.

7. Paraphrased in part from David Orr, *Earth in Mind* (Washington, D.C.: Island Press, 1994), 1.

8. Peter Singer, "The Singer Solution to World Poverty," *New York Times Magazine*, September 5, 1999.

9. Fritz Stern, "The Importance of 'Why,'" *World Policy Journal*, Spring 2000, 1–8.

10. William Rathje and Cullen Murphy, *Rubbish* (Tucson: University of Arizona Press, 2001), 142.

11. Rathje and Murphy, *Rubbish*, 143.

12. Alan T. Durning, *How Much Is Enough: The Consumer Society and the Future of the Earth* (New York: Norton, 1992).

13. Bill McKibben, "A Special Moment in History," *Atlantic Monthly*, May 1998, 55–78.

14. Eugene Odum, *Ecological Vignettes* (Australia: Harwood, 1998).

15. David Korten, *When Corporations Rule the World* (West Hartford, Conn.: Kumarian Press, 1995).

16. Adapted from John C. Ryan and Alan T. Durning, *Stuff: The Secret Lives of Everyday Things* (Seattle, Wash.: Northwest Environmental Watch, 1997), 62. Contact information: Northwest Environmental Watch, 1402 Third Ave., Suite 500, Seattle, WA 98101.

17. Ryan and Durning, *Stuff*, 4–5.

18. Alan T. Durning, "Are We Happy Yet?" in *Ecopsychology*, edited by Theodore Roszak, Mary E. Gomes, and Allen D. Kanner (San Francisco: Sierra Club Books, 1995), 68–76.

19. David Gilmore, *Manhood in the Making* (New Haven: Yale University Press, 1990) as described in Fritjof Capra, *The Hidden Connections* (New York: Doubleday, 2002).

20. Capra, *The Hidden Connections*, 264.

21. Robert Settle and Pamela Alreck, *Why They Buy: American Consumers Inside and Out* (New York: Wiley, 1986).

22. Sharif Abdullah, *Creating a World That Works for All* (San Francisco: Berrett-Koehler, 1999), 53–54.

23. Derrick Jensen, *A Language Older Than Words* (New York: Context Books, 2000), 40.

24. The Stanley Milgram quote is taken from Howard Zinn, *Declarations of Independence* (New York: HarperPerennial, 1990), 38.

25. David Whyte, *Crossing the Unknown Sea: Work as a Pilgrimage of Identity* (New York: Riverhead Books, 2001), 43–44.

26. The "givens" are from the following: J. N. Abramovitz and A. T. Mattoon, *Paper Cuts: Recovering the Paper Landscape*, Worldwatch Paper 149 (Washington, D.C.: Worldwatch Institute, 1999); Mathis Wackernagel and William Rees, *Our Ecological Footprint* (Gabriola Island, British Columbia: New Society, 1996); and Dr. James Findley, personal communication. Note: The newspaper footprint would be reduced by about one-third if raw pulp came from plantation sources.

27. Nicky Chambers, Craig Simmons, and Mathis Wackernagel, *Sharing Nature's Interest* (London: Earthscan, 2000).

28. M. Wackernagel, et al., "National Natural Capital Accounting with the Ecological Footprint Concept," *Ecological Economics* 29 (1999): 375–390.

29. Wackernagel and Rees, *Our Ecological Footprint*.

30. Chambers, Simmons, and Wackernagel, *Sharing Nature's Interest*.

31. Ellen J. Langer, *Mindfulness* (Cambridge, Mass.: Perseus Books, 1989), 9.

32. Jeremy W. Hayward, *Letters to Vanessa* (Boston: Shambhala, 1997).

33. For more information, see www.rockymountain.org.

34. Pam Chang, "Jeepless in Seattle," *Yes! A Journal of Positive Futures*, Winter 2002, 10.

35. David La Chapelle, *Navigating the Tides of Change* (Gabriola Island, British Columbia: New Society, 2001), 26.

7

COURAGE: PLANETARY DESTABILIZATION

It is obvious that something is not working properly. How else can we explain our mismanagement of resources and of our own population, the pollution and the destruction of our environment, or the mass murder of our own species? We cannot see the bigger picture and how we fit into it. We are no longer in our bodies; we are not in our right minds.

—Wes Nisker[1]

This is the last chapter of part II, Assessing the Health of Earth. Our field checks on trees, birds, water, and soil (chapter 5) revealed that all is not well on planet Earth. In chapter 6 we saw that human numbers and, especially, high levels of resource consumption are leading to planetary breakdown. In this chapter we explore the long-term consequences of continuing on our present path by examining three dilemmas now confronting humanity:

1. the web of species that humankind depends on for life support is unraveling as people expand their control over more and more of Earth's surface;
2. Earth's climate is becoming increasingly unstable as humans continue to release enormous quantities of greenhouse gases into the atmosphere;
3. human health and reproductive functioning is increasingly in jeopardy, as humankind persists in introducing new synthetic chemicals into workplaces, homes, and communities.

This chapter's practices center on courage. It takes courage to squarely face up to the ecological crisis that is unfolding around us—courage to speak and courage to act. This word, "courage," comes from Old French, and it means "heart." Our courage comes forth when our hearts are cracked open by compassion for the condition of the world.

FOUNDATION 7.1: DESTROYING THE WEB OF LIFE

The great American conservationist Aldo Leopold wrote more than a half-century ago of the importance of saving all the pieces—all the species of Earth's ecosystems. Leopold warned that if we, through our arrogance or ignorance, drive other species to extinction, we may face unexpected consequences.[2]

The Extinction of the Dinosaurs

Species extinction is nothing new. Throughout Earth's history, new species have evolved into existence while other species, unable to adapt to changing conditions, have gone extinct. Only a tiny fraction of all the species that have ever existed on Earth are still alive today. Examinations of the fossil record reveal that "mass extinction" events occurred five times over the last six hundred million years. The most recent mass extinction was at the end of the Cretaceous period, sixty-five million years ago, when the dinosaurs disappeared.

An asteroid from space is what likely precipitated the extinction of the dinosaurs: Some sixty-five million years ago a massive object, traveling at a speed of at least fifty thousand miles per hour, crashed into the Gulf of Mexico (which at the time was a shallow sea), creating a crater more than one hundred miles across. Paleoecologist Tim Flannery reconstructs the gargantuan crash:

> The great smoking pit left by the impact must have been an incomprehensible sight. How long, I wonder, did the ocean continue to pour into it. So big was the hole that one could not have seen from one side to the other, and so deep that no cliff on Earth today could produce such vertigo.[3]

The sea water that rushed into the giant crater likely later rushed back out across the sea and land as a colossal tsunami. The immediate physical damage was devastating—for instance, almost all plant species disappeared as far north as present-day North Dakota—and it seems that the asteroid exerted an equally severe impact by polluting the atmosphere, creating what astronomer Carl Sagan once called "impact winter." Again, Flannery creates the scene:

> The ejecta from the impact made the atmosphere opaque, preventing energy from reaching the Earth's surface. Just how much of the sun's energy was lost is difficult to estimate, but a loss of one fifth for a decade seems reasonable. This would have produced ten years of freezing or near-freezing temperature across the globe. Even if it did not become that cold, the rock ejected into the atmosphere may have created a twilight sufficiently dim to prevent photosynthesis. In effect, the whole world may have been plunged into a long polar night, thereby starving its plants.[4]

According to recent studies, many dinosaur species—among them, triceratops and tyrannosaurus—went extinct at about the time of the asteroid impact. Although not in complete agreement, most scientists generally believe that the dinosaur's nemesis was closely connected to this errant asteroid.[5]

The Extinction of the North American Megafauna

Significant extinction events have also happened in more recent times. For example, you may find it a surprise to hear that North America, as recently as fourteen thousand years ago, was populated by a bizarre array of mammals—the so-called North American megafauna.

Take a step back, in your mind, to a time 14,000 years ago. Imagine that you find yourself in what Americans today refer to as the Ohio Valley. You note with relief that the vegetation is familiar to you—the grasses and herbs and trees. However, many of the animals living in the grassy plains and forests are completely alien. In the open grassland, you see a huge tusked creature with a humped head and sloping back, its body covered in dense black hair; this is a woolly mammoth. On the edge of the forest, you spot a slow-moving elephant-like creature, ten feet high at the shoulder: the American mastodon. You enter the forest and behold a three-ton animal covered with thick fur, reaching up twenty feet to feed on canopy leaves: a giant sloth. Off to the side you spot several long-horned bison with their enormous straight horns. Given the abundance of large-bodied herbivores, you begin to wonder (and worry) about carnivores that might be lurking about, so you climb a tree at the edge of the grassland and sit motionless through the late afternoon and early evening. Slowly the predators reveal themselves: the dire wolf, the short-faced bear (largest meat-eating predator known to have ever existed), the American lion (almost twice the weight of today's African lion), the saber-toothed cat (an ambush predator with short, powerful legs), and the cheetah (distinct from today's African species). This is the North America biota you would have encountered before the Clovis people appeared in North America about 13,200 years ago.[6]

Many (but not all) archeologists believe that the Clovis people were the first modern *Homo sapiens* to enter North America. They entered from Asia and quickly spread from the Pacific to the Atlantic coast and down to present-day Guatemala. Archeological remains suggest that the Clovis people hunted big game, like woolly mammoth, using large fluted spearheads called "Clovis points." That no tools for grinding nuts or grains were found in archaeological digs at Clovis camps indicates that these early Americans may have relied almost exclusively on meat.

Using radiocarbon dating, archeologists have discovered that the Clovis people only made these lethal spear points for three hundred years (between 13,200 and 12,900 years ago). Then the manufacturing of Clovis points suddenly ceased—perhaps, as Flannery has suggested, because the very large creatures the Clovis points were designed to kill had been largely exterminated.[7] The archeological evidence to date reveals that most of the common species making up this North American megafauna disappeared about 13,000 years ago, coincident with the heyday of these Clovis people. In a few isolated areas, the megafauna persisted for longer. For example, while the giant ground sloth disappeared from North America about 13,000 years ago, this sloth held on in Cuba until 6,000 years ago, which is when humans first migrated to Cuba.[8]

The rapid demise of the North American megafauna may have occurred, in part, because most of these giant creatures had been living in North America for more than a million years prior to human arrival and presumably had no fear response to the diminutive newcomers. In addition, the use of ruses and tricks by Clovis hunters may have resulted in massive slaughters. As Flannery points out:

> Analysis of the creatures' age from kill sites reinforces the idea that whole herds of mammoth were massacred—not just young, old or isolated individuals. In this the Cloves may have acted much like modern African elephant cullers who kill complete family units. . . .[9]

So, just a short time ago extraordinary creatures like woolly mammoths, mastodons, cheetahs, and giant sloths roamed the wilds of North America. It was as if the Serengeti of Africa existed right in the heartland of America. Now this megafauna is gone. Our Clovis ancestors of thirteen thousand years ago certainly did not fully grasp that their actions would cause the extinction of an entire fauna; they were acting opportunistically, abiding by natural impulses. Today, humankind has access to real-time scientific assessments of species population declines, and thus, we are in a position to grasp the consequences of our actions.

The Situation Today

If you had walked into an elementary school classroom a half-century ago and called out "extinction" and then asked students for the first word that came to mind, chances are good that they would have unanimously called out "dinosaurs!" As recently as the 1960s, "extinction" was regarded as something that happened far back in time. But try this in a classroom today, and students, in addition to calling out "dinosaurs," are likely to volunteer the names of a raft of rare and endangered species: California condor, blue whale, snow leopard, American alligator, Florida panther, monk seal, gorilla, gibbon, and so on.

Biologists now concur that humans are precipitating a new "mass extinction" episode. Whether it is clearing swaths of forest for farming in the Amazon Basin or replacing a woodland with a subdivision in New Jersey or damming a river in Oregon, the outcome is the same: the habitats of plant and animal populations are being destroyed. Excessive utilitarianism leads humans to see the forest as a "woodlot" and the stream as a "hydropower generator"; but, of course, forests and streams are first and foremost homes for other species that inhabit Earth together with human beings.

When scientists talk about the magnitude of the current extinction crisis, they often contextualize their remarks by referring to "background rates" of extinction. Based on the fossil record, species survive, on average, for a million years or so before going extinct. Hence, for each million species on Earth, only one, theoretically, goes extinct each year; this is the background rate of extinction. But Earth is now losing species at a rate that is estimated to be somewhere between a hundred and a thousand times faster than these "normal" background levels.[10] Data from the World Conservation Union's *Red Book* reveal that "39% of the mammals and fish, 30% of amphibians, 26% of reptiles and 20% of bird species" on Earth are now in danger of extinction.[11]

What does this mean for the future? Stuart Pimm, an ecologist at the University of Tennessee, has used a combination of field observations, published data, and computer modeling to address this question; and he concluded that "we might lose between a third and a half of the life on Earth as a consequence of our actions"[12] (see box).

SO WHAT IF WE ARE LOSING SPECIES?

Taking a completely utilitarian perspective, are the high rates of species extinction that scientists now report and project for the near future really anything for humans to concern themselves with? After all, many species seem to do pretty much the same thing. Why do we need ten different tree species in a forest? Wouldn't one be good enough? Here is a helpful way of thinking about this question. Imagine that you are boarding an aircraft and that you notice a guy up on the wing busily popping out rivets. You walk over and yell up, "What the heck are you doing?"

He replies nonchalantly, "Our airline company has discovered that we can sell these rivets for two dollars apiece, and we need the money. If we don't pop rivets, we won't be able to keep expanding and offering our customers more services and flight options."

If you were to encounter such a situation, you would probably head back to the terminal, report the rivet popper to the FAA, and book passage with another carrier. The point of the story is that all of us are riding on a very large carrier, spaceship Earth, and we don't have the option to switch to another craft. The parts of our "aircraft" are Earth's ecosystems, and our craft's "rivets" are akin to the species in Earth's ecosystems. It is probably true that removing a few rivets from an airplane wing or driving a few species to extinction in an ecosystem will not compromise the functioning of the airplane or the ecosystem. However, it could well be that the ninth rivet popped from a wing flap or the extinction of a key species involved in the cycling of nitrogen in the forest could spell doom for the airplane and the ecosystem.[13]

Saving All the Pieces

What would be required to save all the species, or "save all the pieces" as Leopold counseled? To consider this question, imagine that you have been invited to a birthday party for all the species on Earth. You are the human representative among ten million other invitees—one individual from each of Earth's estimated ten million species. When it comes time to cut the cake, these ten million representatives from the earth's flora and fauna, including all the microscopic species, ask you to come forward first to cut your species' piece of the cake. You step forward and take the knife. But then, as you turn to look at the magnificent assemblage of Earth's species, you stand, immobilized, uncertain of your "fair share" of Earth's cake.[14] You consider the needs of your species, and then you consider the needs of the ten million species that have gathered together with you on this day. And you ponder. . . .

This very question is one that many biologists are now attempting to answer from a strictly scientific perspective. In effect, they are trying to figure out how much of earth (i.e., "cake") humans should set aside to avoid further species extinctions. William Newmark is one of the scientists who has worked on this question. Newmark came to ecology in a roundabout way.[15] He was a political science major in college, but he took a couple of ecology courses on the side and eventually decided to get a second bachelor's degree in biology and then study ecology in graduate school. For his dissertation Newmark decided to figure out if national parks in western North America were big enough to protect the species living within their borders. At the outset he knew that the parks were of different sizes: the biggest parks are more than one hundred times larger than the smallest ones (figure 7.1).

Newmark began his research by visiting each park and talking with park personnel. He soon learned that the parks had records of faunal sightings dating back to their founding. These records revealed which mammal species had been seen each year over the park's history. At the same time, the records had the potential to provide clues on species extinction. For example, if a park's records all of a sudden revealed no further sightings of a species beyond a certain date, this would implicitly reveal when that species had gone extinct within the park boundaries.

Newmark discovered that over time there had been few, if any, extinctions in the larger parks, but as park size decreased, the number of extinctions increased. For example, the red fox, spotted skunk, and white-tailed jackrabbit have been lost from Bryce Canyon National Park; at Mount Rainier, the lynx, the fisher, and the striped skunk have blinked out; and Crater Lake National Park has witnessed the local extinction of the river otter, the ermine, the mink, and the spotted skunk.

Newmark concluded that many parks, although seemingly huge to their human visitors, are in fact too small to maintain populations of all of their original mammal species. Of course, when these parks were first formed, their effective size was much bigger than their present-day boundaries would suggest because the surrounding lands were largely undeveloped; but now this has changed, and many parks are too small to maintain their original faunal diversity.

Biologists now concur that to "preserve all the pieces," as Leopold enjoined us to do, will require a significant increase in the amount of land set aside for wild nature. In other words, it will require that humans leave a generous helping of the planetary "cake" for Earth's myriad nonhuman species. If, however, humans continue to usher fellow species into oblivion—extinguishing,

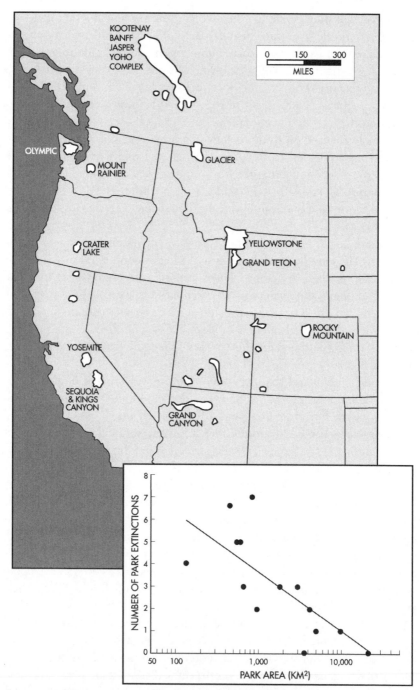

Figure 7.1. The size and location of selected national parks in western North America and the relationship between local mammal extinctions and park size[16]

one by one, the threads of life's fabric—we may ultimately lose our own life-support system and invite our own demise. The point is this: As we proceed with our "rivet popping," we are sailing uncharted waters, engaged in an experiment with an uncertain outcome.

Reflections

It seems to me that we have little likelihood of stemming the extinction crisis unless we experience a fundamental shift in how we humans see ourselves in the context of the larger community of life. At present, human actions in the world are usually grounded in homocentrism—the belief that the earth is here for us to use as we see fit. This worldview has provided us with a convenient justification for the elimination of plant and animal populations from areas that we covet. Even in those instances where we have been moved to protect wild species, our rationale is often imbued with homocentrism. For example, we might decide to protect certain tropical ecosystems, believing that their plant and animal species contain compounds or genes that will help cure human diseases or enhance human crop-breeding programs. Even when we set aside park lands, we do it for our enjoyment—to serve our recreational needs. In short, we save things to further a human agenda.

Ecocentricism—valuing and respecting all species and ecosystems to the same degree that we value and respect humans—stands as a counterpoint to homocentrism. An ecocentric orientation would lead humans to embrace the Noah principle, which states, plain and simple, that a species' long standing existence on Earth carries with it unimpeachable rights to continued existence. Although humans have not yet adopted the Noah principle, we have taken some important steps in this direction. For example, over its short history, the United States has been gradually extending rights and acknowledging moral obligations to the disenfranchised human segments of society—to children, women, blacks, minorities, and so forth. And, through legislation such as the Endangered Species Act, the United States has begun to consider its moral obligations to other species.

This same historical progression toward an ever-wider and more inclusive sense of moral obligation, ideally, is recapitulated during an individual's lifetime. For example, as a child, one's sphere of awareness and concern is limited to oneself and one's immediate family. Then, in grade school, awareness expands to friends and neighborhood residents; later in high school, the purview of concern extends to the local community; and by college, the young adult begins to experience himself as a citizen of a nation. With further maturity, a human being will come to see himself as connected, not just to other humans, near and far, but to all life on Earth, as well as to the solar system, the Milky Way galaxy, and the universe (figure 7.2).

Questions for Reflection

- How big a slice of Earth's "cake" do you believe humans should take?
- What connections, if any, do you see between the Noah principle and ecological consciousness?

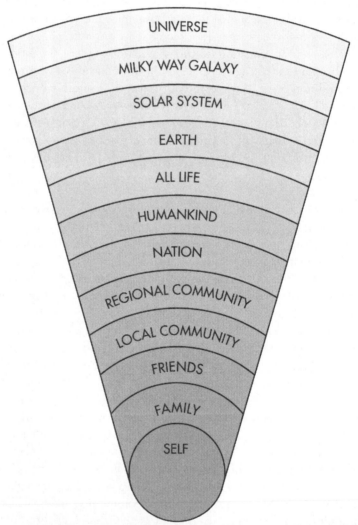

Figure 7.2.   The expansion of human consciousness from childhood through late adulthood[17]

Practice: Mustering Courage to Speak for Beings Who Have No Voice in Human Affairs

Responsibility comes with knowledge. If, for example, I hear on the news that a tornado is headed toward my town, I believe that I have a responsibility to share this knowledge with fellow citizens. In a similar vein, if our best science reveals to us that we are in the early stages of a mass extinction crisis, I believe that we all have a responsibility to share this information—which means finding ways to speak for all the other-than-human beings that have no voice in human affairs. Learning to speak on behalf of other beings is a practice that can expand ecological consciousness.

How might we speak for the biota? A clue comes when we consider what gives some people the courage to speak for and act on behalf of minorities who are voiceless. Their courage is born of compassion. An important step in cultivating compassion for all beings—not just other humans—is to allow the extinction crisis to fully enter our consciousness. Right now one or more of Earth's species is going extinct each day. Environmental educator Mitchell Thomashow laments: "Who bears witness to extinction? Who feels the emptiness, the sadness, and the incomprehensible sense of loss when the last individual of a species expires? Typically there are no human witnesses."[18]

And yet, each species is a creative unfolding. So it is that losing a species is the loss of creativity; it is the end of evolutionary possibility; it is a diminishment of life's creative force. As we come to fully acknowledge the present extinction crisis, life's multitude of remaining species become more precious. It is these species that need our voice.

Although it does not come naturally to us, we do have ways in which humans can speak for Earth's biota. When we truly care, our courage and creativity bubbles up; and we find a way. For example, in the 1980s two activists, Joanna Macy and John Seed, created a ritual, the "Council of All Beings," designed to awaken humans to the suffering that their actions cause in other species.[19] For two decades people all over the world have been participating in these councils.[20] In fact, anyone can create one: start by inviting a group a people—fifteen to thirty is a good number—to join you for an experience in consciousness expansion.

Once everyone is present, the host explains that the first step in the council is to put aside our human identity and open to the possibility of speaking on behalf of another life-form. The participants are then invited to walk outside in silence for a time, all the while being open to being "chosen" by another life-form. The idea isn't to consciously choose what to become, but instead, to allow yourself to be chosen by a life-form that "wants" to speak through you. Some may recoil at this idea, forgetting that an important part of what makes us human is our imagination and intuition. Spirit guide Jesse Wolf Hardin offers this advice:

> It is not your favorite animal, nor the animal you would most like to be. It's your personal totem. You do not pick it—it picks you. It lives you. Know it yet or not, you are a permanent part of its clan. Despite your rational self's best efforts to suppress it, you will react in life as it does.
>
> You! The bear people—heavy footed, clinging hugs, a grizzly in the office briskly shoving piles of paper aside! Squirrel people—furtive, excitable, stowing away conversations like seeds in a hollow log. Snake people, confesssss!—low profile, eye for details, unwavering stare, hidden defensive fangs, the happy wiggle, the craving for hot sand beneath a smooth belly.[21]

It might be a plant or an animal or even part of the landscape—a mountain, stream, or marsh. Allow yourself to be surprised; you might be chosen by a frog, dragonfly, maple tree, or swamp. It may be that you do not know much about the "other" that selects you. No matter. Use your powers of imagination to enter into the body of this "other." For example, if you have been chosen by spruce tree, become spruce. Place yourself on a particular mountain top, give yourself a particular age and form. Register how it is to be a mountain-top spruce tree.

Once all the participants of the council have an "other" in mind, the host invites them to make a mask with materials that are on hand—for example, markers, scissors, cardboard, wood, bark, leaves, string, fabric scraps. This is done in silence. The mask is a symbolic representation of the being that has selected each person. Participants are encouraged to allow this being to guide them in their mask making.

After the masks are completed (with eye and mouth holes), the participants go behind them and begin to move around, placing themselves in the body of this "other" that has chosen them. Council beings then gather in threes and fours, looking out through their masks and speaking, one at a time, about their qualities and characteristics and how it feels to be their being.

> I am spruce. I grow in these mountains. Birds feed on my seeds and build nests in my branches. My fungal friends festoon my roots; microbes, feeding on my old needles, return nutrients to me. In winter the snows cover me and my fire grows dim, but in spring the Sun's warmth and energy revive me and I rejoice; and in summer I thrive in the cool mountain air.

Next, the beings come together in one large circle, and with this, the formal Council of All Beings commences. The host asks for a roll call, and each being discloses his or her identity: "I am bear, and I speak for the bear people"; "I am ocean and I speak for the oceans of the world." The beings then proceed to consider how things have changed for them as a result of the spread of humans over the planet.

I come from a long line of spruce. My ancestors grew here by the tens of thousands long before you two-leggeds walked this land. For many generations your people and my "people" lived in peace. We didn't mind when, from time to time, you took a few of our elders to build your houses. Nor did we mind when you removed our dead parts and used them in your cooking fires. Always we heard your words of gratitude for our gifts and we felt honored.

But now it is different. Now you drive your skidders and chippers up into our ancestral mountain home and you snatch us from the ground when we are still young. And the spring rains which fall on us are no longer clean and pure but tinged with acid, and this makes life hard for us. And the winter snows, which once blanketed us, no longer come with their familiar abundance and the cool summer, our favorite time, is no longer cool as we like it. Things are changing and we are no longer thriving in our homeland.

After a time, the host observes that the suffering that the council beings speak of seems to be caused by one upstart species, *Homo sapiens*; and so it is important to have the presence of human ones in the council. The host asks a third of those gathered to remove their masks and move to the middle of the circle facing outward. In so doing, they assume the role of humans. The host then speaks directly to the human ones:

Hear us humans. This is our world too. And we've been here a lot longer than you. Yet now our days are numbered because of what you are doing. Be still for once, and listen to us.[22]

And again, the council beings speak:

We spruce have always lived by cooperating; we share Earth's bounty with the fungi and the bacteria and the birds of the sky. We know that there is enough of life's goodness to go around. You, too, once knew this but you seem to have forgotten.

Finally, council beings are asked to consider what gift each of them might want to give the "human ones" so that they might become more fully conscious of Earth's plight:

I speak now for all plants—the grasses of the prairie and herbs of the meadow and algae of the sea and vines of the forest. We watch as you wage your wars over oil and we know—though you have forgotten—that you are literally fighting "over our dead body," for the coal and oil and gas that you covet is our ancient remains.

In this, your time of turmoil, know that we are the key to your survival. It is only we, the plant ones, that have the capacity to directly convert the energy of Father Sun to the food energy that sustains you.

We enjoin you to cease fighting each other and to overcome your addiction to fossil fuels. You are a clever people. It is time for you to learn to live on present-time sunlight as we plants do. In this, we invite you to study our leaf "factories" and to use them as templates to design clean, renewable energy systems for your homes and industries. You have the ingenuity. May you have the will and the wisdom!

The council might end with each participant thanking the being that chose them for the privilege of speaking for them. Then, all the participants might kneel, placing hands and forehead on the ground, returning the energy that has passed through them back to the earth for the healing of all beings. As Macy observes, the Council of All Beings is an excellent practice for "growing the ecological self, for it brings a sense of our solidarity with all life, and fresh appreciation for the damage wrought by one upstart species."[23]

As you practice extending consciousness to other beings, through whatever means, write down your thoughts and experiences in the form of a short letter or poem; and carry this with you. An opportunity will come—during a conversation with a friend, a special dinner, a pause in a meeting, a church gathering—when you muster your courage and speak for the voiceless—sharing the words you carry, bringing compassion into the world.

FOUNDATION 7.2: DESTABILIZING EARTH'S CLIMATE

The colossal scale of the human enterprise now significantly affects the atmosphere enveloping Earth. If your great-great-grandfather had climbed to the top of a remote mountain in 1850 and collected a bottle of air, and if you were to go to that same remote mountain today and collect a new sample, the two air samples would be substantially different. Their basic chemistry would have changed. We need to let this fact register. "The air around us, even where it is clean, and smells like spring, and is filled with birds, is *different*, significantly changed."[24]

Analysis of the two air bottles, collected from that remote mountain, would reveal that since 1850 the earth's atmosphere has become enriched in carbon dioxide (CO_2), along with other gases such as nitrous oxide, methane, and chloroflorocarbons (CFCs). In 1850 the atmosphere contained about 280 parts CO_2 per million, but this was before humans started burning oil, coal, and gas in significant quantities. Fossil fuels are mostly composed of carbon, and when they are burned, this carbon combines with oxygen in the air to produce CO_2. The worldwide cutting and burning of forests—for example, to make room for farms and human settlements— also result in the release of CO_2, albeit on a much smaller scale than the release associated with fossil-fuel burning. By 2002 the CO_2 concentration of Earth's atmosphere had risen to 373 parts per million and was going up by about 0.4 percent per year. If humans continue on their present path, the air could test out at close to 500 parts per million CO_2 by 2100. This number matters because changes in carbon dioxide concentrations can alter the heat balance of the earth (see box).

HOW COULD THIS BE?

It seems like an outrageous proposition to suggest that we, mere humans, could actually change the climate of something as grand as planet Earth. But the earth's atmosphere doesn't extend endlessly out into space; it is only a wafer-thin protective layer. How thin? Imagine the earth scaled down to the size of a basketball. Dip this ball in water and pull it out. That gossamer-thin sheen of water enveloping the ball represents, to scale, the extent of Earth's atmosphere. Human activities can and do affect this narrow, life-sustaining band.

If the earth's temperature is to remain more or less constant, the energy entering the earth system must remain similar to the energy leaving Earth. Energy enters Earth as sunlight and leaves in the form of infrared radiation (heat) that escapes back to space. Greenhouse gases like carbon dioxide absorb infrared radiation but not sunlight. Hence, when the concentration of greenhouse gases increases, some of the infrared radiation, which previously escaped to space, is intercepted and reflected back toward Earth, warming its surface.

Some trapping of infrared radiation is a good thing. Without it, the earth would be brutally cold—with the oceans frozen from top to bottom. At the other extreme, if none of the solar radiation entering the earth's atmosphere was able to escape as long-wave heat, Earth would quickly heat up to unbearable temperatures. Maintaining the right balance is important. Humans upset this balance when they pump too many "greenhouse gases" into the atmosphere, causing more heat trapping than would otherwise occur under "normal" conditions.

The growing worldwide concern over climate destabilization has prompted governments to provide scientists with hundreds of millions of dollars to study Earth's climate. Researchers analyzing air bubbles in ice cores taken from Greenland have discovered that the concentrations of greenhouse gases on Earth are now higher than they have been at any time during the past 160,000 years. These studies also reveal that when the concentrations of carbon dioxide in Earth's atmosphere increased in the past, Earth's temperature rose, and vice versa.

In addition to this paleoclimate research, scientists have been studying contemporary climate in great detail. Recent temperature records reveal that the earth's average temperature fluctuates from year to year, and researchers are getting better and better at explaining the reasons behind these short-term fluctuations (figure 7.3). For example, atmospheric scientists attribute the drop in Earth's average temperature from 59.8 to 59.4 Fahrenheit between 1990 and 1992 to the eruption of Mount Pinatubo in the Philippines. When this volcano erupted, it ejected a fine mist of smoke, ash, and sulfate aerosols into the atmosphere. These "pollutants" mixed with the upper atmosphere and were dispersed around the globe. Due to their relatively large size, these pollutants actually blocked out, to a degree, the short-wave solar radiation coming into the earth, thus causing the observed cooling. As these volcanic ejecta gradually fell out of the atmosphere, the earth's temperature began to rise again between 1992 and 1993.

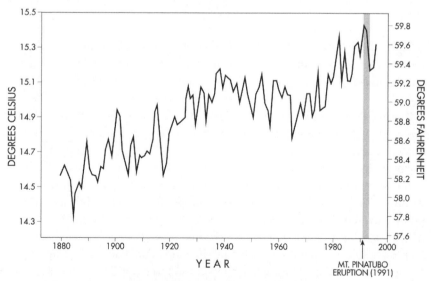

Figure 7.3. Changes in Earth's temperature since 1880[25]

Figure 7.3 also reveals that the earth's temperature underwent a slight cooling and then a stabilization from the 1940s to the 1960s; but since then, it has experienced a strong warming trend. Indeed, the eleven warmest years in recorded history have occurred since 1980, and the decade of the 1990s was the hottest since the 1860s.[26] The oceans have also been warming in recent years. As water warms, it expands, and this expansion has contributed to a four- to eight-inch rise in sea level over the last one hundred years. Other indications of climate change include the following:

- The thickness of polar ice has declined by 40 percent in the last forty years.
- The world's nonpolar glaciers have lost nearly half their ice over the last one hundred years.
- The frequency of "severe" storms (i.e., those involving more than two inches of precipitation within twenty-four hours) has increased by 20 percent in recent years.

Overall, it appears that humankind is in the midst of a giant planetary experiment. The science driving the experiment is, in many ways, straightforward: On a hotter Earth, there is more evaporation of water, which means more rain; land and water surfaces are also hotter, and this means more air movement—more wind—which in turn contributes to the clashing of cold and warm fronts, and the resultant chaotic weather.

Climate Change—So What?

Atmospheric scientists still can't predict with precision the magnitude of climate change that lies ahead, but their climate models, running on super computers, are becoming ever more sophisticated. And the more they learn, sometimes the more grave their predictions become. For example, in the mid-1990s, the United Nations Intergovernmental Panel on Climate Change predicted a temperature increase of several degrees Fahrenheit for Earth over the twenty-first century; but in their most recent report, released in 2001, this panel announced that the Earth might experience a temperature rise as great as ten degrees Fahrenheit over the present century. This prediction represents the general consensus of atmospheric scientists from around the world (see box).

The consequences of a ten-degree change in the earth's temperature are not pleasant to contemplate. This rate of climate change would be many times faster than anything humans have ever lived through. What might it be like to live in this "greenhouse" world?

BUT WHAT ABOUT THE NEWS REPORTS
DENYING CLIMATE CHANGE?

A small number of scientists doubt that human activities are in any way responsible for the current warming trends on Earth. These dissenting voices sometimes, though not always, receive their funding from fossil-fuel conglomerates that have a vested interest in making the public believe that great disagreement exists among scientists regarding climate change. The media—always on the lookout for controversy because it attracts readers and viewers—often gives significant air time and print space to the voices of a few dissenting scientists, making it appear as if the scientific community is deeply divided. In the case of climate change, the virtual consensus among scientists is that Earth's atmosphere and climate are changing. This said, it is important to recognize that lots of healthy disagreement among scientists exists concerning the particulars of climate change. For example, scientists are not entirely sure if humans are solely responsible for the present-day changes in climate. Other factors—like changes in Earth's rotation, orbit around the Sun, and natural variation in ocean circulation patterns—are also known to influence global climate. Scientists also share great uncertainty concerning the medium and long-term impacts that climate change will have on human welfare and how we should respond to potential threats.

- Contemplate significant changes in rainfall patterns, which in turn might cripple agriculture in some areas.
- Consider the possibility that the native plants and animals in some national parks would no longer be adapted to the altered climate within park boundaries and that they might have little or no access to protected refuges offering suitable climate and habitat.
- Imagine that diseases such as malaria, yellow fever, and cholera might expand into temperate areas rendered semitropical by global warming.
- Contemplate the possibility of hundreds of billions, perhaps trillions, of dollars in economic loss in the form of farm-sector collapse, rising sea levels, elevated storm damage, and increased illness and death due to heat stress.
- Imagine that this biophysical instability could cause social instability and lead to the declaration of martial law, perhaps propelling some nations down the slippery slope to totalitarianism.

It would be naïve, however, to assume that climate change only brings deleterious effects. For example, perhaps as Earth's surface warms, frigid northern areas, previously unsuitable for agriculture, will become available for farming. Similarly, global warming might lead to reductions in the use of fossil fuels for heating and declines in cold-related health problems.

Although scientists might disagree about many of the particulars surrounding climate change, one thing they do agree on is that Earth's climate will become more and more unpredictable in the coming years (see box).

LINK BETWEEN GLOBAL WARMING AND THE ONSET OF A NEW ICE AGE?

Unpredictability is now the hallmark of Earth's climate. According to some climatologists, the warming now occurring on Earth could even lead to an abrupt cooling—a climate flip-flop in the not-too-distant future. The thinking goes like this: The warming of Earth that we are now experiencing could lead to increases in high-latitude rainfall and the melting of ice caps,

- which in turn could place huge amounts of fresh water into the world's oceans,
- which in turn could suppress the natural movements of warm surface waters from the tropics to the cold north,
- which in turn could cause the cooling of northern continental land masses,
- which ultimately could trigger the onset of a new ice age.

Even though there are a lot of "coulds" in this story, the scenario is not as farfetched as it may sound. Paleoclimatologists have discovered that many such climate flip-flops have occurred in the past, even in the absence of human intervention.[27]

Given all these uncertainties, insurance companies are particularly nervous about climate change. Insurers base their fee structure on probabilities, but since the 1980s, these probabilities no longer reflect reality, especially given the dramatic rise in severe storm events.

In sum, humans have changed the atmosphere, and it is contributing to changes in Earth's climate. The earth's temperature and rainfall are no longer entirely the work of some separate, natural force; instead, they are partially a product of our actions. According to McKibben:

If we can bring our various emissions quickly and sharply under control we can limit the damage, reduce dramatically the chance of horrible surprises, and preserve more of the biology we were born into. But do not underestimate the task. The UN's Intergovernmental Panel on Climate Change projects that an immediate 60% reduction in fossil-fuel use is necessary just to stabilize climate at the current level of disruption.[28]

Here is the UN Climate Change Panel's mandate in numerical terms: Humankind now emits approximately 8.5 billion tons of CO_2 each year. To stabilize climate, we will have to bring our emissions down to 3 to 4 billion tons per year. If we are given a peak world population of eight billion (the rosiest projection), the average per capita CO_2 emissions would need to drop to less than a half-ton per year. To appreciate the magnitude of this challenge, note that per capita CO_2 emissions in the United States presently stand at approximately 10 tons per year.

Although the challenge is daunting, it is within the realm of the possible to dramatically reduce greenhouse-gas emissions. In the United States, the solution lies in moving away from heavy dependence on fossil fuels while switching to renewable-energy technologies based on wind, solar, geothermal, and hydrogen power. Technologies already exist for these alternative energy sources, but they are generally opposed by the coal, oil, utility, and automobile industries. However, 80 percent of Americans now believe, along with the majority of the scientific community, that climate change is a real phenomenon; and an even higher percentage of Americans would prefer to get their energy from renewable resources rather than from fossil fuel or nuclear sources.[29]

Reflection

Imagine that it is evening and the lights suddenly go out in your house. What would it take, really take, to get those lights to come back on? Where does the electricity for your lights come from? What's the fuel? Where's the power plant?

In my case (central Pennsylvania), virtually all the power I consume for electricity comes from coal. A few years back I decided to figure out how much coal a typical family of four needs to keep the lights on in its home for an evening. Here's what I discovered:

- Number of 100-watt bulbs burning per night:
  8 bulbs per household
- Average hours lights turned on per night:
  3 hours per light bulb
- Kilowatt-hour energy consumed per day:
  2.4 (8 bulbs × 100-W × 3 hours × 1 kW/1000W)
- Kilowatt-hour energy consumed per year:
  (2.4 kWh/day × 365 days)

- Money spent on lighting-related energy per year:
  $54 (876 kWh $\times$ $.0615/kWh)
- Pounds coal burned per year for lighting:
  716 (876 kWh/yr $\times$ 0.817 pounds coal/kWh)
- Pounds CO_2 emitted per year for lighting:
  1,848 (876 kWh/yr $\times$ 2.11 pounds CO_2/kWh)

This exercise makes the normally *invisible* connection between lighting and climate change *visible*. Working through the numbers, I discovered that a household that has eight one-hundred-watt bulbs turned on for three hours each evening uses approximately seven hundred pounds of coal in a year, which results in the release of almost one ton of carbon dioxide to the atmosphere. Fortunately, we can participate in simple exercises to dramatically reduce our lighting "footprints" (see chapter 9).

Questions for Reflection

- What happens when you allow yourself to imagine life in a greenhouse world?
- What feelings arise?
- What fears, if any?
- What resolve, if any?

Practice: Mustering Courage to Talk to One Another about What Really Matters

We live in a mythic time. Never before, it seems have humans wielded so much power for both good and evil. We can annihilate one another in war; we can embrace one another in peace. We can poison Earth; we can restore Earth to wholeness. This should be a time of big talk and big ideas; but what predominates in many quarters is small talk and small ideas. People's reluctance to engage in meaningful conversation—conversation that honors their spirit and intellect—is captured is this anecdote from psychologist Marshall Rosenberg:

> Once at a cocktail party I was in the midst of an abundant flow of words that to me, however, seemed lifeless. "Excuse me," I broke in, addressing the group of nine other people I'd found myself with, "I'm feeling impatient because I'd like to be more connected with you, but our conversation isn't creating the kind of connection I'm wanting. I'd like to know if the conversation we've been having is meeting your needs, and if so, what needs of yours were being met through it."
>
> All nine people stared at me as if I had thrown a rat in the punch bowl. Fortunately, I remembered to tune in to the feelings and needs being expressed

through their silence. "Are you annoyed with my interrupting because you would have liked to continue the conversation?" I asked.

After another silence, one of the men replied, "No, I'm not annoyed. I was thinking about what you were asking. No, I wasn't enjoying the conversation; in fact, I was totally bored with it."

At the time, I was surprised to hear his response because he had been the one doing most of the talking! Now I am no longer surprised: I have since discovered that conversations that are lifeless for the listener are equally so for the speaker.[30]

Rosenberg's story reminds us that good conversation is alive. It involves speaking with one another about what really matters—what has life for us. If we want big talk instead of small talk, we need to be willing to introduce questions with heart into our conversations.

Vicki Robin and Monica Wood of Seattle, along with a team of others, are calling on Americans to rediscover the power and importance of conversation. This duo has spearheaded "conversation cafes" in several dozen coffeehouses in and around Seattle. Why conversation cafes? Robin and Wood respond, "Because when you put strangers, caffeine, and ideas in the same room, brilliant things can happen. The British Parliament banned coffeehouses in the 1700s as hotbeds of sedition."[31]

Seattleites now know that any evening of the week they can drop into a cafe and participate in good conversation centered on juicy questions, such as

- What is working well in our city? What general lessons can we draw from this?
- When have you experienced good listening?
- What breaks your heart about living in these times? How do you cope with the pain?
- What constitutes meaningful work? How can we make all forms of work meaningful?

Part of what makes conversation cafes successful is that participants honor a few simple "agreements," like "seek to understand rather than persuade"; "invite and honor diversity of opinion"; and "speak what has personal heart and meaning."[32] Each conversation is hosted so someone is always present to welcome people, read the agreements, and commence and conclude the gathering.

The conversations often begin by lighting a candle as a way of symbolically bringing light to the circle. The host then invites those gathered to spend a moment writing down a question or concern that is alive for them. People share what they have written, and from among the various contributions, a topic or question is selected for the conversation. Then, each person shares a personal thought or feeling around the topic. A stone, or other "talking object," is passed from speaker to speaker; the person with the talking stone has the attention of all. There is then a second round, where those gathered, one by one, respond to what was said in the first round. After this, the conversation opens up. During this open period, the talking object is placed in the middle of the table. If the cross-talk becomes too frenetic, anyone can pick up the object to signal their need for space to speak or simply for a moment of reflection. As the conversation draws to a close, the host invites all the participants to reflect on what they might be taking away from the evening.

When this type of more deliberate communication is practiced, the overall quality of exchange is often refreshingly different from what passes for ordinary talk. No one is trying to "win" the conversation. The pace is slow and respectful, and silences are no longer awkward.

I recently had the opportunity to experiment with this communication form while in Seattle. I sat with seven others in a circle. We were ethnically diverse, and we had only just met. The question that we decided to discuss was "In what ways do you contribute to the suffering in the world?" We sat in silence thinking. Eventually our host picked up the talking stone and offered her personal response to this question. She then handed the stone to the man next to her. He spoke slowly, knowing that no one would interrupt him. We all knew that we would have a turn to speak (or pass) and that now was the time to listen to one another's words and to attend to what these words stirred in us—judgment, curiosity, confusion. . . . The stone went from person to person, each speaking and then repeating the question as he or she relinquished the stone. Suspending any need to "solve" the problem, we were able to go deeply into the heart of the question and into our own hearts. This was a very different experience for me. I felt heard, and I was able to hear things that normally I would have blocked out. It was surprising to discover that wisdom did not lie with any one person, but between and among the perspectives of all of us. It was as if we were all listening to one another and thinking: to know myself, I have to know you.

You don't need to go to Seattle to talk about what matters. You can do it in your local gathering places: the dinner table, the church basement, the classroom, the laundromat, the park bench; as well, you do not need to convene a group of people to have a meaningful conversation (see box).

THE FIVE-WORD RESPONSE GAME

A way to get into conversation that has heart and meaning when you are with just one other person is to ask questions back and forth, limiting responses to five words. This structure ensures that neither person dominates, and it allows the participants to introduce provocative questions. For example, while having dinner with my daughter in a restaurant, I asked, "In three to five words, what do you most enjoy about being alive?" Genny become very quiet and after a time said, "The times I am alive." Then, she asked me, "What do you believe will happen to you after you die?" I relaxed and waited for the answer to surface: "I will merge into space-time." When our waitress came, we told her what we were doing, and she joined in, posing a question for us and offering responses to our questions as she visited our table throughout the evening. Later in the meal, after Genny and I had exchanged a handful of questions and answers, we began to elaborate on some of our more opaque responses. So it was that this simple five-word response format drew us into meaningful conversation.

Summing up, it is through conversation that culture reflects on itself and changes. All the great social movements in history were seeded in conversation. In this vein, author and communication specialist Margaret Wheatley writes:

> I've seen that there's no more powerful way to initiate significant change than to convene conversation. . . . It's easy to observe this in our own lives, and also in recent history. Poland's Solidarity began with conversation—fewer than a dozen workers in a Gdasnk shipyard speaking to each other about their despair, their need for change and their need for freedom. And in less than a month, Solidarity had grown to 9.5 million workers. There was no e-mail then, just people talking to each other about their own struggles, and finding their needs were shared by millions of fellow citizens. At the end of that month, they acted as one voice for change. They shut down the country in a general strike.[33]

Breaking the silences that we keep as a people and speaking from the heart about what really matters are radical acts. Humankind's very survival may depend on it.

FOUNDATION 7.3: JEOPARDIZING HUMAN HEALTH AND REPRODUCTION

Commentators sometimes refer to the time in which we live as the "Information Age," but future historians might just as readily dub it the "Age of

Chemicals." More than eighty thousand different man-made chemicals are now in circulation on planet Earth. These include solvents, pesticides, preservatives, detergents, lubricants, emulsifiers, extractants, as well as reagents of all sorts.[34] Almost all of these chemical formulations have been "invented" over the last hundred years, and most have been developed since 1950 (see box on page 212).

Many modern chemicals have brought humanity benefits. But, while chemicals solve some problems, they often create other problems—that is, they have a shadow side that is often only fully revealed after the chemical has been in use for a number of years.

Because literally billions of tons of synthetic chemicals have been produced and released into the environment in recent decades, most humans have several hundred different foreign chemicals in their body tissues. A significant body of scientific evidence now links many man-made chemicals to cancers and reproductive disorders in both wildlife and humans. The incidence of many human cancers has been going up in recent years in all ethnic groups and both sexes. In some cases, this higher incidence rate may be partially the result of increased early-detection success—for instance, prostate cancer can be detected more effectively now than in the past; but in the case of childhood cancers, which have had a 30 percent increase since 1970, early detection technologies are not employed. Early detection is also not a factor in the dramatic rise in testicular cancer for young men between nineteen and forty years old (threefold increase since 1940); nor is it a factor in the rise in brain cancers among both children and the elderly.[35]

The rise in human cancers has been paralleled by a rise in cancers among animals, especially since 1960. These animal cancers—for example, fish with tumors, clams with gonadal cancer—are almost always associated with chemically contaminated environments, such as polluted bays. Results from these animal studies are actually easier to interpret because animals don't smoke, eat fatty food, experience work-related stress, and so forth. In other words, animals don't introduce the confounding variables that make it so difficult to pin down causality in human cancer studies. Furthermore, the animal studies are highly relevant to humans because many of the genes that govern cell division (i.e., those associated with cancers) are very "conservative"—that is, they are the same (or very similar) in humans and animals.[36]

Foreign Chemicals and Human Reproduction

For many years biologists have been concerned with the ways in which man-made chemicals might increase the incidence of cancers in adult hu-

FOREIGN CHEMICALS IN OUR BODIES

Throughout history humans have been plagued by pests such as flies and mosquitoes. Crop plants have also been regularly attacked by insect pests. Hence, the development of synthetic pesticides was regarded as a monumental scientific breakthrough. Unfortunately, we now know that some of these chemicals that kill "pests" can also find their way into the human body and cause harm. They do it through a process known as "biomagnification."

Here is how biomagnification works. Imagine that the insecticide DDT has been sprayed repeatedly on a lake over a period of years to control mosquitoes. A biologist studying the lake finds that the water has a very low DDT concentration—only 0.001 parts per million. This quantity doesn't seem like anything to worry about. However, if the biologist were to sample the lake's plankton (small microscopic organisms), she would find that they have a much higher concentration of DDT—perhaps 0.1 parts per million—than that found in the lake water. And if she were to examine the fish that feed on the plankton, she would probably discover that their body tissues have DDT concentrations in the vicinity of ten parts per million. Meanwhile, the flesh of birds feeding on these fish could have DDT concentrations as high as one thousand parts per milliom.

DDT and many other synthetic chemicals become progressively more concentrated along a feeding sequence because they are fat soluble. Hence, DDT passes from the water across the cellular membranes of plankton and becomes incorporated in lipids (fat) inside plankton. Then, when fish feed on plankton, a significant portion of the planktonic DDT becomes incorporated into fish fat, instead of being excreted. A similar thing happens when a bird or human eats the fish. Hence, the higher up in a food web that an organism feeds, the more fat-soluble synthetic chemicals it will accumulate in its tissues, with all else being equal.

---

mans. But it wasn't until the 1980s that a significant number of researchers began to entertain the possibility that foreign chemicals might also have impacts on humans at the very beginning of life—that is, during the development of the human embryo in the mother's womb. We now know that a surprising number of synthetic chemicals can inadvertently scramble the hormone signals that direct human development and reproduction.

Hormones play a particularly crucial role before birth, during embryogenesis. As the human fetus develops in the womb, hormones orchestrate critical events, such as sexual differentiation and the proper construction of the brain. Normal development depends on getting the right amount of the right hormone to the right place at the right time. If the hormone messages

don't arrive or if they arrive in the wrong amounts, the embryo's development can be derailed. For example, from conception until the human fetus
is almost two months old, it is impossible to tell if it will be male or female.
Then, on a certain day, small amounts of hormones kick in, "instructing" a
specific patch of the embryo's tissue to begin to develop into either testicles
in males or ovaries in females. This normal unfolding of gender can be confused if foreign chemicals that mimic or block natural hormone expression
are present in the fetus (figure 7.4).

Scientists have already identified more than seventy individual chemicals
or chemical families that disrupt natural hormone function in humans, either
by mimicking hormones or blocking their action. They refer to these chemicals, collectively, as "endocrine disrupters" because they disrupt the normal
functioning of the specialized endocrine system organs—such as the pitu-
itary, thyroid, ovary, and testes—that produce, store, and secrete hormones.
Some endocrine disrupters are found in everyday household products, such
as detergents, cosmetics, plastic toys, and food containers. To date, only a
tiny fraction of the tens of thousands of synthetic chemicals now free in the
environment have been screened for their hormone-disrupting potential.

Numerous studies of amphibians, birds, and mammals have documented
many types of severe developmental abnormalities that are linked to endocrine disrupters—for example, extra limbs, shrunken penises, eyes developing in mouth cavities.[38] The possible consequences of scrambled hormone messages in human development include lowered IQ, testicular
cancer, attention deficit disorder, reduced sperm count, breast cancer, and
genital and reproductive-tract deformities.

Of all these possible consequences, reports on reductions in the human
male sperm count (i.e., the number of spermatozoa per milliliter of semen)
have garnered special attention. Biologist Michael Zimmerman explains why:

> A frightening collection of recent studies has begun to suggest that emascula-
> tion is taking place. . . . A massive study published in 1992 by the British Med-
> ical Journal . . . found that average sperm counts have decreased 42% from 113
> million to 66 million sperm cells per milliliter, within the past fifty years. The
> study was able to factor out a host of extraneous factors, like the once held view
> that men's tight-fitting briefs elevated body temperature to the point that
> sperm production is jeopardized, and concluded that the decrease is due to
> hazardous chemicals in the environment. . . . What is novel and truly frighten-
> ing about the results reported in the British Medical Journal is that the decline
> in sperm production was not limited to men with particularly high occupation
> risk. Rather, the declines were found in the general population, demonstrating
> that we have reached the point that ambient levels of toxic chemicals are high
> enough to affect our health.[39]

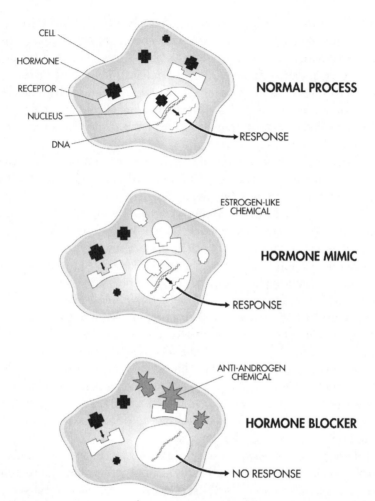

Figure 7.4. Hormones trigger cellular activity by hooking up—lock-in-key fash-ion—to "hormone receptor" sites, thereby activating genes in the nucleus (top cell). Foreign chemicals in the developing embryo can bind with hormone recep-tor sites (hormone mimics; middle cell) or block these sites (hormone blocker; bottom cell), thereby disrupting the normal unfolding of embryo development.[37]

The consequences of synthetic chemical proliferation in the environment are perhaps experienced most poignantly by women. Take the case of a young mother who has just given birth to her first child. She longs to expe-rience the joy and wholesomeness of breast feeding. In addition, she has heard of the advantages of breast feeding—how it will develop her new-born's immune system, helping to ensure that her breast-fed infant will have fewer bouts with sickness. But what she hasn't heard very much about are the chemical contaminants in her breast milk. Of all human food, breast milk is now among the most contaminated. Consider: When this young mother nurses, fat globules throughout her body are mobilized and carried to her breasts, where they are transformed into milk under the direction of the hormone prolactin. Toxins, like dioxin, that have slowly accumulated in her body fat over the years will be part of those mobilized fat globules that will be transformed to milk and passed on to her infant (see box).

What is the young mother to do? Forego the many advantages of breast milk for her child by feeding her inferior infant formula? Or fortify her child's immune system with breast milk while simultaneously contaminating her baby with the synthetic chemicals stored in her own body fat? This is not a decision any woman should be asked to make. Biologist, cancer patient, and mother Sandra Steingraber offers this counsel:

> Breast feeding is a sacred act. It is a holy thing. To talk about breast feeding versus bottle feeding, to weigh the known risks of infectious diseases against the possible risks of childhood or adult cancers is an obscene argument.[40]

As is the case with species extinctions and greenhouse-gas emissions (foundations 7.1 and 7.2), the unprecedented production and release of synthetic chemicals throughout the biosphere in recent times constitutes a unpremeditated planetary experiment that humans have set in motion. Here is the experiment: For millions of years the human lineage has lived in an environment of naturally occurring chemicals; then about a hundred years ago *Homo sapiens* began creating tens of thousands of new

---

THE SOURCE OF BREAST-MILK DIOXINS

What is the origin of the dioxins that are now found in mother's breast milk? A common source is polyvinyl chlorides (PVCs), a category of everyday plastics used in such things as food packaging, toys, and vinyl siding. When PVC plastics are thrown away, they often end up in huge trash incinerators. Such incinerators are like "laboratories" for the inadvertent manufacture of dioxins, with PVCs serving as one of the main raw materials in this process. The dioxins emitted from incinerator stacks settle on land and water bodies; then they enter into the food web, eventually finding their way back to humans. So it is that a plastic toy a mother buys her firstborn, now trashed and incinerated, finds its way back into that same mother's breast milk as dioxin (via contaminated food), which she then passes onto her secondborn.

chemicals, many of which were completely outside of the evolutionary experience of primates. Given the magnitude of this chemical experiment, perhaps we should not be surprised that we are experiencing a proliferation of previously rare cancers, that the human sperm count is apparently declining, and that developmental abnormalities are on the rise.

Fortunately, not everyone is standing idly by. Among nations, Sweden has taken the lead in slowing down this global chemical experiment by banning from commerce any chemical that "biomagnifies" in organisms or that persists in the environment (i.e., that doesn't naturally break down to harmless constituents in the environment). Coupled with this, Sweden is using the precautionary principle with regard to the release of new chemicals, which means that before any new chemical is licensed for sale, chemical manufacturers must demonstrate that it is totally safe.[41]

Reflection

The load of foreign chemicals that I carry in my body tissues—and, by extension, the probability that I might suffer from an environmental-related cancer—depends in part on where I have lived during my lifetime. Cancer mapping studies often reveal a correlation between the abundance of certain categories of chemicals in the environment (pesticides, industrial solvents, radioisotopes) and certain types of cancers. For example, breast, bladder, and colon cancer are most common in regions that are highly industrial: eastern seaboard, Great Lakes Basin, Baton Rouge area. Non-Hodgkin's lymphoma is most concentrated in the Great Plains up to Minnesota and Wisconsin; it is associated with agricultural regions and with occupations that involve the application of pesticides.[42]

What about the chemical load in the place where you live? With a computer you can visit www.scorecard.org, a website that provides an inventory of the types and amounts of chemicals in use in every county in the United States. Just type in your zip code, and this site will tell you which chemicals are in your county's air, water, and soil—and, by inference, probably in your body.

When I visited this website, I learned that, living in Centre County, Pennsylvania, I face a cancer risk more than a hundred times higher than the goal set by the Clear Air Act. I also learned that local industries annually release more than forty tons of zinc into the environment; and that zinc is a suspected developmental, immunological, reproductive, and respiratory toxicant.

Questions for Reflection

- How do you feel knowing that there are hundreds of foreign chemicals in your body tissues?
- Why do you think that people are, for the most part, passive with regard to the chemical contamination of their very own bodies?

Practice: Mustering Courage to Live Divided No More

As people begin to learn about, and open their minds and hearts to, the battered condition of Earth, some common responses are:

- Why didn't anyone ever tell me about this before?
- I want to cry when I think about what humans are doing to the earth.
- I care about Earth and what we are doing to it, but what about other people?

Actually, tens of millions of people share these concerns and sensitivities, but most remain silent.

Ecopsycologist and activist Joanna Macy believes that it is fear that is at the root of our silence. First, we simply fear the pain we will experience if we open our hearts to the suffering that is occurring on Earth. Western culture teaches that pain is to be avoided at all costs. Second, observes Macy, we fear guilt. At some level, we know that we share in the responsibility for ecological problems, but we don't want to be reminded because it induces guilt. Third, we fear appearing ignorant; we remain silent because we fear, if we speak out, that we might get embroiled in a debate when we don't have all the facts. We also fear appearing unpatriotic. Criticizing our country, even when its policies harm the earth, seems almost un-American. Together, these and related fears often leave us silent, immobilized, and disempowered (see box).[43]

We avoid facing our fear through many channels: We seek comfort in shopping, entertainment, and travel; or by indulging in specialty foods, fancy cars, therapy, Prozac, and so forth; but still the fear is there.[45] Macy emphasizes that the despair and pain that people feel concerning the damaged state of Earth is not a sign of neurosis. To the contrary, the capacity to feel pain stems from compassion; it is a measure of one's humanity and sense of connectedness to the earth. "Pain is the price of consciousness in a threatened and suffering world," writes Macy. "[Feeling pain] is not only natural, it is an absolutely necessary component of our collective healing. As in all organisms, pain has a purpose: it is a warning signal, designed to trigger remedial actions."[46] Seen in this new light, perhaps the incapacity to feel pain for the condition of the earth is the true sign of neurosis.

FEAR AS BIOLOGICAL LEGACY

The fear that humans experience when confronted with things that are strange and new may actually be an adaptive response. As Peter Seidel points out in his book *Invisible Walls*:

> There is a good reason that evolution has made higher animals, humans included, fearful and ready to flee from what they perceive to be different. What is different can be dangerous, and fear and suspicion are important survival mechanisms that warn us of possible dangers so that we can react appropriately.[44]

But fear can also be counteradaptive. In a state of fear, we may be too scared to act and, thus, fail to take advantage of worthwhile opportunities. Today, for example, humans need to respond with decisive action to address the deepening ecological crisis; but our natural, biological inclination is to be cautious and resist change.

By embracing the anger and sorrow that one feels for the condition of the earth, one gains the power to speak and act on behalf of the world. It bears repeating that the pain or the sorrow that you may feel for the condition of the earth in these times is a measure of your compassion. When one weeps for the desecration of life on Earth, love burns clear; and in this love there is great power. When we face what *is*, tremendous intelligence, courage, and creativity come forth. On the other hand, when we cap our heart, we cap our life.

An essential practice in these times, then, is to give voice to our feelings and thoughts regarding the condition of the Earth. To do so takes courage, but it can be done a step at a time. One place to start is with a trusted friend. Sit with this friend at a time when neither of you is feeling rushed and check if it would be okay to ask her a serious question. Assuming that she agrees, ask: "What makes you feel sad or angry or despairing about the condition of the world?" Listen in silence, and receive your friend's words with openness and respect. When she has finished, thank her and then ask her permission to repeat the question: "What makes you feel sad or angry or despairing about the condition of the world?" By repeating the question, you give your friend the space to explore it more fully. Tell her that you are content to sit in silence as she ponders the question. In this way, you offer your friend the space to crawl down into the corners of her heart and to speak things that perhaps she has never spoken before. Asking the same question a third, fourth, or even fifth time often deepens the inquiry. Eventually, ask your partner if she would switch roles and allow you to openly reflect on this same question.

Sometimes, in helping colleagues and students make contact with and express their capped-over feelings regarding the condition of the planet, I select objects that symbolize different emotional states and place them in the middle of our gathering space. For example, if the destruction of the earth causes your heart to feel cold and numb, you might want to literally pick up a rock and, holding it, speak of the numbness you feel. If it is anger that you feel, then you might pick up a stick and, holding it fiercely in both hands, give voice to your anger. If you are feeling sadness, you might want to pick up dried leaves and speak of your aching heart. Finally, cradling a large empty bowl in your lap, you might be better able to speak of any emptiness, confusion, or hopelessness that you are feeling.[47]

The purpose here is not to inflict new pain. Rather, it is to bring to consciousness all the capped pain that we all carry. We carry this pain because we are a total expression of the living, and in these times wounded, body of Earth. Hence, our pain is the world's pain; it is feedback for the world and, as such, should not be privatized.

Ultimately, the challenge before us is to live divided no more—which means quite literally "coming out," and speaking and acting with compassion for what we personally know to be right and true. This is the path that awakens spirit and brings us to full consciousness.

CONCLUSION

It is the destruction of the world in our own lives that drives us half insane, and more than half. To destroy that which we were given in trust: how will we bear it?

—Wendell Berry[48]

If humankind continues to barrel ahead—growing, consuming, and expanding its footprint—we will jeopardize the health of our grandchildren and leave them a tattered planet with a rumbling climate. It is even possible that there may be no future generations for the human species; humankind may not make it through this difficult period. Just as other species have gone extinct, humans may now be moving toward their own extinction. It is important that we acknowledge this possibility.

Fortunately, we humans do have some say in this matter. We can help to avoid the grim specter of our own extinction by, among other things, adopting the precautionary principle. This principle reminds us that—in situations where there is both scientific uncertainty and a reasonable suspicion of harm—we should stop what we are doing.

In all three of this chapter's foundations—the sharp rise in species extinctions, the disruption of Earth's climate, and the proliferation of potentially harmful chemicals in the environment—there is, clearly, "reasonable suspicion of harm." Adopting the precautionary principle would therefore involve taking forthright action to do the following:

- Stop destroying the habitats of Earth's species so that the biological diversity composing the web of life is protected.
- Stop the burning of fossil fuels by making a concerted effort to switch quickly to renewable energy resources so that further disruption to Earth's climate can be avoided.
- Stop producing chemicals that are potential carcinogens and endocrine disrupters so that human health and the well-being of the biosphere can be protected.

This chapter's practices focus on cultivating courage. After mulling over all the numbers and considering all the arguments, there comes a time when we all need to move from our heads to our hearts and therein find the courage that gives rise to action. In an effort to ground myself in courage, I sometimes place a spoonful of fine silt—dust—on my desk and spread it around. This is the dust that came from distant star systems billions of years back in time—the dust that formed the earth and that forms my body—your body. This is the dust that we will all return to when we die. We are dust; we are Earth; our voices are dust come alive. Our hearts—our courage—are spirit come alive:

What am I here for?   What are you here for?
What have I been afraid to say?   What have you been afraid to say?
What I am being called to do now?   What are you being called to do now?

NOTES

1. Wes Nisker, *The Essential Crazy Wisdom* (Berkeley, Calif.: Ten Speed Press, 2001), 133.

2. Aldo Leopold, *A Sand County Almanac* (Oxford: Oxford University Press, 1949).

3. Tim Flannery, *The Eternal Frontier* (New York: Atlantic Monthly Press, 2001), 16.

4. Flannery, *The Eternal Frontier*, 20.

5. Flannery, *The Eternal Frontier*.

6. Flannery, *The Eternal Frontier*.

7. Flannery, *The Eternal Frontier*.

8. Flannery, *The Eternal Frontier*.

9. Flannery, *The Eternal Frontier*, 197.

10. Staurt L. Pimm, *The World According to Pimm: A Scientist Audits the Earth* (New York: McGraw-Hill, 2001).

11. Nicky Chambers, Craig Simmons, and Mathis Wackernagel, *Sharing Nature's Interest* (London: Earthscan, 2000), 44.

12. Pimm, *The World According to Pimm*, 231.

13. I received the idea for this analogy from Paul Ehrlich and Ann Ehrlich, *Extinction* (New York: Random House, 1981).

14. Mathis Wackernagel and William Rees, *Our Ecological Footprint* (Gabriola Island, British Columbia: New Society, 1996).

15. This description of Newmark's work is from David Quammen, *Song of the Dodo* (New York: Simon & Schuster, 1996).

16. Map adapted from Quammen, *Song of the Dodo*, 490. Graph adapted from William D. Newmark, "A Land-Bridge Island Perspective on Mammalian Extinctions in Western North American Parks," *Nature* 325 (January 29, 1987).

17. This figure was inspired by Michael L. McKinney and Robert. M. Schoch, *Environmental Science: Systems and Solutions* (New York: West, 1996), 596–597.

18. Mitchell Thomashow, *Bringing the Biosphere Home* (Cambridge, Mass.: MIT Press, 2002), 53.

19. "Council for All Beings" is fully described in J. Seed, J. Macy, P. Fleming, and A. Naess, *Thinking Like a Mountain* (Gabriola Island, British Columbia: New Society, 1988).

20. Joanna Macy, *Coming Back to Life* (Gabriola Island, British Columbia: New Society, 1998).

21. Jesse Wolf Hardin, *Kindred Spirits: Sacred Earth Wisdom* (Columbus, N.C.: Swan-Raven, 2001), 15.

22. Macy, *Coming Back to Life*, 163.

23. Macy, *Coming Back to Life*, 161.

24. Bill McKibben, *The End of Nature* (New York: Anchor Books, 1989), 18.

25. This graph is adapted from Michael L. McKinney and Robert. M. Schoch, *Environmental Science: Systems and Solutions* (New York: West, 1996); B. W. Pipkin, *Geology and the Environment* (St. Paul, Minn.: West, 1994); and L. Brown, N. Lenssen, and H. Kane, *Vital Signs 1995* (New York: W. W. Norton, 1995), 65.

26. Ross Gelbspan, "A Global Warming Crisis," *Yes! A Journal of Positive Futures*, Winter 1999–2000, 12–18.

27. W. H. Calvin, "The Great Climate Flip-Flop," *Atlantic Monthly*, January 1998, 47–74.

28. Bill McKibben, "A Special Moment in History," *Atlantic Monthly*, May 1998, 55–60, 62–65, 68–73, 76.

29. Denis Hayes, "Earth's Energy Future," *Yes! A Journal of Positive Futures*, Winter 1999–2000, 35–39.

30. Marshall Rosenberg, *Nonviolent Communication: A Language of Compassion* (Encinitas, Calif.: Puddledancer Press, 1999), 128–129.

31. Vicki Robin and Monica Wood, "Mini-Manual for Conversation Cafe Hosts" (Seattle, Wash.: New Road Map Foundation, 2001). Contact information: New Road Map Foundation, P.O. Box 15981, Seattle, WA, 98115; www.newroadmap.org.

32. Accessed at www.conversationcafe.org.

33. Margaret Wheatley, "Some Friends and I Started Talking," *Shambhala Sun*, March 2002, 15.

34. Theo Colburn, "Toxic Legacy," *Yes! A Journal of Positive Futures*, Summer 1998, 14–18.

35. Sandra Steingraber, Penn State University Public Lecture, October 23, 2000.

36. Sandra Steingraber, Penn State University Public Lecture, October 23, 2000.

37. This figure was inspired by Theo Colburn, Diane Dumanoski, J. Peter Myers, *Our Stolen Future* (New York: Dutton, 1996), 72.

38. Colburn, Dumanoski, Myers, *Our Stolen Future*.

39. Michael Zimmerman, *Science, Nonscience, and Nonsense: Approaching Environmental Literacy* (Baltimore, Md.: Johns Hopkins University Press, 1995), 117–118.

40. Sandra Steingraber, "PVC and the Breasts of Mothers," *Rachel's Environment and Health News*, no. 658 (July 8, 1999); accessed at www.rachel.org.

41. Diane Dumanoski, "Hormonal Havoc," broadcast on Alternative Radio (Boulder, CO), April 18, 2000.

42. Sandra Steingraber, Penn State University Public Lecture, October 23, 2000.

43. Macy, *Coming Back to Life*.

44. Peter Seidel, *Invisible Walls* (Amherst, N.Y.: Prometheus Books, 1998), 71.

45. Robert Sardello, *Freeing the Soul from Fear* (New York: Riverhead Books, 1999).

46. Macy, *Coming Back to Life*, 27.

47. Macy, *Coming Back to Life*.

48. Wendell Berry, *A Timbered Choir* (Boulder, Colo.: Counterpoint Press, 1998).

Part II: Summary

ASSESSING THE HEALTH OF EARTH

> In order for us to maintain our way of living, we must, in a broad sense, tell lies to each other, and especially to ourselves. . . . The lies act as barriers to truth. These barriers to truth are necessary because without them many deplorable acts would become impossibilities.
>
> —Derrick Jensen[1]

In part II we examined the earth as a medical doctor might, looking for signs and symptoms of well-being as well as disease. The diagnosis emerging from this exploration is that Earth is sick. Consider its many signs of ill health: forest decline, migratory bird declines, aquifer depletion, soil degradation, species extinction, climate destabilization, hormonal upsets. These are not isolated problems—they are all connected to the explosive growth in consumption of Earth's resources and the concordant generation of waste. Just as a human being who is experiencing stress slowly sickens, so it is that earth now sickens, ailing under humankind's expanding ecological footprint.

When the earth is not well, we, its human inhabitants, are not well. How could it be otherwise? When we contaminate our water, we are quite literally contaminating ourselves because our bodies are mostly water; when we pollute the air, we are polluting ourselves because air is what we breathe into our bodies moment by moment; and when we taint our food with pesticides, we are tainting ourselves because food is what maintains and rebuilds our bodies day by day.

In addition to physical suffering, untold psychological suffering occurs when human beings lose their connection to place, meaning, and purpose. Religious writer Thomas Moore refers to this as loss of soul:

> The great malady of the 20th century implicated in all our troubles and affecting us individually and socially is 'loss of soul.' When soul is neglected it doesn't go away, it appears symptomatically in obsession, addictions, violence and loss of meaning.[2]

The anomie that Moore refers to is poignantly reflected in the increasing suicide rate among young Americans. When Sharif Abdullah taught at Oregon's prestigious Governors School, he discovered that more than half of the mostly middle-class teens in the school's programs had "either attempted or seriously contemplated suicide." Although these young people had everything that society could provide, their lives were bereft of meaning. Sharif suggests that:

> Suicide is an extreme vote of no confidence in a society that doesn't work. Suicide is a radical opting out. We have created a society our children want to leave. They want connection and instead get materiality. They want meaning and instead get a life devoid of cultural and spiritual riches, a life ripped free of context—historical, social, spiritual, communal. . . . America's middle-class children face spiritual starvation on a mass scale.[3]

Spiritual teacher Eckhart Tolle, in his recent book *The Power of Now,* observes that at this moment in history, "humans are a dangerously insane and very sick species. That's not a judgment. It's a fact. It is also a fact that the sanity is there underneath the madness." As a testament to "insanity," humans killed a hundred million fellow humans in the twentieth century alone. No other species violates itself on such a grand scale. As Tolle observes, "Only people who are in a deeply negative state, who feel very bad indeed, would create such a reality as a reflection of how they feel."[4]

A nerve is touched when we summon the courage to open our eyes and look straight on at the suffering, abuse, and violence now being inflicted on people, other-than-human beings, and whole landscapes and ecosystems. But it is painful to face these things, and so it is not surprising that many choose to remain asleep. They become walking dead. Life washes over them like a mush of garbled letters:

*—nIroerdrfos otnmiaaitnrouywafoglniivi,ewtmsu,niadbaro
essen,ltlesleiothecaroeth,danyelslpaeicotsoeuvrle. . . .
—eThsleitacsasbraerirothttru*

But we can say "yes" to life by remaining open and curious and unafraid to feel the Earth's suffering. If we do so, that same garbled mush of letters will begin to sort itself out into clusters with glimmers of meaning:

—*nI roerd rfo su ot nmiaaitn rou ywa fo glniivi, ew tmsu, ni a dbaro essen, ltle slei o theca roeth, dan yelslpaeic ot soeuvrle. . . .*
—*eTh slei tac sa sbraerir ot httru*

If we persist in our quest for truth and understanding—by questioning assumptions and by thinking critically about our contributions to contemporary problems—we will discover one day that those clustered letters, once transposed, become words that blend into an important message:

—*In order for us to maintain our way of living, we must, in a broad sense, tell lies to each other, and especially to ourselves. . . .*
—*The lies act as barriers to truth.*

Part II has been about overcoming "barriers to truth" by our looking fearlessly at the endangered state of Earth and by allowing ourselves to feel the sorrow and despair of these times. When we refuse to allow ourselves to feel, we close the door to caring. The ecological crisis is, at base, a crisis of caring. Once we open to our feelings, our caring nature is awakened; and with this awakening, our capacity for action is engaged. In sum, this journey of discovery transforms confusion, denial, and numbness into understanding and purposefulness. It is the pathway to expanded consciousness.

NOTES

1. Derrick Jensen, *A Language Older Than Words* (New York: Context Books, 2000), 2.
2. Thomas Moore, *Care of the Soul* (New York: HarperCollins, 1992), xi.
3. Sharif Abdullah, *Creating a World That Works for All* (San Francisco: Berrett-Koehler, 1999), 43.
4. Eckhart Tolle, *The Power of Now* (Novato, Calif.: New World Library, 1999), 67.

Part III: Introduction

HEALING OURSELVES, HEALING EARTH

The voyage of discovery lies not in seeking new vistas but in having new eyes.

—Yeats

Residents of southern India have a clever way of capturing monkeys. They drill a hole in a hollow coconut, place rice inside, and then chain the coconut to a tree. Here is the clever part: The hole in the coconut is large enough for a monkey to insert his hand inside, but too small to extract it once a handful of rice has been grasped. Instead of letting go of the rice, the monkey often holds on greedily, only to be captured in a net by the villagers.[1] Just as the monkey is captured because of his refusal to let go, we as a society are now trapped because we are clinging, tenaciously, to counterproductive ways of thinking and acting in the world.

Our present situation is part of a long process of cultural and technological unfolding. In our striving for knowledge, control, and power, we have come to equate progress with conquest—using brute force to fell forests, mine mountains, and tame rivers; and relying on powerful machines, weapons of mass destruction, and unfettered capitalism to exercise control over the earth and one another. But that time may now be passing. As Americans look around at the fruits of their "progress," they sense dis-ease. It is apparent in the troubled eyes of mall shoppers, the defiant swagger of high school students, the detached personas of young professionals, the stressed psyches of single mothers.

This pain is apparent everywhere—that is, if we care to look—and I believe it is a prelude to a massive waking-up. Throughout America—in businesses, in churches, in schools, in government—many people are beginning to ask each other, "Is this it? Is this what it means to be alive, to be a human being?" The answer is often, "No, this is not what I am here for." This "no" is important. Saying "No, I don't believe this . . . No, I won't do this any more . . ." is a prelude to creating the world that deserves our "Yes."

The third and final part of this book focuses on creating this world of "yes." In chapter 8 we will see how humankind's old story—based on control, consumption, and domination—is no longer working. These times are tumultuous precisely because our old story is dying. Humankind is now in the midst of a grand process of social change, and out of this process a new story is striving to emerge. The new story is based on partnership, not domination; sustainability, not exploitation; and connection, not separation.

This journey toward wholeness requires radical change, both at the individual and societal levels. As we will see in chapter 9, certain technical, economic, and political tools are available for making this journey; and in chapter 10, we will learn about the strong conceptual models for societal transformation—models that demystify power and map the process of social change. Equipped with these tools and models, we can serve as change agents within our communities and workplaces to bring foward a new, life-sustaining society.

NOTE

1. Daniel Chiras, *Lessons from Nature* (Washington, D.C.: Island Press, 1992).

8

STORY: CREATING MEANING IN A TIME OF CRISIS

Tell me the story of the river and the valley and the streams and woodlands and wetlands, of shellfish and finfish. Tell me a story. A story of where we are and how we got here and the characters and roles that we play. Tell me a story, a story that will be my story as well as the story of everyone and everything about me, the story that brings us together in a valley community, a story that brings together the human community with every living being in the valley, a story that brings us together under the arc of the great blue sky in the day and the starry heavens at night.

—Thomas Berry[1]

We all yearn to hear a good story. Even in the sciences, if someone has written a brilliant research paper, my colleagues and I may refer to the paper as "a great story." And when my graduate students prepare to present their research findings at a symposium, I invariably challenge them to find the "story line" in their data.

A good story ignites the imagination; it enlarges us. Depth psychologist Clarrisa Pinkola Estes suggests that the "mother tongue" that unites all peoples is symbolic language—the language of art, music, poetry, and story. Symbolic language allows us to grapple with mystery, awe, and anguish.

Life's biggest mysteries and sources of existential angst center on questions of origins: How did we come to find ourselves in this world? . . . and in this destiny? What are we to make of this life? In our time, the way that

people and society as a whole answer these fundamental questions determines, to a significant degree, the ecological health of Earth.

An essential first step in addressing the ecological crisis that is now unfolding on Earth is to see clearly how human culture—especially modern Western
culture—shapes human consciousness (foundation 8.1). An important next
step is to understand the particular ways in which Western culture's "story"
about life and its purpose is leading humankind as a whole ever deeper into
ecological crisis (foundation 8.2). The final step is to recognize that it is possible to create a new "story"—one that can engender a more enlightened consciousness and lead humanity out of crisis, toward a just and life-sustaining
world (foundation 8.3). Paralleling these foundations, this chapter's practices
center on the power of story to create meaning and to give purpose and
direction to our personal lives.

FOUNDATION 8.1: OUR STORY
SHAPES OUR CONSCIOUSNESS

Since the mid-1990s my students have been raving about Daniel Quinn's
book *Ishmael*. Quinn's book offers meaning in a time of confusion by seeking to explain how humanity's fractured relationship with Earth is a product
of Western culture's overarching story.

A culture's story is a living mythology that explains how things came to be
and how we are to act. Quinn offers Nazi Germany as a vivid example of
how a people's story can have disastrous consequences. Hitler offered the
German people a story that told how the Aryan race had been discriminated
against and abused by mongrel races over history. His story went on to describe how the Aryan race would rise up and wreak havoc on its oppressors
and then assume its rightful place as the master of all races. Many Germans
didn't see Hitler's story as simply a misguided attempt at making meaning;
they embraced it as destiny. Similarly, many of the citizens of Greece in the
time of Homer probably didn't regard their stories as "Greek mythology,"
but simply as the way things were.[2] Our times are no different. We live
within a story that attempts to give meaning to our actions:

> Like the people of Nazi Germany, [we, too,] are the captives of a story. Of
> course, we don't even think that there is a story for us. This is simply because
> the story is so ingrained that we have ceased to recognize it as a story. Every
> one knows it by heart by the time they are six or seven. Black and white, male
> and female, rich and poor, Christian and Jew, American and Russian . . . we all

hear it. And we hear it incessantly, because every medium of propaganda, every medium of education pours it out incessantly. . . . It is always there humming away in the background like a distant motor that never stops.[3]

Once we become aware of a culture's story, we see how it is employed to explain and justify behaviors. The Nazi Germany story, for example, provided a justification for the creation of a world that would benefit Aryans. One hundred and fifty years ago, white Americans created a similar story of racial superiority to justify slavery.

Today, we in America continue to live within a story. Think back to your history and social studies textbooks in grade school and high school. An underlying theme in most of those texts was humanity's inexorable march of progress. We learned that our distant ancestors were impoverished and backward but that today, thanks to the development of science and the application of technology, we have become enlightened and advanced. Nowhere is this march toward progress more prevalent, we were told, than in the United States of America, the land of freedom and abundance. For proof of U.S. superiority, we were given statistics emphasizing the phenomenal productivity and growth of the American economy. As we absorbed this story, it was only natural that many of us were led to want to support and defend our culture's values and accomplishments. Most of us never realized that our story was laden with a particular set of values.[4] All national cultures mark their citizens. If we had been born in France, our story would have valued European cultural traditions much more strongly than production and growth.

One way to appreciate the power of our culture's particular story is to imagine how it would be for you if you had been born into a society or civilization with a very different story. For example, imagine that your culture's story was centered on the belief that "this life is merely a preparation or test for the life to come"; or "there is nothing you can do to change your condition—it is your fate or karma from past lives"; or in a more positive vein, how about "everything you do has an effect on the unfolding cosmos—through your life you cocreate the universe." Each of these beliefs would clearly lead you to a different stance and response regarding the future and your individual planetary responsibility. As futurist Barbara Marx Hubbard points out: "As we see ourselves, so we become." Herein lies the colossal power—for both destruction and salvation—of a culture's "story."[5]

A Story of Domination

The overall, big-brush story of Western civilization has been predicated upon the implicit, but often unstated, assumption that the world was "made" for humans. Embedded in this story is the belief that combat and war are often the means by which goodness overcomes evil. This story elevates humans and implies that the emergence of "man" was the central event in the history of the cosmos. For example, the Christian Bible teaches that the world was unfinished without man. It needed a ruler. Man had to subdue the world. Even today, we hear it over and over: Man is conquering the deserts; man is subduing the oceans; man is conquering the atom and the human genome; man is taming outer space. According to Western mythology—the Western story—man was born to control, manage, and exercise dominion over the earth; this is man's destiny.[6]

Of course, this homocentric notion that man is the climax, the final objective, of the whole cosmic drama of creation is simply a story that Westerners have, for the most part, unconsciously absorbed. But these days it takes a fair measure of naiveté to imagine that the entire cosmic unfolding of the universe, extending back some thirteen billion years, was finally accomplished when man appeared on a little planet we now call "Earth." After all, since the appearance of man, the universe has continued to expand; new stars have continued to be born; and the principles and processes undergirding biological evolution and speciation have continued, just as if man had never appeared. In fact, to say that "man is the pinnacle of the evolutionary process" is no less shortsighted than imagining a time far back in evolutionary history—say when photosynthetic bacteria were the most complex life form on Earth—and thinking that the appearance of these amazing bacteria was the final objective of the entire cosmic unfolding. The creative unfurling of the cosmos didn't stop with the photosynthetic bacteria, nor has it stopped with us.[7]

A Story of Partnership

As should be clear by now, our Western story is rooted in assumptions. We have assumed that humans are the endpoint of evolution and that humankind's highest purpose is to exercise control over Earth. But perhaps it is not human destiny to be aggressive and controlling. In her provocative book *The Chalice and the Blade*, Riane Eisler differentiates between societies with "dominator" versus "partnership" characteristics. In dominator societies, interactions and negotiations tend to be hierarchical. A male god often sits at the top; men, women, and children, in descending order of importance, are below; and nature is at the very bottom. The heroes in dominator cultures are often violent; power is often determined by the ability to control nature and dominate one another.

In partnership societies, giving birth and nurturance are valued more than control and domination; cooperation is more important than competition. These societies are attuned with nature; daily activities are graced by art, ceremony, and celebration; capacities for empathy, intuition, and imagination are nurtured.

Anthropologists have documented partnership characteristics in some contemporary societies; and Eisler believes (despite her critics) that, based on evidence from archaeological excavations, our ancient past included long periods of peace and prosperity—in some places, at least—where societies were neither violent nor strongly hierarchical. According to Eisler, partnership societies existed in areas of Europe and the Near East that we know today as Palestine, Lebanon, Syria, Turkey, Greece, Romania, Bulgaria, Cyprus, and Crete.[8]

If you were able to go to these places seven to eight thousand years ago (during the Neolithic period), you would encounter a people who had script, trading boats, and complex religious and social institutions. These people, Eisler contends, were led to see the life-giving and sustaining powers of the world more in female form than in male. The remains of female figurines are common in the archeological ruins from this period, and a significant amount of evidence suggests that complex religions centered on the worship of a Mother Goddess.

The extensive art of this period—in the form of shrines, wall paintings, religious statuary, decorative motifs on vases, pictures on seals, and engravings on jewelry—reveals a reverential posture toward the beauty and mystery of life. Unlike the art of later dominator periods, partnership art is devoid of imagery that idealizes armed might, warriors, or scenes of battles.

According to Eisler, the decline of these partnership societies was tied to the Kurgan invasions. These nomadic invaders had a dominator model of social organization. Successive waves of invasion led to the progressive impoverishment of partnership cultures. The archeological record reveals a steady loss of shrines, finely crafted artifacts, and works of art. At the same time, fortifications appeared everywhere, as dominator societies supplanted partnership cultures.[9]

Although the dominator mentality is still very strong today, partnership consciousness is also present. A creative tension exists between these two mind-sets both in our individual psyches and in Western culture at large. This tension provides the energy for change. We will evolve as a culture not by returning to the mythic partnership societies of the past, nor by placing heavy emphasis on dominator characteristics, as we have done in more recent times. Rather, the challenge for our times appears to lie in evolving beyond this dualistic fixation to a new consciousness grounded in inclusiveness, creativity, shared leadership, and compassion.

Reflection

When my son was growing up I would tell him a story each night. It all began one day when we were walking along the beach. Jake was two and a half at the time, and he was tired. Instead of carrying him, I decided to tell him a story in hopes of diverting his attention. The story was about a lad named Johnny who woke up very early one morning and tiptoed downstairs so as not to awaken his parents. Johnny quietly prepared a lunch, grabbed his backpack, and slipped out the back door. Then, feeling free and ready for adventure, little Johnny walked down to the corner where he encountered Sam, a dump-truck driver and fellow adventurer. Johnny hopped up into the cab of Sam's truck, and off they went to fix a bridge that had been partially damaged by a flood. Sam and Johnny worked together all day until the bridge was fixed. Then Sam drove back to town and left Johnny off at the same corner where they had met in the early morning.

That was the gist of the story. Of course, I embellished it with some details: the clothes Johnny and Sam were wearing; the traffic jam they encountered; the discussion they had about which tools to use to fix the bridge; and so forth. But the story itself was simply about Johnny and Sam's fixing a bridge. As I told the story, my son was transfixed; he walked along for a half hour never once indicating that his legs were tired.

That night, when it was time for bed, Jake said, "Dad, tell me a 'Johnny' story." At first, I didn't know what he was talking about, but then I remembered. And so I repeated the formula: Johnny gets up early, tiptoes down the stairs, fixes his lunch, slips out the back door, meets Sam, and they set out to solve a problem; the problem is not easy to solve, but with determination and hard work Johnny and Sam figure it out.

Year after year I repeated the same story line, only the problems changed. One night it was a forest fire; then a traffic signal that didn't work; then a leaky roof at the airport; then a farmer who needed his tractor fixed; and on and on. As Jake got older, the problems that Johnny and Sam confronted changed: an exotic species invading a natural area; a rare disease that can only be cured by a plant growing in Amazonia; a young man who is struggling with nicotine addiction; a midget who abandons the local circus and wants to start a new life. . . . Each time, Johnny and Sam figured out a solution—sometimes technical, sometimes psychological, sometimes political.

In the process of telling these stories, I came to understand how stories can bring meaning to our lives and shape our view of the world. The stories

I told Jake were ostensibly about fixing bridges and putting out fires, but they carried deeper messages, such as "bad things happen, but we don't have to be victims"; "it is important to think before acting"; "when you work as a team, you can solve any problem"; and so forth.

Since that time I have come to understand that when I was growing up, I too received a story about the world—where it came from, how it works, and my role in it. Now I realize that it was largely my inheritance from Western culture that I was passing on to my son through story.

Questions for Reflection

- What are the stories that you were told when you were growing up, and how have these stories shaped your understanding of the world?
- What are the underlying themes and messages in the stories you receive from the news media each day? How do these messages shape your understanding of the world?

Practice: Discovering the Power of Story

The Inuit Indians have a saying that the Great Spirit must have loved stories, for why else would he have created so many people? Each of our lives is indeed a story, with each day a chapter—which is why, when we meet a friend or loved one at the end of the day, they ask, "So, what happened to you today?" When we respond "Not much" or "You know, same old stuff," we may have just "slept" through our day.

We can adopt simple practices to awaken to the drama of our lives. A good starting place is to learn to look at our daily life as an anthology of stories. For example, see the morning newspaper as a storybook; regard all your daily encounters and activities as having a beginning, middle, and end; see each meal as a story involving food preparation, eating, and cleaning up; listen to the eleven o'clock news as a "goodnight story"; and so forth. Consider the characteristics of these daily stories. What effects, for better or worse, do they have on you? What is memorable about them?

To formalize this practice, consider doing an "evening review" of the events of your day. In other words, run the movie of your day through your mind. As you do this, pay particular attention to the activities that you engaged in, the interactions you had with others, and the thoughts and conversations surrounding any decisions you made during your day. Think of this evening review as a way of harvesting the stories of your life.

Some people make it a practice to take a bit of time before bed to write about the events of their day in a journal. The journal becomes a place where the stories of their days are recorded: the difficult encounters as well as the sweet moments; the recurrent conundrums as well as the fresh insights. If you take up this practice, you may find it helpful to have several questions to guide your journaling. Possible questions include: What happened today that I am grateful for? When was I timid today? When was I courageous? How much of my day was spent in a state of worry, fear, and inadequacy? How much was spent in equanimity, confidence, and sufficiency? When was I fully present to others? (see box).

Taking time to reflect on the stories and lessons embedded in each day cultivates self-esteem. When we do this, we are saying, in effect, "I care enough about myself to want to pay attention to the daily unfolding of my life journey."

HARVESTING THE STORIES OF OUR LIVES

I was waiting in line to board an airplane. The line was long and moving slowly. As I stood in a state of impatience, an elderly couple pushed past me. They were both large people, and it appeared as if the man had suffered a stroke because his speech was slurred. A few minutes later, when I approached my seat at the back of the plane, I noticed that the same couple was seated in my row. I confess, with some embarrassment, that I felt irritated that I would be sitting next to them, but as it turned out, my seat was on the other side of the aisle.

About halfway through the flight, a stewardess came down the aisle holding up a twenty dollar bill and asking for change. It occurred to me that I might have change, but I decided that it would be too big of a hassle to undo my seatbelt, dig my wallet out of my back pocket, and search. Just after I reached this decision, I heard a man with garbled speech say that he thought he had change. It was the man I had earlier wished to distance myself from. For several minutes he wrestled with his seatbelt and then his wallet; finally, he produced change for the twenty, and the stewardess thanked him.

It wasn't until later, when I paused to review the events of my day, that I was able to fully bring this story and its significance to consciousness. I remembered the slight irritation I felt when the couple pushed their way past me; I remembered my chagrin when I thought that I would have to sit with them; and then I recalled my deep humiliation and embarrassment because the man helped the stewardess when it would have been so much easier for me to have done so. That man had been my teacher that day, and I was grateful to him. To the extent that we are alive and engaged in life, each day is bristling with stories. When we fall into bed at night without reflecting on our day, we fail to harvest life's lessons and deprive our lives of meaning.

Just as it is important to harvest the stories of our life, it is also worthwhile to take the time to tell our life stories. In the exchange of stories with friends, we offer them the opportunity to empathize with our circumstances, whether happy or sad. In this way our friends become co-participants in our lives, helping us see how our stories might offer us moral guidance and wisdom. As English professor Scott Russell Saunders points out: "Stories gather experience into shapes we can hold and pass through time, much the way DNA molecules in our cells record genetic discoveries and pass them on."[10]

The best storytellers are people who are authentic, entirely comfortable in their skins, and fully awake to their lives. Yet we all have the capacity to become storytellers. Natalie Goldberg, in her book *Writing Down the Bones*, describes how she sometimes calls friends together to form "story-telling circles." She invites her guests to sit on the floor in a circle with a candle in the center to "create a sense of magic." Then Goldberg invites stories to come forth by saying: "Tell us about a time you were really happy"; or "Tell us about a place you really love"; or "Describe a time when you were really down"; or "Give us a magic moment that you remember from last week."[11] Any one of these questions is sufficient to elicit a rich round of stories. In one of the circles, a man named Lauchlan described a magic moment:

> There was one summer that I was a forest ranger in Oregon for four months. I was alone for that whole time and I hardly ever wore any clothes that summer, because there was no one around. I was deep in the woods. By the end of the summer I was very tan and very calm. It was late August and I was squatting, picking the berries off a berry bush and eating them. Suddenly I felt a tongue licking my shoulder and I slowly turned my head. There was a deer licking the sweat on my back! I didn't move. Then she moved next to me and together we silently ate berries off the bush. I was stunned. An animal trusted me that much![12]

In sum, we don't need to search for stories in books or on the television. Our lives are filled with story, and we can all be storytellers. As we engage in practices like the evening review, journaling, and story-telling circles, we build community and create culture from the bottom up.

FOUNDATION 8.2: OUR PRESENT STORY AND THE ECOLOGICAL CRISIS

The general ideological thrust of Western culture's story has continued unchanged for the last several thousand years. Today, Westerners still mostly see themselves as separate and superior beings, charged with taking dominion over the earth. However, in recent history, movements such as Renaissance humanism, the scientific revolution, and the Enlightenment have worked together to give the Western story a distinctly secular character.

In the Middle Ages, prior to the Renaissance, Western life was organized primarily around religion: the basis for governance was formed around religion; works of art were frequently created to glorify God; and armies were assembled and wars were waged mostly to further religious agendas. The emergence of Renaissance humanism (fifteenth century) was a revolutionary reaction against the constraints imposed by the church–state power of the medieval world. Humanism posited that "man," by virtue of his rational mind, had the capacity to understand the workings of nature. By the sixteenth and seventeenth centuries, scientists such as Copernicus, Galileo, and Newton were demonstrating that it was possible to ascertain reliable knowledge of the workings of the physical world. Gradually, the medieval story—about salvation through divine grace—was replaced by the doctrine of material progress. Man, not God, gradually became the central focus of life; and reason triumphed over dogma.

The Enlightenment (eighteenth century) represented the consolidation and extension of the humanist–scientific orientation to all spheres of human endeavor. The "medieval Christian world, wherein all creation was infused with God's presence and direction, was replaced by a sense of a clockwork-like universe that had been set in motion by God in the beginning but otherwise operated autonomously, according to Newtonian laws."[13] By the end of the eighteenth century, modern man had emerged as the detached manipulator of the world, with a strong secular–rationalist orientation centered on the pursuit of self-interest. Once scientists had made it abundantly clear that Earth was not at the center of the universe, the Enlightenment philosophers elevated the individual to that exalted position.

In sum, the expansion of secularism, since the 1500s, isn't so much a new story as it is a new spin on Western culture's old story. If anything, modern secularism has reinforced the tendency for humans to see themselves as the culmination of evolution and to believe that it is their destiny to control and manipulate nature. At the same time, secularism is a welcome historical development insofar as it signals an expansion of human consciousness. The ascendancy of science and its decisive separation from religion has allowed mankind to shed medieval superstitions and gain a much broader understanding of life and the cosmos.

Economism: The Modern Permutation of Secularism

Imagine that you have been hired to make sense of human culture as it is enacted in the United States. To do so would require that you assume the role of a detached observer as you visit homes, churches, workplaces, and schools throughout the United States. Listening to what people talk about and noting how people spend their time, you discover that, aside from family and friends, much of what occupies human consciousness is linked to money, work, possessions, production, efficiency, communication, and technology.

Social change activist Barbara Brandt has used the word "economism" to characterize the story or belief system undergirding contemporary U.S. culture. Economism is an approach to life that places great meaning on money, work, and possessions. Most Americans are undeclared adherents of economism insofar as they believe that their primary purpose in life is to work hard (or study hard) so that they can make money to buy the things that will bring them personal happiness. Like any powerful story, economism is so fully a part of American thinking and culture that most people simply see it as the way things are (see box).[14]

A CRICKET AND SOME COINS

Gerry was walking down a sidewalk in Washington, D.C., with a Native American friend who worked in the Bureau of Indian Affairs. It was lunchtime in Washington. People were husslin' and busslin' along the sidewalks, and car honks and hurried engine noises filled the streets. In the middle of all this traffic, Gerry's friend stopped and said, "Hey, a cricket!"

"What?" said Gerry.

"Yeah, a cricket," said his friend. "Look here," and he pulled aside some of the bushes that separated the sidewalk from the government buildings. There in the shade was a cricket chirping away.

"Wow," said Gerry. "How did you hear that with all this noise and traffic?"

"Oh," said the Native man, "It was the way I was raised. . . what I was taught to listen for. Here, I'll show you something."

The Native man reached into his pocket and pulled out a handful of coins . . . nickels, quarters, dimes . . . and dropped them on the sidewalk. Everyone who was rushing by stopped . . . to listen.[15]

A measure of economism's effectiveness has been its ability to reframe the original purpose of many societal institutions and functions. Many Americans have now come to believe that the central purpose of schools is to teach children the skills they will need at work so that U.S. businesses can remain competitive in the global economy; the main purpose of government is to promote policies that will help our economy flourish; and the primary purpose of the natural environment is to provide resources that will fuel our economy.[16] Under economism, the needs and values of business have come to dominate society. Activities that generate a profit or bring a high return on investment are judged as desirable, often irrespective of whether they are wise, wholesome, or morally defensible.[17]

Economism has become so fully integrated into U.S. culture that it now serves as a kind of pseudoreligion. Strange as it may sound, economism has the equivalent of deities, high priests, missionaries, places of worship, and commandments. Among economism's deities are

- money, the ultimate source of security;
- science, the ultimate source of knowledge;
- technology, the ultimate source of power.

This is the Trinity that many in Western culture now seem to trust in for earthly salvation. It follows that the "high priests" of economism—those who are closest to these gods—are economists, scientists, and technologists. The "missionaries" are the minions from transnational corporations seeking to convert all nations to "free-market capitalism." Economism's "churches" are the outlets of the corporate-controlled media, especially television— that is, the places where people hear about the power of money and hard work to satisfy their "needs," and where they hear about all the new products and cures generated by science and technology.[17] Finally, just like a bonafide religion, economism has commandments:

Work hard to earn money!
Watch television to stay hip!
Buy on credit!
Keep up with the Joneses!

These commandments have become so institutionalized that most people see them simply as the way life is lived in America. Although somewhat simplistic, this metaphor does offer a way to begin making sense out of contemporary American values, beliefs, and behaviors.

Cracks in Our Story

Religion, at its best, offers us a way of seeing ourselves that summons up what is most noble in us; it appeals to our highest selves—our adult selves— and it inspires us to be kind, courageous, and loving. However, the pseudoreligion of economism tends to appeal to our lowest selves—our adolescent selves—drawing out our greed and small-mindedness. Students of U.S. culture suggest that Americans have created an adolescent culture that thrives on consumption because promoting perpetual consumption is the way to make money; and under economism, money is the principal measure of progress, success, and happiness.[18] But there is a growing tension in America between values and appetites—between our will and our wants—and this tension is leading to cognitive dissonance (see box).

The strain that economism is now putting on the human psyche was revealed in an article entitled "Yearning for Balance," which was a comprehensive analysis of American perspectives on consumption. According to this study, when Americans—irrespective of gender, age, or race—look at the condition of the world today, they come to a similar conclusion:

> Things are seriously out of whack. People describe a society at odds with itself and its own most important values. They see their fellow Americans growing

---

THEY DIDN'T SKIP A BEAT

Recently, I asked a gathering of college freshman to reflect on what really mattered to them. At first they were taken aback by this question, but after some thought, they began to speak. For many, what mattered most was family, friends, the places where they grew up, their religion, health, learning, personal freedom, and natural beauty. It was wonderful to hear these freshman speak in public about what they cherished. Next, I asked them to reflect on what really mattered to society at large. This time they didn't skip a beat. "This country is all about image," they said. "It's about convenience and competition and getting ahead; it's about business and making money." I was surprised by undertones of anger and cynicism in some of their responses. Here were young Americans who had personal values that were generous and life-affirming; yet they were also young Americans who were living in a culture that they perceived as crass. When I introduced the idea of "cognitive dissonance" (i.e., the tension that ensues when two belief systems or mental models clash), many students were grateful to finally have a word (a diagnosis!) to describe their divided mental state.

increasingly atomized, selfish, and irresponsible; they worry that our society is losing its moral center.[19]

Cognitive dissonance often prompts a change in attitudes and behavior. For instance, significant numbers of Americans are less and less lured by the fancier car, the second house, the luxury vacation. Just like the college freshmen mentioned above, these people somehow have come to know that true wealth lies not so much in financial assets as in

- social relationships—friends and family;
- cognitive capacities—abilities to read, learn, and reason;
- wholesome, natural, and man-made surroundings—beautiful buildings and parks; healthy oceans, rivers, and forests;
- cultural legacies—traditions, literature, artistic expression;
- political liberties—civil rights, open civic exchange, and local governance.

Material wealth is not irrelevant for these people, but its role is largely in terms of how it facilitates their ability to access other, more meaningful forms of wealth.[20]

Summing up, a society behaves as it does because of a deeply ingrained, yet largely invisible, story. Economism, the present permutation of Western culture's story, is based on the notion that the earth is primarily a lump of resources waiting to be transformed into products for humankind. This story's main plot revolves around corporations, markets, resources, manufacturing, capital transfers, and advertising. The story's central character is the human, a being with a seemingly insatiable appetite for material goods. But a characteristic of human cultural evolution is that our old stories eventually no longer supply enough meaning to satisfy us. The cognitive dissonance now manifest in America is a sign that we are in the throes of birthing a new story.

Reflection

I find this concept of "economism" intriguing and disturbing at the same time. I can see that it offers a useful description of contemporary life. Many of us do spend vast amounts of our waking life working or studying so that we can make money to acquire the things—many of which are not "necessities"—that, we assume, will bring us ease and happiness. But economism, as a worldview, presents a depressing characterization of the human enterprise, suggesting as it does that work, money, and material goods are the central reason for—and significance of—our lives.

At the same time I can see how, over evolutionary history, the human's appetite for more—more territory, more food, more status, more mates—has helped to ensure our survival as a species. Cognitive psychologist Timothy Miller has suggested that the success of all species, including humans, results from the fact that they lack an "enough" switch. According to this view, if a species were to say to itself, "This is good enough. I really don't need more territory, food, status, mates. I'll just take it easy now," it would soon be out-competed by other species and rendered extinct.

It may well be that having no "enough" switch ensured the survival of our species in the distant past, but today this inability to define "enough" may be putting humanity's survival in jeopardy. Fortunately, we do have the ability to consciously cultivate an awareness of how much is enough and then learn to live within those bounds. I have come to believe that economism—with its emphasis on constant growth and consumption—is not our destiny, but simply a stepping stone on the path to fuller human consciousness.[21]

Questions for Reflection

- In what ways is "economism" similar and dissimilar to other "isms": nationalism, sexism, pluralism, and so on?
- If economism is Western culture's present "story," what is the story that is waiting to be born?

Practice: Becoming Aware of Our Personal Life Story

Just as consciousness can change and evolve over time within a society, we as individuals have the opportunity during our lifetimes to evolve in our understanding and wisdom. We can sit back and passively react to what comes along in our lives, or we can proactively create our life story. When we choose the latter path, our lives become stories about the deepening and widening of consciousness.

Certain exercises and practices can help bring awareness and agency to one's unfolding life story. A good place to begin is to figure out what has contributed to your present understanding of your life and its purpose. While I was working on this book, I was invited to participate in an exercise with just this purpose. We gathered in Richard's living room, and he asked us to pick a specific time in the past when we were going through a transition—for example, graduation from high school, starting our first job, beginning a special friendship. Next, he invited us to bring to mind the people, experiences, places, activities, and institutions that had influenced us up to that point and also to recall what we understood to be true about the world at that time.

After allowing us to mull this over for a few minutes, Richard placed a box of magic markers in the center of our circle and gave us each a transparency sheet. He then asked us to put pictures and symbols (avoiding words if possible) on the transparency sheet and, in so doing, to describe what influenced us and what was true for us at that particular time in the past.

When we finished our drawings, Richard invited each of us to talk about the big themes that shaped our worldview up to the historical marker that we had selected from our past. For example, I chose the transition between high school and college. My worldview up to that time was heavily shaped by the Catholic Church: authority, morality, ceremony, self-sacrifice.

Then Richard asked us to hold our transparencies in front of our faces and to look through them. This was an "aha!" moment. What each of us had on our transparencies was what we had absorbed from our families, schools, authorities, natural surroundings, and culture. Each of us looked out onto the world through lenses tinted and modified by our particular past.

Someone in the group observed that when we forget that we have a lens on, we make the mistake of thinking that the way we see the world is the way the world really is. We see things not as they are, but rather as we are. In reality we are like the proverbial blind men describing the elephant, each operating from his own limited viewpoint.

Of course, we are not stuck with our lenses. Some people noted that certain things on their transparencies from the past were no longer present. Others spoke of their new, "corrective lenses," noting that in recent years they had incorporated new things into their lenses (e.g., more compassion), which had improved their ability to be present more skillfully in the world. Richard was particularly interested in these positive changes, asking about the circumstances surrounding them. In many cases, the insights arose during periods of trauma or loss.

The phrase "no pain, no gain" does seem to describe the setting of most personal growth. Noting this, I sometimes ask my students to prepare a "failure resume" to go along with their standard success-oriented resumes. I tell them to fill it up with all their worst failures. I then ask them to note the lessons encoded in each of their failures and how their failures have changed their lives. The upshot is that students come to see that their failures reveal much more about themselves than a simple list of their accomplishments. Our responses to disappointment and failure are to a significant degree a measure of our self-acceptance, determination, flexibility, and integrity.

It is often during periods of failure in our lives that we are most fully alive because these are the times when we are living on the edge, taking risks. As Gregg Levoy notes in his book *Callings:*

It's almost axiomatic that the important risks we don't take now become the regrets we have later. In fact, I was once told that if I'm not failing regularly, I'm living so far below my potential that I'm failing anyway.[22]

Another helpful practice for discovering one's life story involves drawing. Drawing is a way of gaining access to those parts of ourselves that, more so than words, are grounded in images, sounds, sensations, and feelings. Psychologist Molly Brown encourages the use of drawings to explore major life questions. I was a bit skeptical about this, but I decided to give it a try.[23] So, feeling a bit sheepish, I went to my local bookstore and bought a box of crayons. When I sat down with a clean sheet of paper, I felt unsettled and decided to take a few minutes simply to breathe, calm my body, and quiet my thoughts. Once I was relaxed, I asked myself, "Where am I now in my life story?" Then, I simply paid attention to the thoughts, images, sensations, and feelings that arose. After a time, I began to work with the crayons, giving form, color, and shape to my thoughts and feelings. I found, to my surprise, that I enjoyed using crayons to express my nonverbal responses to this question, especially knowing that I was not performing for anybody. Afterward, I examined the drawing, then I shared it with a friend, and we talked about it. Brown recommends hanging these "life" drawings on your bedroom wall and examining them from time to time until you have fully absorbed their messages.

As we gain the capacity to see our life as an unfolding story, we remember to become full participants in it. Levoy recommends pausing from time to time to embrace your life story as a grand myth—an epic story of your destiny and travails. Why do this?

[Myths] get at the heart of human behavior, at profound truths, universal truths, ageless patterns. They are . . . stories of transformation: from chaos to form, sleep to awakening, woundedness to wholeness, folly to wisdom from being lost to finding our way. They describe the stages of life, the initiations we all go through as we move from one level to another: child to adult, young to old, single to married, cowardly to courageous, life to death, death to life.[24]

I remember a June morning many years back when I looked out my kitchen window and saw my son bending over a large plastic dump truck in the sandbox. Jake was three at the time. As I watched him, I realized that he was "studying" the mechanics of movement. With excruciating slowness, he pushed the dump truck up a sand mound. His head was bent down, his face six inches from the truck's rolling wheels. Then, he painstakingly guided the truck down to the bottom, studying the mechanics of "reverse."

This went on for a full ten minutes—very slow, very focused. Jake is now a young man; his heart sings when he is engaged in mechanical work—fixing his car, building a house.

David Whyte believes:

> Each of us, somewhere in the biography of our childhood, remembers a moment where we felt a portion of the world calling and beckoning to us. . . . Somewhere inside us, the child is still running enthusiastically toward a horizon it once glimpsed. Our future life depends on finding this original directional movement in our lives. . . .[25]

Seeing ourselves as mythic characters on the hero's journey of transformation is not loony or far-fetched. It is a form of self-respect. We see our lives as unique endeavors with noble purpose. Conveyed as myth, the struggles in our lives become tests of initiation in an epic adventure; our earthly friends become our guardians and muses; our heart's calling or vocation becomes the sacred treasure that we are willing to risk all for.[26]

FOUNDATION 8.3: OUR NEW STORY AND THE SUSTAINABILITY REVOLUTION

Culture, more than anything, shapes both personal life stories and societal belief systems. Thus, a lasting solution to the ecological crisis now confronting humanity lies—of necessity—in the gestation and emergence of a new life-affirming belief system: a new story.

Humankind needs, in effect, a revolution on the order of the agricultural and industrial revolutions of our past. These earlier revolutions were also likely catalyzed by crisis. In the case of the agricultural revolution, the central crisis appears to have been food scarcity. Prior to ten thousand years ago, humans were nomadic. As human numbers grew, edible wild plant and game resources became less available; and it is believed that this scarcity prompted humans to migrate out of Africa and the Middle East. However, some human groups hit on an entirely different strategy: They settled down in one place; they domesticated animals; and they cultivated plants. This was a radical idea, and it subsequently changed the face of Earth in ways that could never have been imagined at the time.

Staying in one place, rather than constantly moving from place to place, meant that humans could accumulate possessions for the first time in their history. Some people apparently were more adept at accumulating things than others. Over time this accumulation of possessions—of wealth—

helped create the conditions for the emergence of money, crafts, trade, cities, governments, and armies.

The increased food availability made possible through agriculture was a key factor in allowing the human population to grow from five to ten million at the time of the agricultural revolution to almost one billion by the late 1700s. But then, once again, scarcities began to cause stress; this time it was energy and land, especially in the case of Europe. As the availability of trees as a fuel source declined, Europeans—first in England then elsewhere—began to use coal to satisfy their energy needs. The mining, transport, and combustion of coal presented many technological challenges, which in their solving triggered many changes: "Coal led to steam engines. Machines, not land, became the central means of production."[27]

The spectacular success of the industrial revolution, along with improvements in sanitation and medicine, led to a further explosion in human numbers. Now, in the early twenty-first century, humanity faces yet another suite of scarcities—namely, diminishing supplies of fuels, metals, and land, as well as a reduced capacity of the environment to absorb the polluting byproducts of industrialization.

Our species is now, quite literally, living in a "middle time." The old stories that brought meaning to life for our ancestors no longer work, and a compelling new story has not yet fully taken hold in our psyches. Those alive today have the opportunity to literally act as midwives in the birthing of a new story. Alan AtKisson, in his book *Believing Cassandra*, summarizes the challenge as follows:

> To . . . prevent global collapse, we need an idea that is both visionary and profitable, a solution that can appeal to both the ardent altruist and the hardened venture capitalist. We need a source of hope that is also a business opportunity, a hot investment that is also intensely idealistic. We need something that will challenge our higher natures and attract our baser instincts, coaxing us into the game of transformation without polarizing society or fomenting [violent] revolution. We need something that has not been seen since humans first began plowing up dirt, building skyscrapers, and messing around with atmospheric chemistry. We need something that has the power to command a lifetime of allegiance. . . .[28]

This sounds intimidating. Fortunately, though, there is already a word to describe the new "something" that AtKisson speaks of: "sustainability." This word began to be used in academic circles in the 1980s; by the 1990s, the general public was hearing the word; and now in the twenty-first century, schoolchildren are introduced to the concept of sustainability. Entire magazines are devoted to exploring its many facets; prizes are awarded for sustainable buildings; cities have developed sustainability indicators; and food labels tout sustainable farming practices.

"Sustainability" is not an easy word to say, nor is its meaning crystal clear. But in spite of its seeming awkwardness, "sustainability" is popping up everywhere precisely because this word is so necessary for our times. Again, AtKisson provides a useful perspective:

> History is full of examples of new and complex ideas overturning the old order, often against seemingly long odds. An example is 'democracy.' Today, most people throughout the world take it as a given that governments should be elected by the people. But this is a fairly recent idea and not a simple one (nor is the word particularly beautiful). Before a rudimentary form of democratic government took hold in the late 1700s in the newly formed United States—inspired in part by the ancient Greeks and the Iroquois Confederacy of Nations—democracy was not exactly a household word. Nor was this form of social organization widely understood, accepted, or practiced.[29]

Sustainability as the Foundation for a New Story

Sustainability's power as an organizing concept for modern life has come about, in part, because of three remarkable lessons that humans slowly absorbed over the second half of the twentieth century. The first was the lesson of exponential growth, most stunningly illustrated by the J-form of the human population growth curve, but exhibited just as dramatically in scores of production, consumption, and waste trends. The second lesson was that Earth, in terms of materials, is a closed system. Resources are finite, and there are physical limits to growth. The third lesson was that humans by sheer force of their numbers and appetites could outstrip the earth's carrying capacity. Taken together, these three remarkable lessons are gradually sensitizing humanity to the fragility of Earth and the need to practice more sustainable lifeways.

Those championing sustainability recognize that the human economy is embedded in the living earth. If Earth were eliminated, the human economy would cease to exist. Hence, in a sustainable society, the central focus is no longer on the money cycle but on the life cycle. The concept of sustainability is most fully understood by referring to the core principles or values that undergird all sustainable enterprises (see box).

Though the concept of sustainability may be new, the substance of its principles is clearly embedded in widely shared human values. For example, "living within limits" embodies the traditional American values of frugality and thrift; "accounting for full costs" is a call to remember the value of honesty and complete disclosure; and "sharing power" is (theoretically) what we

THE FIVE SUSTAINABILITY PRINCIPLES

Governments, organizations, and households seeking to become sustainable behave in accord with the following core principles:

Respecting life and natural processes. Sustainability commits us to explicit consideration of the effects of our decisions and actions on the health and well-being of the entire community of life.

Living within limits. Sustainability involves an awareness that the natural resources upon which all life depends—forests, fertile soils, fisheries, pure water, and clean air—are finite endowments to be used with care and prudence, at a rate consonant with their capacity for regeneration.

Valuing the local. Sustainability commits us to show respect for the natural components of our neighborhoods and bioregions; to preservation, restoration, and use of local knowledge; and to creation of strong, self-reliant local economies.

Accounting for full costs. Sustainability requires that we become aware of the costs generated by our products—from "source to sink"—to the environment and society. Product prices must reflect this awareness.

Sharing power. Sustainability demands we recognize that we are all interconnected—people, biota, and physical elements. Problems are solved by each individual assuming a share of the responsibility.[30]

believe American democracy is all about. It would seem that sustainability, as embodied in these core principles, offers a richer, deeper, more generous value system than the worldview based on economism (table 8.1).

Contemporary mass culture, grounded as it is in economism, does not live in accord with sustainability principles. For example, this culture often fails to respect life: it frequently regards the earth's biodiversity as raw material for human ends; and it fails to live within limits by consuming the natural stocks of soil, ocean fishes, and forests more rapidly than they are able to regenerate through natural processes. Furthermore, this consumer culture fails to account for full costs: it sells things cheaply and in the process violates the rights of workers, the environment, and future generations. In addition, it fails to value the local by harming local economies, traditions, and cultures in a rush for short-term profits. And, finally, this culture in many instances fails to share power to any significant degree by increasingly leaving its citizens in a state of disempowerment and dependency.[31] Although there are some exceptions to these generalizations in aggregate, the human attitudes and behaviors spawned by economism are far from ennobling.

Table 8.1. The Values Undergirding Two Worldviews: Economism and Sustainability

Economism	Sustainability
1. Life on Earth is for our use. —Humans are separate from nature. —Earth is a static system. —Concern for this generation only	**1. Life on Earth supports us.** —Humans are part of nature. —Earth is a living, evolving system. —Concern for future generations
2. We can expand forever. —Resource supplies are infinite. —Emphasis on consumption and constant growth —Happiness through acquisition	**2. There are limits that we must live within.** —Resource supplies are limited. —Emphasis on conservation and steady state —Happiness through relationship
3. The market will guide society. —Humans are only motivated by self-interest. —De-emphasis on government regulation —Economic growth is more important than environmental protection.	**3. The market is amoral—it is not a good guide.** —Humans have the capacity to act for the common good. —Government regulations are necessary to protect the public. —Environmental protection is more important than economic growth.
4. We must globalize everything. —We need *free* trade. —Emphasis on mass media —Materials and food come from far away.	**4. We must accord respect to the local.** —We need *fair* trade. —Emphasis on face-to-face interaction —Materials and food come from local sources when possible.
5. We must impose control from above. —Hierarchy: "power over" —Society built around competition —Wisdom resides at the top. —Strength in separateness	**5. We must share power and wealth.** —Equity: "power with" —Society built around cooperation —Wisdom resides in the network. —Strength and mutual well-being through partnership

Sustainability Means Defining a New Bottom Line

Citizens can take an active role in redefining societal values, and in so doing they can transform culture. Indeed, sustainability is a radical concept precisely because it invites citizens to comprehensively redefine the "bottom line" in all sectors of society. In this vein, Rabbi Michael Lerner calls on citizens to get together in their offices, factories, schools, churches, and businesses to consider what their workplace, professional endeavor, or daily routine would look like if it had a new bottom line—one that promoted personal self-esteem, caring human relationships, and ecological integrity (table 8.2). For example, imagine what it would be like if the primary goal of a university education were to cultivate a human being who is curious and intellectually alive, loving and able to show deep caring for others, awake to the spiritual and ethical dimensions of being, ecologically attuned, and creative.[32] How would this be different from the present experience of university education? Or what about rethinking the bottom line for business? A commitment to sustainability means literally expanding the traditional business bottom line beyond *profit* to include the well-being of the *planet* and the planet's *people*. Triple bottom-line companies strike a balance among the three Ps: profit, planet, and people.

Science, Sustainability, and the Evolution of Human Consciousness

The ecological crisis offers humans the opportunity to understand themselves in a new, more expansive way. As was true five hundred years ago in the time of Galileo (and more recently in the time of Darwin), so it is today that the discoveries in the sciences are playing a central role in catalyzing the expansion of human consciousness. For example, new discoveries—born of modern physics, astronomy, and biology—concerning the nature of matter, the origins of the universe, and biological evolution allow us as never before to see ourselves as participating in a most extraordinary story. The entire story of the universe is the story of the evolution of consciousness within an expanding universe—from the flaring forth thirteen billion years ago, to the creation of the first atoms of hydrogen, to the birth of our Sun and Earth, to the appearance of bacteria, and to the emergence of the human and onward. We humans are not the pinnacle of evolution, nor is it in our best interest to rule the earth. Rather, it appears that we are co-participants in a process leading to ever greater complexity and perhaps ever greater consciousness.

Table 8.2. Redefining the Bottom Line in Alignment with Sustainability Principles in Different Sectors of Society

Sector of Society	Status Quo Bottom Line	Redefinition of Bottom Line
Education	Passive: something that one submits to in order to "get ahead."	Active: something that one willingly participates in for self-actualization.
Business	Single bottom line: Success defined only in terms of profit; only growth is rewarded.	Triple bottom line: Success defined in terms of profit, people, and planet.
Work	Occupation: A job defined by someone else that one agrees to do in exchange for money.	Calling: A passionate pursuit that springs from within a person and engages mind, body, and soul, while contributing to the common good.
Government	Federal: Concentration of power at the top; role of citizens is to work, consume, and obey the laws of the land.	Bioregional: Power vested in citizens at the regional level; all voices valued; citizens work collectively to promote the common good.
Church	Salvation: Emphasis on personal salvation, good vs. evil, punishment, separation, shame.	Transcendence: Emphasis on expanding consciousness, insight, interconnection, acceptance, forgiveness, transformation.

Western science over the past five hundred years has focused on understanding nature's parts. By reducing the world to its parts—taking a so-called reductionistic approach—scientists believed that they could fully understand it. In the reductionistic view, the world is composed of clearly defined objects; and to the extent that they occur, relationships among objects are secondary. The reductionistic approach has been extremely fruitful but, in the end, not wholly satisfactory. Scientists now know that we can't fully understand the essence of things by simply taking them apart. The whole is *more* than the sum of its parts. For example, when iron and nickel are blended, they produce a material, steel, with a tensile strength far greater than the combined strengths of iron plus nickel. Likewise, the combination of hydrogen and oxygen produces a substance, water, with properties that aren't predictable based on a separate knowledge of each element. The case of hydrogen is particularly instructive: A "complete" knowledge of the hydrogen atom would not allow one to predict that hydrogen could self-organize to produce galaxies and planets and apple blossoms, but this is precisely what has happened over the past thirteen billion years. The properties of any "whole" are the result of the interactive relationships among the parts, and these interactions produce "emergent properties." This phenomenon is particularly true of living systems.

The relatively new field of systems science serves as a kind of counterpoint to exclusively reductionistic and mechanistic approaches to science. Scientists with a "systems" orientation are primarily concerned with understanding the patterns of interactions—the relationships among the parts—and this orientation leads to different ways of speaking about and seeing the world (figure 8.1).

In the reductionistic mind-set, each human being has sharply defined boundaries (figure 8.1). In this view, the way to ensure personal well-being is to make one's boundaries strong. Hence, the emphasis is on separation, often expressed as hyperindividualism. However, in the systems view, human beings are seen as participating in larger patterns of flows. Rather than being *mostly* separate, people are *mostly* connected through flows and interactions of matter, energy, and information. In this systems view, relationships are primary, and the way to ensure well-being is to soften one's boundaries and become permeable to the whole.

This systems view of reality, though difficult to grasp, appears to be a more accurate portrayal of living systems than the old mechanistic view. This new way of seeing gives more emphasis to interdependence rather than independence, spirit rather than ego, fluidity rather than rigidity, wholes rather than parts, union rather than division, connection rather than separation, synthesis rather than dissection (see box).

SYSTEMS THINKING AND LANGUAGE

The way we use words has an immense, though frequently unrecognized, effect on how we see and experience the world. People schooled in reductionistic thinking (i.e., most of us) often think in "thing" language. For example, in referring to Earth, we might say:

1. Earth is a rock in space.
2. Earth is a planet in the solar system.
3. Earth is a bundle of resources.

However, someone schooled in systems thinking would see Earth as a process alive with relationship and say:

4. Earth is a cosmic happening.
5. Earth is our living body.
6. Earth is allurement, relationship, potentiality. . . .

Reflection

I confess, I am both terrified and exhilarated by this present moment in history. On the one hand, humanity has never experienced such a dangerous time; we live in the shadow of nuclear annihilation, widespread chemical poisoning, and catastrophic climate change. On the other hand, we live at a moment when we can finally see ourselves as part of a grand cosmic unfolding—a time when it is possible, at least theoretically, to build a life-affirming, sustainable society.

We are, as Joanna Macy says, at the time of the "Great Turning." It is an in-between time. We may not make it; the dangers are real and daunting. But what a time to be alive! If we do pull through, our descendants will likely look back at our time with envy, knowing that we lived at a time of high adventure. They will say, "They were the ones alive at the Great Turning. At first, they were burdened with despair; their actions were paltry; their words tentative. But bless them for they rose above their despair; they found their voices; they acted; they were the people who took part in the Great Turning."[33]

Questions for Reflection

- Which of the five sustainability principles might you consider embracing? In what ways might you manifest these values in your daily life?
- How might you consider participating in the Great Turning?

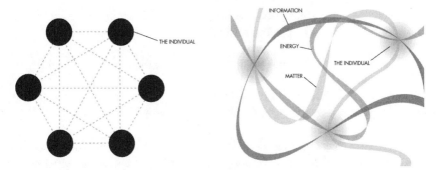

Figure 8.1.   Mechanistic versus systems view of reality

Practice: Creating Our Culture's New Story

Culture is not static. It changes and evolves as a result of the things that we human beings do—the ways that we behave, the stories that we tell, the conversations that we have, and the values we espouse during our lifetimes. Each day, humans participate in the process of making culture. In this vein, the efforts throughout the planet to give birth to a new story—one grounded in sustainability—can be viewed as a grand social-change movement.

Not long ago I participated in a practice called the "Double Circle," which shed light on this cultural process.[34] There were twenty of us, and we sat facing one another in two concentric circles, ten people per circle. Those in the inner circle, facing out, were humans from a hundred years in the future; they had come to listen. The people in the outer circle, facing in, were simply themselves, beings of the present. I was in this outer circle. My partner and I sat facing each other, knee to knee.

Our guide welcomed us into a "middle world"—a space outside of time where the generations can meet. She said that the future ones (in the center) had four questions for those of us in the outside circle. The future ones were to convey their questions telepathically to the present ones through our guide's voice.

The first question from the future ones was:

Ancestor, I have been told about the terrible times in which you lived, wars, and preparations for war, hunger and homelessness, the rich getting richer, the poor getting poorer, poisons in the seas and soil and air, the dying of many species. . . . It is hard to believe. Was that really true? Tell me.

Each of us present-time beings responded, speaking out of our personal experience. The future ones, sitting across from us, simply listened to our words. After a time, when we present ones had no more to say, the future ones were asked to rotate one position to the right. Then, using the guide as a mouthpiece, the future ones asked their new partners:

> Ancestor, what was it like for you in the midst of all that? How did you feel?

After we present ones had finished responding to this question, the future ones again moved to the right for the third question:

> Ancestor, we have songs and stories that still tell of what you and your friends did back then for the Great Turning. Now what I want to know is this: How did you start? You must have felt lonely and confused sometimes, especially at the beginning. What first steps did you take?

And, finally, after another rotation, the fourth question came forth through the guide:

> Ancestor, I know you didn't stop with those first actions on behalf of Earth. Tell me, where did you find the strength and joy to continue working so hard, despite all the obstacles and discouragements?[35]

After listening to our responses to this final question, the future ones were then given an opportunity to speak. They addressed us present ones, expressing what they were thinking and feeling after all that they had heard.

Our guide concluded this ceremonial practice by inviting us to come back to present time. Then, we formed a large circle and shared our thoughts and insights. Many of those who had been in the outer circle—the present-time people—found this experience ennobling; they felt heard and appreciated, and their lives seemed to take on added meaning. Those from the future experienced great admiration and empathy for the present ones and, by extension, compassion for themselves as flesh-and-blood present-time beings. Exercises like this can be empowering insofar as they encourage us to see ourselves as participating in an epic moment in history.[36]

The birth of the life-affirming new story, necessary to usher in the Great Turning, will not occur without midwives—people who understand their interconnectedness with all life, who feel the pain of the earth's suffering, and who realize that it is both necessary and possible to create a sustainable world.

A simple beginning practice for people wishing to participate in this Great Turning is to change their "news diet." Why? Because what we take in through the media affects not only our mind-sets but also who we become. Start with a "news fast": Turn off the tired messages from mainstream newspapers, television, and radio. This is not a shutting down so much as an opening up to a whole new world of power and possibility. Insofar as the mainstream media is financed and controlled by the keepers of the old story, it is largely incapable of seeing and communicating the emergence of a new story based on different values, principles, and beliefs.

Next, invite new messages—new sources of news and inspiration—into your life. Consider subscribing to *Yes!*, *Resurgence*, *Orion*, *Sun*, or *Earth Light*. These are some of the magazines of the sustainability revolution; many of their articles offer stories about people and communities awakening to the wonder of life, adopting sustainable practices to reduce their ecological footprint, and acting for the common good. The enactment of these individual life stories is giving rise to a grand "new story"—a story about connection, not separation; about awe and wonder, not business as usual; about partnership, not domination. Ultimately, this new story transforms our understanding of what it means to be human.

Becoming midwives for the new story also means learning to use a new vocabulary, a vocabulary of solutions. In recent years this new vocabulary has been popping up everywhere. People are coming to hear words and phrases such as *living machine, organic agriculture, wind power, ecological footprint, growth boundary, natural capitalism, car-share, green buildings, voluntary simplicity, community-supported agriculture, co-housing, watershed stewardship, holistic health, integrated pest management, restorative justice, socially responsible investing, community land trust, micro-radio*, and on and on. This is the vocabulary of the sustainability revolution.

As we will see in the next chapter, sustainable solutions are now available to meet humankind's needs for energy, food, transportation, housing, water, and education. What is lacking isn't political will so much as its antecedent—individual will made manifest through personal decisions and actions. In our buying decisions in particular, each of us has countless opportunities to make choices that foster sustainability. In this vein, a powerful practice is to jot down the five sustainability principles (see box on page 249) on a card and carry this card with you in your wallet. Then, every time you take out your wallet to buy something, you can ask, in making this particular purchase, if you are

1. respecting life,
2. honoring planetary limits,
3. helping your local economy,
4. supporting ecologically responsible businesses,
5. contributing to social justice.

The new life-sustaining culture will emerge as individuals adopt sustainable practices and tell the stories of these practices. I was reminded of this recently while attending a presentation on voluntary simplicity. Toward the end of the evening, Beth stood up and said she had a personal story to tell. This wasn't a scheduled part of the program, but the host consented. Beth was nervous, but her determination to share her story allowed her to overcome her jitters. She told us that after many years of commuting to work alone, she had decided to give carpooling a try. Beth confessed that at first she was apprehensive because she didn't have much natural affinity for the two women she had arranged to carpool with. However, she told us that to her surprise she was enjoying the presence of two people who were quite different from herself.

Beth went on to mention two unexpected benefits of carpooling. The first was that, with two other people in the car, she was no longer able to do "impulse shopping" on the way home. This saved her money. Second, knowing that her two colleagues were waiting for her meant that she could no longer dawdle about in her office at the end of the day. When 5 P.M. came, she had to be heading for the door to be on time for her colleagues. So, Beth decided to carpool because she originally wanted to do a good turn for the planet, but in the process she made new friends, saved money, and gained time. Just as important as the particulars is the fact that she shared her story.

In sum, each of us can help to create the new culture of sustainability first by fully appreciating that we live in an epic time and then by tuning into the new story of the sustainability revolution. The practices for this movement consist of switching our news antenna from the old signals to the new frequencies, learning the new vocabulary of solutions, adopting sustainable lifeways, and telling the stories of our actions.

CONCLUSION

It's all a question of story. We are in trouble just now because we do not have a good story. We are in between stories. The old story, the account of how the world came to be and how we fit into it, is no longer effective. Yet we have not learned the new story. . . . We need a story that will educate us, a story that will heal, guide, and discipline us.

—Thomas Berry[37]

In this chapter we have entertained the possibility that Western culture's story—a story founded on control, expansion, and separation—is at the root of the ecological crisis. In its modern guise of "economism," this story proclaims: Humans are essentially economic beings; reductionism is the primary pathway to knowing; and the subjugation of nature is necessary for progress.[38] These are the "truths" that we are all taught—American, Russian, Japanese, and Swede—but they are in fact only half-truths, unable to fully support and sustain a healthy and just world. But the solution is not to shun modernity. This would be folly. As philosopher Ken Wilber reminds us:

> The rise of modernity . . . served many useful and extraordinary purposes. We might mention: the rise of democracy; the banishing of slavery; the emergence of liberal feminism; the differentiation of art and science and morality; the widespread emergence of empirical sciences, including the systems sciences and ecological sciences; an increase in average life span of almost three decades; the introduction of relativity and perspectivism in art and morals and science; the move from ethnocentric to worldcentric morality; and, in general, the undoing of dominator social hierarchies in numerous significant ways. These are extraordinary accomplishments, and the anti-modernist critics who do nothing but vocally condemn modernity, while gladly basking in its many benefits, are hypocritical in the extreme. . . .[39]

The solution for we modern ones is to take the next step in human cultural evolution. Humankind has moved from hunting and gathering to plant and animal domestication to industrialization. Now we need to take the good that modernity offers and leave behind its harmful effects, attitudes, and behaviors. Religion, science, and economics need to be integrated into a larger whole that is grounded in enduring values, spirit, and truth-seeking.[40]

The essential practices for participating in this Great Turning center on story. Story is a key catalyst in both personal and cultural change. At the personal level, our fidelity to our unique calling (i.e., personal story) is forged by our passion and sense of purpose. At the societal level, what is necessary to ensure the survival of our species is a mythic story with the power to excite our imagination and call forth that which is good and noble in us. If we, as a people, can change our story, we can change our destiny.

NOTES

1. Thomas Berry, *The Dream of the Earth* (San Francisco: Sierra Club Books, 1988), 171.
2. Daniel Quinn, *Ishmael* (New York: Bantam, Turner, 1992).
3. Quinn, *Ishmael*, 35–36.
4. Quinn, *Ishmael*.

5. Barbara M. Hubbard, *Conscious Evolution: Awakening the Power of our Social Potential* (Novato, Calif.: New World Library, 1998), 178.

6. Quinn, *Ishmael.*

7. Quinn, *Ishmael.*

8. Riane Eisler, *The Chalice and the Blade* (New York: HarperCollins, 1987).

9. Eisler, *The Chalice and the Blade.*

10. Scott Russell Saunders, "The Most Human Art," *Utne Reader*, September/October 1997.

11. Natalie Goldberg, *Writing Down the Bones* (Boston: Shambhala, 1986), 147.

12. Goldberg, *Writing Down the Bones*, 147–148.

13. Charlene Spretnak, *The Resurgence of the Real* (New York: Routledge, 1999), 56.

14. Barbara Brandt, *Whole Life Economics* (Gabriola Island, British Columbia: New Society, 1995).

15. This story is directly quoted from Susan Strauss, *The Passionate Fact* (Golden, Colo.: North American Press, 1996), 9.

16. Brandt, *Whole Life Economics.*

17. Jerry Mander, *In the Absence of the Sacred* (San Francisco: Sierra Club Books, 1991).

18. Benjamin Barber, "McWorld and the Free Market's Threat," broadcast on Alternative Radio (Boulder, CO), October 7, 1998.

19. Harwood Group, "Yearning for Balance," *Yes! A Journal of Positive Futures*, Spring/Summer 1996, 17.

20. Jerome Segal, *Graceful Simplicity: Toward a Philosophy and Politics of Simple Living* (New York: Henry Holt, 2000).

21. Mark A. Burch, *Stepping Lightly* (Gabriola Island, British Columbia: New Society, 2000).

22. Gregg Levoy, *Callings: Finding and Following an Authentic Life* (New York: Three Rivers Press, 1997), 264.

23. Molly Y. Brown, *Growing Whole* (Center City, Minn.: Hazelden Educational Materials, 1993).

24. Levoy, *Callings*, 139.

25. David Whyte, *Crossing the Unknown Sea: Work as a Pilgrimage of Identity* (New York: Riverhead Books, 2001), 65, 70–71.

26. Levoy, *Callings.*

27. Donella Meadows, Dennis Meadows, and Jorgen Randers, *Beyond the Limits: Confronting Global Collapse, Envisioning a Sustainable Future* (Post Mills, Vt.: Chelsea Green, 1992), 220–221.

28. Alan AtKisson, *Believing Cassandra* (White River Junction, Vt.: Chelsea Green, 1999), 130.

29. AtKisson, *Believing Cassandra*, 145.

30. Christopher Uhl, "Green Destiny: Universities Leading the Way to a Sustainable Future," *BioScience* 51, no. 1 (2001): 36–42.

31. Uhl, "Green Destiny."

32. Michael Lerner, *Spirit Matters* (Charlottesville, Va.: Hampton Roads, 2000).

33. Joanna Macy, personal communication.

34. Macy, *Coming Back to Life.*

35. Macy, *Coming Back to Life*, 146–148.

36. For other exercises, see Macy, *Coming Back to Life.*

37. Thomas Berry, *The Dream of the Earth* (San Francisco: Sierra Club Books, 1988), 123–124.

38. Spretnak, *The Resurgence of the Real.*

39. Ken Wilbur, *A Brief History of Everything* (Boston: Shambhala, 1996), 69.

40. Wilbur, *A Brief History of Everything.*

⑨

CREATIVITY: DESIGNING
A SUSTAINABLE SOCIETY

> We live in two interpenetrating worlds. The first is the living world, which has been forged in an evolutionary crucible over a period of four billion years. The second is the world of roads and cities, farms, and artifacts, that people have been designing for themselves over the last few millennia. The condition that threatens both worlds—unsustainability—results from a lack of integration between them.
>
> —Sim Van der Ryn and Stuart Cowan[1]

The metaphor for the industrial revolution was the "machine." Humans imagined that by reducing matter to its constituent parts, they could understand the mysteries of the whole. The metaphor for the sustainability revolution, by contrast, is the "organism."

The first organisms (bacteria) appeared on Earth nearly four billion years ago. Starting six hundred million years ago, the creative, self-organizing life force began to bring forth shellfish and lizard, tree and bird, rodent and human. Humankind is not outside the family of life but a co-participant in an evolutionary unfolding. The pathway to sustainability is encoded, to a significant degree, in the larger economy of nature.

The sustainability revolution is already underway. This is a peaceful revolution; it will not destroy the old but, rather, reinvent it—shaping, recasting, reforging. First and foremost, this is a design revolution. Humankind now needs new approaches to the design of everything from household appliances, through manufacturing processes, to towns and cities (foundation 9.1). Our old economic system, with its exclusive fixation on profit, also needs to be reforged into a sustainable economics that values people and the planet just as much as profit (foundation 9.2). Finally, the creation of a sustainable society requires that we reinvent democracy from the bottom up, recognizing that politics, at its best, is a collective striving to serve the common good (foundation 9.3). This chapter's practices offer personal ways of exploring the political, economic, and design dimensions of sustainability.

FOUNDATION 9.1: DESIGN DIMENSIONS OF SUSTAINABILITY

Activist Nancy Todd tells a story about a conversation she had with a Native American woman. The woman observed that the blade of grass, the tree, the spider, the elk all live by their "instructions" and that her people try to do the same. Then, the native woman looked at Todd and said, "Our people don't know what your 'instructions' are." In that instant, Todd realized that her people, the white European settlers in North America—in their excitement over what they could create in America—neglected to concern themselves with the "instructions," with the basic ecological and planetary rules; and they unwittingly designed a world that is not sustainable.

Pay attention the next time you are out walking through your town. Almost everything you see is the result of a design decision. Much of what has been built throughout America, particularly over the past fifty years, is not sustainable—shopping centers, industrial parks, subdivisions, and apartment complexes are all snapped together with a massive network of highways, power lines, and storm and sanitary sewers (see box).[2]

A Design Challenge

The task of creating a sustainable society is, in large part, a design problem. To convey the magnitude of the challenge, green-design experts William McDonough and Michael Braungart invite people to imagine that they have just been presented with the industrial revolution as a retroactive design assignment. Your assignment might sound something like this—design a system that

1. puts billions of kilograms of toxic material into the air, water, and soil each year;
2. measures prosperity by the speed at which things are made, used, and thrown away;

EXAMINING BUILDINGS THROUGH THE LENS OF SUSTAINABILITY

The five sustainability principles, introduced in the last chapter, are a useful filter for examining the sustainability of our buildings. I sometimes challenge my students to examine our classroom building by responding to questions relating to each sustainability principle.

> *Principle 1: Respect life and natural processes.* Was the building created in a way that minimized damage to soil, air, and biota?
>
> *Principle 2: Live within limits.* Is the building designed in such a way that it requires minimal energy for heating and cooling—for example, is it sited to maximize solar gain? Are the building's energy sources renewable and local in origin?
>
> *Principle 3. Value the local.* Is there anything about the building, inside or out, that tells you where you are—for example, does the building embody regional culture and values? Was the local economy supported in constructing the building, and is it currently supported through the building's maintenance activities?
>
> *Principle 4. Account for full costs.* What were the "upstream" environmental impacts associated with the mining and manufacturing of the building's materials? What are the present-day environmental impacts of the waste (including greenhouse-gas emissions) associated with the building's operation? Have all the people who brought the building into existence, as well as those who now maintain it, been justly compensated—for instance, are their efforts on behalf of the building reflected in their health and retirement benefits?
>
> *Principle 5: Share power.* Who was involved in the decision to construct the building and in the decisions about the building's appearance and features? Were decisions made based on principles of inclusivity?

Each of these questions is linked to sustainability. For example, as my students reflect on the "sharing power" questions, they realize that the people who occupy campus buildings rarely have a significant say in determining what is built and how new buildings are designed. Nor do they have any significant control over the operations of the buildings they spend time in. When all building operations are electronically controlled, students learn passivity and the irresponsibility invited by never having to know how things work.

These same questions can be applied to assess the sustainability of homes, churches, businesses, and workplaces. In the case of classroom buildings, the important lesson that emerges is this: Learning takes place inside buildings, but it also takes place as a result of how buildings are designed and by whom; how they are constructed and from what materials; how they harmonize with their location; and how efficiently they operate.[3]

3. produces materials so dangerous that they will require constant vigilance from future generations—for example, radioactive waste;
4. contributes to the extinction of species.

You may be scratching your head until you realize that—from a strictly environmental perspective—the industrial revolution unwittingly accomplished these four design "goals."

McDonough and Braungart point out that humans now have the opportunity to design a sustainable society for the twenty-first century. Our new design assignment is to create a system that

1. introduces no hazardous materials into the air, water, or soil;
2. measures progress by how many buildings have no smokestacks or dangerous effluents;
3. produces no materials or chemicals requiring the vigilance of future generations;
4. celebrates the abundance of biological and cultural diversity and renewable solar income.[4]

The design of such a sustainable world is within reach, but creating it will require a paradigm shift. For starters, the concept of "waste" will need to be eliminated. How?

> To eliminate the concept of waste, all materials and products should be conceived of as nutrition. . . . Humans must design things that, after their useful life, return safely to soil and rebuild it. Other things must be designed to safely reenter industrial production on a continuous basis over many production cycles. Anything that does not fit safely into these two metabolisms (i.e., industrial or biological) must be phased out.[5]

In the 1990s, McDonough and his associates set out to design a totally safe, zero-waste fabric. The team's first idea was to make the fabric with a blend of organically grown cotton and plastic fibers from recycled soda bottles (for durability). Both materials were readily available, market-tested, and cheap. But when the team tested the cotton–plastic fabric, they discovered problems. When a person sat in an office chair and shifted around, the new fabric abraded, setting free tiny particles of plastic fiber that could be inhaled or swallowed by the user and other people nearby. Plastic was not designed to be inhaled. Also, the hybrid fabric would still end up adding junk to landfills because it could not be further recycled.

So the team went back to the drawing board. Eventually, they decided on a fabric blend composed of ramie (plant fiber) and wool (animal fiber). Then, they focused on the finishes (i.e., the dyes and other fabric-processing chemicals). If the fabric was to be safe to humans and able to go back into the soil, it had to be free of mutagens, carcinogens, heavy metals, hormone disrupters, and persistent toxic substances. The team approached sixty chemical companies, and finally one of them, Ciba-Geigy, agreed to screen the thousands of chemicals used in the textile industry. Fewer than one percent were judged totally safe by McDonough's team, but this was an adequate subset to create a new line of chemically benign fabrics. Indeed, when the regulators came to check the fabric factory, they thought their instruments were broken because the water leaving the factory was as clean as the water going in.[6]

A Zero-Waste World

Many companies are beginning to explore and implement zero-waste production systems. For example, in Germany, BMW has built a pilot disassembly plant to recycle its older cars, and new BMWs are being designed with eventual disassembly in mind. Car parts are bar-coded to identify the types of materials they contain, and the number of component materials is being reduced. Design modifications are aiming to yield 100 percent reusability.[7]

Wastewater treatment is another target for waste elimination. When engineer John Todd was asked to design a wastewater treatment facility for an elementary school in Toronto, he used nature as his mentor and began with the principle: waste equals food. Todd's facility, dubbed a "living machine," is part art, part function, and part teacher.[8] Located in the central atrium of the school and consisting of seventeen water tanks arranged in a snail-spiral, the "living machine" looks like a water sculpture. The school uses conventional water-based toilets. Once the waste is flushed, it is rapidly digested as it passes through four tanks—two with oxygen, two without, each filled with microorganisms. Next, the water is pumped to the highest of the seventeen clear plastic tanks. Here, the students may watch as the water flows first through tanks filled with algae, then through tanks filled with aquatic-rooted plants, and finally to tanks with animals, including clams, snails, and fish. By the time the "wastewater" passes through to an adjacent pond, it is clean.[9] Variants on Todd's "living machine" are now being used to process industrial wastes (see box).

RECONCEPTUALIZING WASTE AS PRODUCTS

In their book, *Ecological Design*, Sim Van der Ryn and Stuart Cowan describe how a film studio in Wuxi, China, resolved a difficult pollution problem involving silver-contaminated wastewater. The studio introduced water hyacinths into a series of ponds. These water hyacinths, in effect, "mined" silver from the wastewater and accumulated it in their roots. Silver concentrations in the hyacinths' roots were up to thirty-five thousand times higher than the concentrations in the surrounding wastewater. The studio harvested the hyacinths' roots, extracted the silver, and then reused it; the overall silver retrieval rate was 95 to 99 percent.[10]

It is even possible to create complex industrial food webs modeled on biological food webs, where waste from one industry serves as a "food stuff" for other industries (and vice versa). A working example of this sort of complex "eco-industrial park" is located in Kalundborg, Denmark, where

> a complex web of waste and energy exchanges has developed among the city, a power plant, a refinery, a fish farm, a pharmaceutical plant, a chemical manufacturer, and a wallboard maker. The exchange works something like this: the power company pipes residual steam to the refinery and, in exchange, receives refinery gas (which used to be flared as waste). The power plant burns the refinery gas to generate electricity and steam. It also sends excess steam [for heat] to a fish farm, the city and a biotechnology plant that makes pharmaceuticals. Sludge from the fish farm and pharmaceutical process becomes fertilizer for nearby farms. Surplus yeast from the biotechnology plant's production of insulin is shipped to farmers for pig food. Further, a cement company uses fly ash from the power plant, while gypsum produced by the power plant's desulfurization process goes to a company that produces gypsum wallboard. Finally, sulfur generated by the refinery's desulfurization process is used by a sulfuric acid manufacturer.[11]

Ideally, eco-industrial parks become closed-looped ecosystems where waste from one factory serves as a foodstuff for other enterprises.

Nature as Design Partner

As human technological design challenges grow, scientists and engineers are turning increasingly to nature for solutions. Fortunately, the earth's biota has been involved in a kind of massive research and development

project for millions of years. Hence, elegant solutions to many human design challenges have already been elaborated in the natural world.[12] The
"technological" sophistication of nature is extraordinary, as illustrated by the
following three examples.

Spiders and Kevlar

Kevlar is known for its incredible strength and toughness; but to make it,
humans have to subject petroleum products to high pressure in a boiling solution of concentrated sulfuric acid. The process is nasty and polluting.
However, spiders are able to produce a biodegradable silk—they have been
doing it for over three hundred million years—that is both tougher and
more elastic than Kevlar—and, ounce for ounce, five times stronger than
steel. In addition, spiders do it at room temperature, without resorting to
the use of toxic chemicals or offshore oil drilling. Potential applications for
spider silk include lightweight bullet-proof vests as well as cable for sus-
pension bridges.[13]

Beetles and Fire Detection

Some species of jewel beetles *(Buprestids)* are able to detect fires at
great distances. Entomologists call these critters "fire beetles" because they
lay their eggs in trees that have been recently killed by forest fires. Fire beetle larvae apparently thrive in heat-sterilized trees.

Adult fire beetles have two sophisticated fire-detection devices. First,
they have a pair of organs located at the base of their middle legs that are
extremely sensitive to infrared radiation. These "sensors" are held aloft during flight, and they evidently enable the beetles to hone in on fires set tens
of miles away. These "heat sensors" are coupled with "smoke detectors" on
the beetle's antennae, which allow them to distinguish between the smoke
of pines, their preferred species of trees, and that of other species. By
studying the physiochemical basis of the beetle's fire-detection "technology," human-based fire detection efforts may be enhanced.[14]

Termites and Air-Conditioning

Some termite species that inhabit hot tropical grasslands build mounds,
honeycombed with tunnels, that rise three or more feet above the ground.
As the sun heats these mounds during the day, the hot air rises through the
tunnels and diffuses to the outside through pores in the mound wall. The

termites usually live in a basement cavern directly below the mound. This chamber is naturally air-conditioned by the cool air rising from deep within the soil. But the heat generated by millions of termite bodies would heat their chamber to unbearable temperatures in no time if it weren't for the above-ground, air-transfer tower. Biologists Andrew Beattie and Paul Ehrlich summarize:

> Termite air-conditioning thus uses soil-cooled air to keep the insects comfortable and, as they warm the air, disposes of it up a porous chimney. The air-conditioning therefore takes advantage of a constant and well known law of nature: warm air rises. Not only is this solution ingenious, it is free.[15]

Nowadays termite mounds are serving as blueprints for architects and construction engineers as they seek to design elegant and low-cost natural air-conditioning systems.

In sum, the old design mentality, born of the industrial revolution, blindly optimized with respect to short-term cost and convenience. This short-sighted approach has been wasteful and polluting. Fortunately, human technological prowess now offers the possibility of efficient, nonpolluting production systems; and the biodiversity of the natural world offers, at least in some instances, sustainable design templates.

Reflection

I received a memorable lesson in sustainable design in 1971 when I spent the winter in Tokyo. I was on a shoestring budget and, in searching for a inexpensive place to live, found my way to a student boardinghouse on the outskirts of the city. The proprietor, a genial man in his fifties, greeted me warmly.

"Yes, there is a room available."

"Might I see it?"

"Yes, yes, follow me."

He led me to a room that measured perhaps five feet by ten feet. It had space for a bed, a desk, and a chair—which was really all I needed. After settling the financial arrangements, I returned to my room to settle in. I then noticed that the building had no heat and that the temperature was in the low forties. In the excitement of locating a room, I had failed to consider heating. But then I noticed something that was to be my salvation. Underneath the desk was a pair of slippers. These slippers had an electrical cord attached to them. I plugged them in and slipped my cold feet into them. Hmm. Not bad. Soon I had my Japanese language book out and was studying. Later, I learned an oriental saying that said something to the effect of: "Warm feet—cold head: Best learning way."

After a time, a bell rang—the call for the evening meal. The dining room was as cold as my little room; but as I sat on the tatami-mat floor with the other students and as I slipped my legs under the low-slung dining table, I encountered another pleasant surprise. The table was fringed with a quilt that draped to the floor, and under it was a small electric space heater. So, while the temperature of the room was forty degrees, my lower body was snug and warm. It suddenly struck me as odd that in America we heat up thousands of cubic feet of air throughout our homes, when really what matters are the few cubic feet of air that envelop our bodies.

I now realize that often the first step in designing for sustainability is to reframe our needs. For example, we want to be warm in the winter, but we don't necessarily want or need central heating. We want mobility, but we don't necessarily want or need cars. We want to have access to light in our homes and workplaces, but we don't necessarily want or need incandescent light fixtures. When we reframe in this way, it is sometimes possible to discover other, more benign ways of meeting our needs (see the following practice section).

Questions for Reflection

- What parts of your own life are well designed from an ecological perspective?
- What parts are ripe for redesign?

Practice: Design and Personal Sustainability

Many of the decisions that Americans make have ecological implications—for example, where to live, what form of transportation to use, how to dry clothes, what kinds of food to buy, at what temperature to keep the thermostat, and so on. Each day we have the freedom to adopt behaviors that reduce our personal ecological impact. This point is especially important in the area of energy use.

Reducing Household-Related Energy Use

The principal reason that the per capita ecological footprint is so large in the United States (chapter 6) is that Americans consume large amounts of energy. To reduce their energy footprint in the home, citizens can participate in the following eight practices.

1. *Lower the thermostat in winter.* It is possible to reduce household heating energy use by 10 to 15 percent simply by wearing a sweater and turning down the thermostat from seventy-two degrees to sixty-eight degrees.[16]

2. *Use a ceiling fan instead of air conditioning.* Fifty years ago air-conditioners were a rarity; today they consume up to one-sixth of the electricity in the United States. We have an elegant alternative: the ceiling fan. The air circulation created by a ceiling fan evaporates moisture from the skin and in so doing makes a room "feel" nine degrees cooler than it really is; in addition, at full tilt, a ceiling fan uses less than 10 percent of the energy that an air-conditioner would use in a medium-size room.[17]

3. *Use compact fluorescent light bulbs.* Roughly one-quarter of U.S. electricity consumption goes to lighting. The lightbulbs in most houses are traditional incandescents. If you are not sure what you have, touch the bulb. If it is too hot to hold on to, it's an incandescent. Indeed, 94 percent of the electricity that incandescent lightbulbs use goes to produce heat, while only 6 percent goes to produce visible light.[18] There is an alternative lightbulb design that is much more efficient. It's called a "compact fluorescent." Compact fluorescents, now available in most hardware stores, use only one-quarter of the energy of conventional incandescents and last ten years—ten times longer than incandescents, thereby paying for their purchase price in electricity savings within two to three years.[19]

4. *Devise new strategies for washing and drying clothes.* It may come as a surprise, but 90 percent of the energy used in conventional clothes washers is used to heat the water.[20] Part of the design solution in this case is simple: Wash with cold water. Another part of the solution is to use less water. In fact, one new clothes-washer design, the horizontal-axis washer, reduces energy and water use by roughly 50 percent compared to conventional vertical-axis washers.[21]

When it comes to drying clothes, we have no more elegant design solution than the clothesline—no machines, no electricity, no pollution. When the sun is not shining, clothes can be dried indoors using a wooden clothes rack. An electric clothes drier, by contrast, is expensive to buy and operate. You could work at a laptop computer for a solid week with the energy consumed in one clothes-drying cycle; and feeding the drier electricity will cost more than one thousand dollars over its lifetime. Moreover, clothes driers reduce the lifetime of clothes (check the lint trap of your drier if you are wondering why your clothes wear out so quickly).[22]

5. *Choose an energy supplier offering renewable energy.* Thanks to electric utility deregulation, many Americans can now choose their energy providers, and many utilities now offer "green options" (see www.green-e.org). For example, in Pennsylvania, it is now possible to purchase energy from Green Mountain Power (GMP), an energy provider committed to offering clean and renewable forms of energy—for example, wind, hydro, geothermal. Purchasing energy from GMP costs a bit more, but the extra money goes toward creating a sustainable energy future—for example, by building new wind turbines.

6. *Avoid choosing a house that is larger than necessary.* This is a first-order design decision with built-in multiplier effects. For example, the amount of energy required to heat and cool a house or apartment (as well as to build it) is closely related to the number of square feet in the domicile. So, all else being equal, a house that is 25 percent bigger than it needs to be to meet an individual's or a family's needs will cost 25 percent more to heat and cool, and it will produce 25 percent more greenhouse gases in the process.[23]

7. *Pay close attention to physical location when choosing a dwelling.* Where people choose to live (close to vs. distant from work, shopping, schools) can dramatically affect the amount of energy they use in transit. All else being equal, people living in the suburbs log 50 percent more vehicle miles per year than people living in urban settings.[24]

8. *Reduce car dependence.* Most people rely on cars to get around in spite of the fact that automobiles are inefficient and highly polluting. For example, only 14 percent percent of the energy in a car's fuel is actually involved in moving the car down the road; the rest is lost as engine heat, drivetrain friction, and general inefficiencies. Carpooling is a simple practice that can dramatically increase fuel-use efficiency. The math is simple. Suppose you carpool with three others in a car that gets thirty miles per gallon. Suddenly with four riders in the car, fuel efficiency jumps to roughly 120 miles per gallon per person. As good as this is, it doesn't come close to the bicycle that runs on renewable energy and has a fuel efficiency equivalent to one thousand miles per gallon. Of course, in the case of the bike, the fuel is the food the biker eats, not the polluting gasoline. What's more, the exercise afforded by biking contributes to human health. Remarkably, 40 percent of all car trips in the United States are two miles or shorter—that is, they are within biking distance.[25]

Step by step, we can adopt practices that dramatically reduce household-related energy consumption. A good place to go for guidance is www.HomeEnergySaver.lbl.gov. After typing in your zip code plus basic household characteristics—for example, type of heating/cooling systems, appliances, number of windows—the site presents specific suggestions for reducing energy consumption.[26]

Paying Attention to the Big Numbers

As you consider practices to reduce your energy footprint, remember that it is more important to place your attention on the big-impact categories than the small-impact ones. For example, many people fret over their consumption of plastic shopping bags at the grocery store (usually only a couple of pounds worth per year), all the while failing to notice that in driving an SUV that averages twenty miles per gallon, they are consuming enormous amounts of energy and are releasing roughly fifteen thousand pounds of carbon dioxide into the atmosphere each year—not to mention all the waste and pollution associated with petroleum extraction, refining, and transportation. Hence, if you can redesign your life to reduce gasoline consumption—for example, by carpooling to work or getting rid of your second car—the benefits for the environment will be far greater than if you manage to use a few less trash bags (see box).[27]

As the shower example suggests (see box), what we quantify with numbers has the potential to change our behavior. Dieters know this and so do economists; now car manufacturers are capitalizing on this principle. For example, Honda has come out with a hybrid car that lets the driver know, second by second, exactly how many miles he is getting per gallon. Bill McKibben, who owns a hybrid, says his driving behavior has been changed by this instant feedback: "When I'm behind the wheel, I'm an American—competitive, score-keeping, out to win. As I pull out of the driveway, what I think about is: can I beat my last trip? Will I make it home averaging 60 miles per gallon? . . ."[28]

It seems that it is instinctual for humans to want to improve their "scores." Give us a number, and we will strive to improve on it. If you live in a house, you receive numbers on a regular basis in the form of utility bills. As a practice, you might consider getting some graph paper and tracking your monthly consumption of water (gallons), oil (gallons), and electricity (kilowatt-hours). Place the graphs on your refrigerator and engage in a competition by trying to lower your scores month by month, year by year.

WORKING THE NUMBERS

It is important to be strategic in deciding where to make lifestyle design changes. For example, suppose you and your four housemates are concerned about water waste. Each of you takes an eight-minute shower each day. With a bucket and a stopwatch you discover that water flows from the shower head at a rate of five gallons per minute. Thus, each day you and your housemates consume two hundred gallons of water in the shower alone, not to mention the energy your water heater uses to heat those two hundred gallons. You meet to discuss the situation, and one of your housemates recommends that everyone limit their showers to six minutes, which would reduce daily water consumption to 150 gallons per day. Another roommate suggests installing a water-conserving shower head that releases only two and a half gallons per minute, which would halve water consumption to one hundred gallons per day. If you were only going to do one thing, it would make the most sense to stick with the eight-minute showers and install the low-flow shower head. After some discussion, everyone agrees that it would be best to adopt both and thereby reduce water consumption to just seventy-five gallons per day. But then a truly radical thought occurs to you: What if we were to take a shower every other day? In this case, average water consumption would be knocked back to thirty-seven and a half gallons per day, just 18 percent of previous consumption. In a later phone conversation, you relate this story to your grandmother. She chuckles and observes, "In my day, we took a bath once a week and we managed to stay plenty clean." You do the math and realize that Grandma's suggestion would take you down to roughly ten gallons per day, a twentyfold reduction in your household's water consumption. Few people are ready to cut consumption twentyfold, but many are ready to take small steps in that direction. Working with the numbers points the way.

Turning off a few lights to cut down on electricity use may not appear to make much difference in the big picture. That's probably what the people of Bangkok thought until their government convinced them otherwise. On a weekday evening at 9 P.M., the city government arranged for all major television stations to show a large dial registering the city's minute-by-minute electricity use. Then television viewers were asked to turn off all unnecessary lights and appliances. As viewers watched, the dial went down by 735 megawatts, enough to switch off two medium-sized coal-fired power plants. The message was clear: Collectively the people of Bangkok had the potential to shut down two power plants, saving resources and money, and reducing pollution.[29]

In the not too distant future perhaps houses will come equipped with feedback meters such as that provided by Honda's hybrid. In this vein, McKibben queries:

> So what if your electric meter was mounted in your kitchen where you could watch it spin? And what if your thermostat gave you an updated oil consumption readout every time you went to turn it up? What if your faucet showed you how much water you'd used in the last day, and how it differed from your annual average? Would that change your behavior?[30]

Overall, each of us can do many things to significantly reduce our consumption of energy and materials, but it is hard to make these changes. It is often so much easier to continue along in the same groove. Here are two tips for moving forward. First, don't go it alone; it is important to have support when you attempt to change ingrained habits. So, connect with some other people who share your interest in living lightly on the planet. Second, go slow. Don't try to do a complete makeover of your lifestyle in one week, or one month, or even one year. Make your changes one small step at a time, and celebrate your advances.

FOUNDATION 9.2: ECONOMIC DIMENSIONS OF SUSTAINABILITY

Environmental scientists throughout the world generally recognize that modern society's survival strategy, one based on constant industrial growth, is not sustainable.[31] The living systems upon which humanity ultimately depends for well-being—lakes, woodlands, savannas, oceans, forests, estuaries—are in decline.

Nature as a Model for a Sustainable Economy

Just as nature can serve as a template for the design of sustainable production and waste-management systems (foundation 9.1), the natural world offers a model for a sustainable economy. Nature's economy is based on sustainable modes of production, wealth creation, and exchange. This economy is successful, in no small part, because it is self-organizing, self-regulating, diverse, creative, and exquisitely adapted to its local environment. Think of a forest with its diverse species, efficient cycling of nutrients, complex gas exchange, and channeling of "goods" from "producers" (plants) to "consumers" (animals and microbes). These same components and characteristics are essential for a sustainable human economy.

Development, not growth, is the primary goal of a sustainable human economy. To "develop" means "to realize the potentialities of; to bring to a fuller, or better state." When genuine development occurs, things can become better without necessarily becoming bigger.[32]

A comprehensively sustainable human economy would, first and foremost, serve life. The primary focus would be on maintaining the integrity of the natural environment (i.e., the human life-support system) while meeting the physical, mental, and emotional needs of people. Like a healthy, natural ecosystem, a sustainable economy would be strongly rooted in a specific place: Local resources would be used to make products; many companies would be locally owned; and loans for new businesses would come from local banks (see box).

Business, money, and investments are integral to a sustainable economy. People would still buy and sell things, but the patterns of consumption would emphasize maintenance; and many people would be employed in repairing, restoring, and recycling. Great effort would be expended in designing and producing durable, efficient, aesthetically pleasing, and easy-to-repair products, rather than turning out gimmicky goods designed to last only a short time. The same number of dollars might circulate in a sustainable economy as in our present economy, but the dollars would

---

ECO-BANKS

Banks can foster sustainable development through their loan policies. For example, imagine a bank that preferentially offers loans to projects that seek to

- manage natural resources to restore and maintain biological diversity;
- apply the highest standards of energy efficiency;
- control waste emissions to prevent damage to the environment;
- seek social justice as well as business returns.

These are, in fact, among the ecological and social investment criteria for ShoreTrust located in Ilwaco, Washington. Just like any other bank, ShoreTrust offers savings, checking, CD, and money market accounts at competitive interest rates. ShoreTrust then takes the bank deposits—"eco-deposits"—from its patrons and uses this money to help create a sustainable economy for Willapa Bay Watershed.[33]

affirm life—that is, they would in no way impoverish Earth—while contributing to human well-being.

Finally, a sustainable economy would have a global face as well as a local demeanor. Nature offers a model for how to strike a balance between the local and the global. A living cell has permeable but managed borders. If there were no cell wall, the cell would lack integrity, and its contents would simply disperse into space. On the other hand, if the cell wall did not allow materials to move in and out, the cell would stagnate and die. Human economies also need permeable but managed borders at each level of organization from the local to the global. This means a high degree of local autonomy along with creative protocols for collaboration among regions and nations.

Ultimately, the pathway to a sustainable economy will probably involve the blending of the local and the global in the form of a nested hierarchy. Nested hierarchies are nature's way: atoms are organized into molecules, molecules into cells, cells into tissues, and so forth—all the way to planets, solar systems, and galaxies. There is a remarkable intelligence and capacity for self-regulation at each level of the natural world, and the integrity and creativity of the whole depends on the strength and dynamism of each level. So it is that self-reliant local economies could form the basis for a truly sustainable global economy.[34] This vision is distinct from the present trajectory of economic globalization, which often wrecks havoc on local communities (see box on page 279).

The largest hurdle barring the way to a sustainable future is humankind's high dependence on fossil fuels. The fossil-fuel economy spawned the industrial revolution of the nineteenth and twentieth centuries, which—in spite of its creative outpourings—constituted the most environmentally destructive in human history. The foundation for a sustainable economy will not be fossil fuels but, instead, sustainable-energy technologies like wind turbines, geothermal heat pumps, solar panels, and hydrogen fuel cells.

Although no contemporary examples exist of a completely sustainable human economy, some of the components of this idealized economy have been achieved by individual countries:

> Denmark has banned the construction of coal-fired power plants. Israel has pioneered new technologies to raise water productivity. South Korea has covered its hills and mountains with trees. Costa Rica has a national energy plan to shift entirely to renewable sources to meet its future energy needs. Germany is leading the way in a major tax-shifting exercise to reduce income taxes and to offset this with an increase in energy taxes. Iceland is planning the world's first hydrogen-based economy. The U.S. has cut soil erosion by nearly 40% since 1982. The Dutch are showing the world how to build urban transport systems that give the bicycle a central role in increasing urban mobility

HOW ECONOMIC GLOBALIZATION NOW
HARMS BOTH THE PLANET AND ITS PEOPLE

It has only been in recent years that global free-market capitalism's negative effects on local communities and economies have been fully registered. David Korten's book *When Corporations Rule the World* was a wake-up call in this regard. In this book, Korten explains how corporations seek to take advantage of the differences in wages, environmental standards, taxes, and social-justice standards among localities. This "hostile takeover," per se, usually means configuring their global operations to ensure the following: first, that products are manufactured in places where environmental and social costs are lowest, that is, in locales with few environmental regulations or human rights protections; and, second, that products are sold in places where markets are most lucrative. Profits, meanwhile, are often shunted to regions where taxes are least burdensome. The ability of companies to abruptly shift production from one place to another shifts the balance of power from local-human interest to the global-corporate interest; and, in so doing, it has the potential to severely undermine local community autonomy and well-being.

and improving the quality of urban life. And Finland has banned the use of non refillable beverage containers. The challenge now is for each country to put all the pieces of an eco-economy together.[35]

Tools for Creating a Sustainable Economy in the United States

Even after we acknowledge that the present-day U.S. economy is not sustainable, we still sense that we are caught in a bind. We are told that if we were to stop spending ("stop growing the economy"), people would be out of work by the millions, and America would slide into a severe depression. But transforming the present U.S. economy into a sustainable economy isn't about putting people out of work; it is about rechanneling entrepreneurial energy in ways that lead to ennobling work, to the production of goods and services that meet genuine needs, and to the creation of business opportunities that feed the human spirit rather than human greed.

In the specific case of the United States, the creation of a sustainable economy will be predicated upon the development of two premises:

1. sound indicators that gauge genuine progress—so that we know where we stand and whether we are moving toward or away from sustainability; and

2. methodologies to assure honest pricing—so that the environmental and social costs that are associated with the production of all commodities are included in their prices.

Indicators to Gauge Genuine Progress

At present, the United States relies on metrics like the Dow Jones, NASDAQ, and gross domestic product to measure economic performance. We often assume that when these economic indicators are pointing upward (i.e, showing growth), things in general are getting better. But is this really so? For example, take the case of the gross domestic product (GDP), which represents the dollar value of all income produced by businesses within the United States in a given year. When the GDP is rising, is the economy improving? . . . and are we, as a people, better off? Not necessarily. The GDP is just one aggregate number. It tells us little about jobs—how many new jobs are being created, whether these jobs are either high quality and well paying, or oppressive and low paying. Nor does the GDP pay heed to the natural capital stocks upon which the economy ultimately depends—that is, the stocks of fertile soil, clean water, forest timber, and uncontaminated air. Moreover, the GDP tells us nothing about the distribution (equitable or not?) of the goods and services it tracks. A rising GDP may mean, for example, that the rich now have even more while the poor are growing poorer. And the GDP is entirely capable of registering deterioration when, in fact, the country has experienced meaningful progress. For example, if we as a society were to cut down on waste, thereby reducing expenditures and improving the health of the environment, the GDP would register this as a decline in wealth creation.[36]

While economists are quick to point out that the GDP was never intended as a measure of societal well-being, the GDP nonetheless has taken on this significance because Americans have come to associate well-being almost exclusively with income. The fundamental problem with the gross domestic product as a metric of well-being is that it only adds; it never subtracts. As an alternative to the GDP, researchers at an organization called "Redefining Progress" have developed an index specifically designed to register well-being; it is called the "genuine progress indicator" (GPI). Like the gross domestic product, the genuine progress indicator represents well-being in monetary terms but with some important differences. For example, the GPI recognizes that the mother who is at home caring for children and the citizen who is doing volunteer work are productive members of the economy just as surely as the steelworker and the restauranteur; hence, their contributions are expressed in monetary terms and added to the genuine progress indicator, rather than ignored as in the case of the GDP. Meanwhile, economic activity associated with a deterioration of our environmental and social fabric is subtracted from the GPI (as is logical), rather than added as is the case for the GDP. Examples of deteriorating factors include the money spent either to clean up pollution or to increase the size of urban police forces.

One way to appreciate the difference between the gross domestic product and the genuine progress indicator is to examine a single commodity, such as the automobile. Consider the following sequence of events:

1. When a car is bought, the gross domestic product goes up.
2. Each time the owner buys gasoline and thus pollutes the air, the GDP goes up.
3. As the number of cars on highways increases, we suffer concomitant increases in insurance, road maintenance, and highway patrols—all of which cause the GDP to go up.
4. More cars means more accidents, medical bills, and repair bills; and, again, the GDP goes up.
5. Wrecked cars are scrapped and new cars bought; and the GDP goes up.

Bottom line: There is no distinction between costs and benefits, well-being or harm—everything is added to the GDP.

The genuine progress indicator, on the other hand, aims to measure the net benefit of the automobile by weighing the benefits the car provides against the negative aspects—for example, the costs associated with pollution and accidents. A business or a household would go bankrupt if it measured progress like the GDP—adding expenses plus income, instead of subtracting one from the other.[37] Overall, the genuine progress indicator presents a very different picture of societal and economic well-being as compared to that offered by the gross domestic product (figure 9.1).

For example, while the gross domestic product in the United States has been rising fairly steadily for several decades, indicating continued economic growth and concomitant improvements in well-being, the genuine progress indicator peaked in the mid-1970s and has been running flat, more or less, ever since. Hence, according to the genuine progress indicator, overall well-being in the United States has not improved appreciably since the mid-1970s in spite of the continual growth of the gross domestic product. The overall significance of the genuine progress indicator, in the context of sustainability, is that it recognizes that not all growth is good, which is an essential step in the shift from our present industrial-growth economy to a life-sustaining economy.

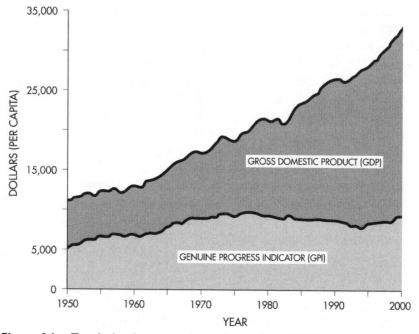

Figure 9.1. **Trends in the gross domestic product (GDP) and the genuine progress indicator (GPI) over four decades in the United States**[38]

Ultimately, though, it is not possible for us to capture movement toward sustainability via one qualitative or quantitative indicator, even with a nuanced number like the GPI. Instead, a suite of indicators will be required. Fortunately, governments, organizations, and cities around the world are beginning to develop ways of measuring genuine progress toward sustainability. One of the pioneers in this movement was the city of Seattle. In the early 1990s, hundreds of Seattle's citizens came together and agreed on forty indicators of sustainability. The Seattle work made it clear that good indicators do the following:

1. reflect something fundamental to the long-term economic, environmental, or social health of a community;
2. are accepted by people as valid signs of well-being (or symptoms of distress); and
3. are statistically measurable.

Seattle's various indicators (www.sustainableseattle.org), when viewed together, provide a kind of sustainability status report for the city (see box).

It is both timely and important to measure sustainability in all realms of society—government, business, education, church. Sustainability is, after

---

**USING INDICATORS TO ASSESS
SUSTAINABILITY WITHIN UNIVERSITIES**

Inspired by the work in Seattle, my colleague Jim Eisenstein and I led a team of students in the assessment of sustainability at Penn State University. We began our study by defining "best" or "sustainable" practices for Penn State's various subsystems. For example, we reasoned that a sustainable energy system would be based on renewable energy, high efficiency, and nonpolluting practices. Hence, our energy indicators measured whether Penn State's energy system was becoming less dependent on fossil fuels, less wasteful, and less polluting over time.

In all, we developed thirty-three indicators for gauging sustainability. Guided by these indicators, we scrutinized Penn State's policies and performance in water conservation, recycling, purchasing, landscaping, energy use, building design, and research ethics; we critically evaluated Penn State's food and transportation systems (and thus asked if the university was moving in a sustainable direction); we checked to see if Penn State's institutional power was being used to strengthen regional economies and promote corporate responsibility; and much more.

The first *Penn State Indicators Report,* released in 1998, depicted an institution whose performance, measured by sustainability indicators, was not exemplary. For category after category (energy, food, materials, transportation, building, decision making), Penn State's practices departed little from the national status quo. Since that time, sustainability has become a major focus, especially within the operations wing of the university; and it is generally acknowledged that our sustainability indicator assessment was catalytic in this shift (www.psu.edu/greendestiny).

---

all, a whole new way of seeing and relating to the world; and just the act of measuring it legitimizes it. It is a good place to start.[39]

Methodologies to Ensure Honest Pricing

Just as a society needs information, such as that offered by the GPI, to gauge genuine progress, citizens need information to guide their consumption (buying) decisions in a sustainable direction. One important tool in economics is called "full-cost accounting," which helps ensure that society charges the right price for its goods and services. Full-cost accounting basically means making sure that the price charged for a commodity reflects all the costs surrounding the commodity's production and disposal.

Although buyers often complain about the high prices of commodities, the prices for many products are below their "full cost." For example, a study by economists at the World Resources Institute (USA), entitled "The Going Rate: What It Really Costs to Drive," concluded that the price of gasoline would increase three- to fourfold if the many hidden costs of car-based transport—for example, highway construction and maintenance, creation of parking facilities, police patrolling, health problems caused by car-generated pollution, lost opportunities on land converted to roads, and so on—were passed on directly to American citizens at the gas pump. Americans now pay some of these costs indirectly through things like taxes and health insurance, but many of the environmental costs associated with driving are being transferred to future generations in the form of an impoverished planet.[40]

It is one thing to do a full-accounting exercise and determine that a commodity like gasoline should be priced at, say, five dollars per gallon; it is quite another to charge this amount at the pump. At this point, an approach called "tax shifting" comes in. The president of the Earth Policy Institute, Lester Brown, describes it this way:

> Tax shifting involves changing the composition of taxes but not the level. It means reducing income taxes and offsetting them with taxes on environmentally destructive activities such as carbon emissions, the generation of toxic waste, the use of virgin raw materials, the use of nonrefillable beverage containers, mercury emissions, the generation of garbage, the use of pesticides, and the use of throwaway products. This is by no means a comprehensive list, but it does include the more important activities that should be discouraged by taxing.[41]

Sweden pioneered tax shifting when it lowered income taxes while raising taxes on carbon and sulfur emissions as a means of discouraging the burning of fossil fuels. Now many countries in Europe—including France, Germany, Italy, and the United Kingdom—are experimenting with tax shifting.

In the United States, President Clinton floated the idea of a carbon tax in the early 1990s, but support for this tax did not materialize. The United States did, however, give a boost to the concept of taxing environmentally harmful activities in 1998 when the tobacco industry consented to reimburse state governments $251 billion for the Medicare costs of treating past smoking-related illnesses. In essence, this retroactive tax was levied on the tobacco industry for all the cigarettes it had sold in previous decades.

In addition to shifting taxes, governments can also shift subsidies. At present, governments throughout the world distribute an estimated $700 billion in harmful subsidies each year. Some governments subsidize the extraction of water in places where water tables are falling; other countries, such as the United States, subsidize the extraction and use of fossil fuels at a time when climate change is a growing concern; still other countries subsidize ocean fishing when ocean fish resources are clearly being overexploited. Subsidy shifting entails taking the hundreds of billions of dollars now spent in harmful subsidies and redirecting these funds to initiatives that would contribute to sustainability efforts in the realms of renewable energy, reforestation, green building technologies, and zero-waste manufacturing.[42]

Once a society starts getting the price right on key commodities and then institutes the right mix of taxes and incentives, its citizens begin to make intelligent choices. For example, if the price of gasoline doubled or tripled (to match its true cost), American citizens would no doubt become very interested in more fuel-efficient forms of transportation, such as the hybrid car, light rail, and bicycle.

In the last analysis, a comprehensively sustainable economy will come into being, or not, to a significant degree as a result of the decisions that individual citizens make:

- Will we be willing to pay a bit more to ensure that our energy comes from renewable sources?
- Will we vote to redesign our cities so that they serve people and not cars?
- Will we refuse to invest in companies that manufacture products that are harmful to human health and safety?
- Will we support innovative tax and subsidy shifting referenda?
- Will we refuse to buy products known to be manufactured in sweatshops?

Each day, we are faced with decisions like these, and the choices we make have the potential to contribute to the emergence of a life-sustaining economy.

Reflection

The humanizing qualities of an economy grounded in sustainability and local resources became evident to me when I began to examine products that meet my most basic needs, such as food. Like many Americans, I have found that the contemporary U.S. food system, while wonderful in many ways, has tended to distance me from the origins of my food. When we select a food package from the supermarket shelf, we usually know almost nothing about its ecological history—nothing about where it came from; the quality of the soil and the chemicals applied in its growing; the working conditions of the people who helped to harvest and process it; and the social responsibility of the companies involved in its growing and marketing.

Something clicked for me when I read that, on average, sixty-five cents of every dollar I spend in the supermarket goes for packaging, delivery, and marketing; and another thirty cents goes to the chemical companies that make the fertilizers and pesticides. Suddenly, I understood why small farmers can't make it—a nickel on the dollar won't do it, unless your farm is very large. Small farmers certainly can't compete with industrial "farms."[43]

They can't compete, that is, unless they offer consumers something more valuable than processed, industrial foods. In this vein, the sustainability revolution is giving birth to whole new forms and structures for the production and marketing of food at the local level. Some small farmers are skipping so-called middlemen and marketing their food directly to local citizens. They invite families to buy a share of their farm's produce for the season. Several of these community-supported agriculture farms (CSAs) operate in the community where I live. To join, a household pays three to four hundred dollars to the farmer at the beginning of the growing season. In return, this "subscriber" household receives fresh produce from the farm throughout the growing season. Early on, the weekly food baskets might be filled with lettuce, spinach, baby beets, turnips, scallions, snap peas; a bit later, in June, families might receive chard, broccoli, zucchini, and green beans, along with bouquets of flowers; next comes, in addition to much of the above, such delectables as carrots, tomatoes, sweet corn, cucumbers, and berries. Finally, at the end of the season, subscribers are likely to receive bounteous supplies of pumpkins, onions, squash, cabbage, parsnips, potatoes, among other foods.

CSAs are great for local folks because they receive fresh, organically grown produce each week. They are good for participating farmers because they know in advance how much to plant and because they have the money up front to buy the seed and tools they need. By cutting out "middlemen," the big chunk of each dollar that normally would be lost from the local economy goes, instead, to compensating the community farmer and reducing food costs for local buyers.[44]

Questions for Reflection

- The U.S. economic system aims to transform resources into more money. In what ways do you, in your own life, transform your personal resources into money?
- What would it mean for you to transform your personal resources into more life?

Practice: Economics and Personal Sustainability

Our lives and households are small-scale economic systems. Through our work we produce products and services—that is, we create wealth—and in return for our labor we receive money. We then exchange this money for other goods and services that meet our "needs." In a very real sense, we are in the business of selling our life energy for money. This life energy—the hours of life we have on Earth—is limited. Hence, it only makes sense to ensure that the money we receive for our life energy is enough to justify our efforts. After all, we really are "paying" for money with our time; and how we spend our time is how we spend our lives.

In their book *Your Money or Your Life*, Joe Dominguez and Vicki Robin offer a simple practice for helping a working person clearly see this trade-off between money and life energy. The first step is to calculate your "real" hourly wage. For example, if you took a standard forty-hour, four-hundred-dollar-per-week (take-home salary) job as, say, an administrative assistant for a small business, you might assume that your hourly wage is ten dollars per hour. Let's examine this assumption.

The first step in this examination is to deduct all work-related expenses from your weekly income. In other words, if your weekly take-home salary is four hundred dollars, subtract your commuting costs, the money spent on clothes for work, the extra cost for work-related meals, the money spent on relaxation after the stress of the work day, job-related illness, and so forth (table 9.1). By subtracting these hidden costs from your wages, you will arrive at your "real" weekly take-home salary. But that's only half of the equation.

After figuring out how much money you are really taking home, now figure out how many hours you are really working. You do this by adding to your formal forty-hour workweek all the hours you spend commuting back and forth to work; the time you spend costuming yourself for work; the time you spend winding down from the stress of the workday; and all other hours indirectly connected with your job. By tallying these hidden hours, you will determine the real length of your workweek. Finally, simply divide your "real" weekly salary by your "real" job hours, and you will get your "real" hourly wage.[45]

The example in table 9.1 reveals how people can trick themselves: imagining that they are working forty hours per week when they are really working more than fifty hours per week; assuming they are earning ten dollars per hour when their "real" salary is less than five dollars per hour.

Once you are clear on how much you are selling your life energy for, you can perform a complementary practice that allows you to evaluate how much satisfaction you are receiving from the things that your life energy buys for you. The practice is simple. It consists of keeping track of all the things you spend your money on. Get a notebook, and keep it in your back pocket; call it your "money log," and use it to record every single penny you spend. This takes discipline. After a month you will have a wealth of information on your spending habits. Now, analyze the data by first placing your expenditures in categories. As Dominguez and Robin observe, "This [practice] isn't just accounting. It's a process of self-discovery, of becoming conscious of how you handle money as opposed to how you think you do."[47]

Next, combine the information on your "real" hourly wage with this spending information. This calculation will reveal how much of your life energy goes to each spending category. For example, if your "real" hourly

Table 9.1. Full-Cost Accounting on the Job[46]

	Hours/week	Dollars/week	Dollars/hour
Nonadjusted Hourly Take-Home Wage	40	$400	$10.00
Adjustments:			
Commuting[a]	+5	−$ 70	
Costuming[b]	+1.5	−$ 10	
Meals[c]		−$ 25	
Decompression[d]	+5	−$ 25	
Vacation[e]	+1.5	−$ 20	
Job-related illness[f]	+1.5	−$ 5	
Total Adjustments	+14.5 hours	−$155	
Real Hourly Wage[g]	54.5 (40 + 14.5)	$245 (400 − 155)	$4.50

[a]Assumes: Workplace is 15 miles from home; daily commute back and forth takes one hour; total cost of car driving = $0.47/mile (based on 1999 estimates from the American Automobile Association).
[b]Assumes: 15–20 minutes/day spent to put on work "costume" and work "face"; $500/year spent for clothes and grooming aids specifically related to job.
[c]Assumes: Expenditures for work-related food (e.g., lattes, snacks, occasional meal out) average $5/day.
[d]Assumes: One hour is spent each day simply decompressing from the rigors and stress of the work day; $5/day in incidental "decompression" expenditures are incurred.
[e]Assumes: Rigors and stress of work make a two-week annual vacation a necessity for recharging; cost of this annual vacation sums to approximately $1,000.
[f]Assumes: Job-related effects result in four days of illness each year at a cost of about $65/day (i.e., cost of medical supplies and health assistance).
[g]The real hourly wage comes to $4.50 ($245/54.5 hours).

wage (after adjustments) is eight dollars per hour and if you spend roughly sixty dollars each week eating out, then you are working one full day each week to pay for your meals out.

The final step in this practice is to bring to mind each of your spending categories and, then, to ask two questions:

1. "In spending money for this product/service, did I receive fulfillment and value in proportion to the life energy I spent to acquire it?"
2. "Is this expenditure of life energy in alignment with my values and life purpose?"[48]

Overall, this is a radical practice insofar as it reminds us that our life energy— our time—is our most precious resource. As such we are wise to allocate that energy only to endeavors that are in accord with our heartfelt values.

FOUNDATION 9.3: POLITICAL DIMENSIONS OF SUSTAINABILITY

Many people today associate politics with scandal, self-interest, and money. They observe the adversarial politics—where most issues tend to be highly polarized and where important decisions are often made out of the public eye—and they are left feeling cynical and disempowered.

Viable democracies have three core qualities, all of which, to varying degrees, appear to be in decline in the United States. First, we need a fair measure of social equality. If a big gap exists between the wealthy and the poor (increasingly the case in the United States), then those lacking wealth cannot participate as equals and democracy is compromised. Second, in a viable democracy, citizens see themselves as part of a community with shared interests and obligations. Democracy is subverted when personal autonomy and personal gain are emphasized at the expense of civic duty and community participation, which increasingly appears to be the case in America today. Finally, a viable democracy depends on an informed populace that actively debates the issues of the day and feels that its participation matters in political deliberations. Here, too, it seems that Americans—insofar as they are recipients of "dumbed-down" news, generated by a largely corporate-controlled media— are increasingly depoliticized, often lacking the knowledge to discuss the complex issues of the day from any but a reactionary posture.[49] Overall, there is legitimate concern that the content and process of U.S. public discourse and decision making may be inadequate for meeting the economic, ecological, and social challenges that America will face in the coming decades.

Just as natural ecosystems offer a compelling template for the redesign of the human-built environment and the human economy (foundations 9.1 and 9.2), they also offer a useful metaphor for sustainable approaches to governance. For example, the web of connections in human communities is just as essential to wholeness and well-being as is the web of life that connects organisms in a natural ecosystem. And just as ecosystems lose their integrity when they are fragmented (isolated) and when certain parts (species) are eliminated, so it is that human communities fall apart when certain voices are silenced and when decisions are made in isolation from those who bear the consequences of those decisions.

Stepping Stones to Sustainable Governance

Democracy is not so much a possession as it is a practice—a practice for nurturing social change. The success of the sustainability revolution hinges to a significant degree on the capacity of people to rediscover the power of democracy. Specifically, this revolution challenges us to recognize that

1. each of us has a public life;
2. vast stores of social capital reside in our communities;
3. immense wisdom lies untapped in our collective intelligence.

Each Person Has a Public Life

Prior to the 1980s, people were referred to on the evening news as "citizens" or "the public." Since that time, these words have been gradually replaced with the word "consumers." So it is that we have come to see ourselves as "consumers." A prerequisite for a sustainability revolution in governance is that people reject the moniker "consumer." After all, we are first and foremost citizens. Along with this, it is essential for us to understand that democracy is not something static that we are born into; rather, it is something that we participate in and cocreate. Whether we know it or not, we all have a public life. As Frances Moore Lappé and Paul Dubois point out in their book *The Quickening of America*: "Everyday—at school, where we work, where we worship, within civic and social groups, as well as at the polls—our behavior shapes the public world."[50] Even the personal choices we make—what we eat, what products we choose to purchase, how far we drive each day—have social, economic, and environmental consequences that reverberate out into the public sphere. We deny our public life when we fall into the trap of imagining that the "market" will solve all of our problems. Analyst Benjamin Barber points out the fallacy in this assumption:

> The market is designed to appeal to our private wants and to create those wants. . . . Democratic and civic institutions [by contrast] give voice to our public side, our civic side, our responsible side; markets only ask us what we think, and want, and desire as consumers.[51]

Insofar as the market asks people to see themselves only as private beings, not public, it may actually undermine the very democratic principles that America has been built upon.

Human Communities Contain Untapped Stores of Social Capital

A vital citizenry generates a unique type of capital known as "social capital" (introduced in chapter 5, page 148). According to Harvard sociologist Robert Putnam: "Social capital refers to features of social organization, such as trust, norms, and networks, that can improve the efficiency of society by facilitating coordinated actions."[52] Just like financial capital, social capital can enable certain economic activities that would not be possible in its absence. Putnam has spent many years documenting the importance of social capital. He has been especially interested in the reasons why some communities flourish and others flounder; and he has discovered that the best predictor of effective governance and economic vitality is not political ideology or wealth, but strong traditions of civic engagement—that is, participation in community affairs, voter turnout, newspaper readership, involvement with civic clubs and associations, volunteerism, and so forth.

Just as financial capital provides the means for business innovation, social capital provides the energy and creativity for social transformation (see box). In communities energized with social capital, citizens take the reigns of governance into their own hands. For example, in Memphis, Tennessee, Shelby County Interfaith (a consortium of fifty-eight congregations and associations) organized 437 house meetings to create its municipal agenda; six thousand people participated in the meetings, and thirty-five thousand people signed a statement outlining what they wanted for their city—for example, affordable housing options, neighborhood boards for dispute resolution, and decentralization of school decision making. These citizens were successful in electing a school board and a city council composed only of people who fully endorsed their agenda. But their goal wasn't merely to get new candidates in office; it was to change the political culture so that all future candidates for political office would respond to questions and agendas set by the people.[54]

SOCIAL CAPITAL IN ACTION

Rotating credit associations provide a wonderful example of citizen empowerment based on social capital. These associations are in operation throughout the world, especially in poor countries. They come into being when a small group of people agree to make regular contributions to a fund that is given, in whole or in part, to each contributor in rotation. In a typical rotating credit association, each member (of, say, twenty) might contribute a monthly sum of five dollars; and each month, a different member receives that month's pot of one hundred dollars. The money might be used to pay for a wedding or a bicycle or a sewing machine. Lacking significant physical assets, the participants in effect pledge their social connections. Thus, "social capital" is leveraged to extend credit, and the credit association becomes a mechanism for strengthening the overall solidarity of the village or town.[53]

In sum, social capital is generated by connections. Weak communities have little social capital; thus, they have few connections. Similarly, the people who are most disempowered are generally those who are most isolated. Strong communities, by contrast, are like healthy ecosystems: they are laced with interconnections.

The Application of Collective Intelligence Leads to Wise Decisions

When people work collectively to solve problems, they often discover that they arrive at more creative and just solutions than when working alone—that is, they experience "collective intelligence." Social capital creates the conditions that allow collective intelligence to flourish in a democracy. Recognizing the power of collective intelligence, the state of Oregon called on its citizens in the 1990s to engage in community meetings so that they could identify the core values that should be used to prioritize state health care decisions, given limited funds. Social change activist Tom Atlee describes the process:

> With experts "on tap" to provide specialized health care knowledge, citizens weighed the trade-offs involved in over 700 approaches to deal with specific medical conditions, and decided which should be given preference. In general, approaches that were inexpensive, highly effective, and/or needed by many people (which included many preventative measures) were given priority over approaches that were expensive, less effective and/or needed by very

few people. Although clearly some people would not get needed care under this system, it was pointed out that some people didn't get needed care under the existing system. The difference was that in the old system, it was poor people who fell through the cracks by default. In the new system Oregonians were trying to make these difficult decisions more consciously, openly and justly.[55]

Atlee contends that by relying on the collective intelligence of its citizenry, Oregon was able to develop health care priorities that were much wiser than anything that could have been produced by an individual or a single group.

We can also tap into collective intelligence at the national level. One way of doing this is to invite citizens to come together, those who represent society's diversity in terms of gender, ethnicity, income, age, values, and political orientation. They could create a consensus vision for the country while the rest of the nation witnesses. Such high-profile "proxy conversations" are already beginning to occur in some countries. As Atlee observes:

These proxy conversations could be used to stimulate ongoing conversations . . . by ordinary citizens in living rooms, schools, churches and bars across the land. This could dramatically change the political environment. Subsequent government decisions would be made in a context of greater public wisdom, sophistication and consensus.[56]

Summing up: When we embrace our public lives and come to see democracy not as a "spectator sport" but as a participatory process, we create the social capital and collective intelligence necessary for positive social change.

New Rules to Create a Sustainable Society

Many modern problems—environmental degradation, poverty, injustice—are, to a significant degree, the result of societal rules. Thus, to create a "new story," one grounded in sustainability, societies will need to use their collective intelligence to devise "new rules." Fortunately, in countries and organizations throughout the world, "new rules" that reinforce sustainability precepts are beginning to be enacted. Here is a small sampling:

- Recognizing that young children lack the sophistication to critically assess advertising claims, Sweden has passed a new rule banning all television advertising targeted at kids under twelve years of age.
- Recognizing that some commodities are produced in ways that unjustly exploit human beings, the state of Maine has a new rule, the

"Maine Anti-Sweatshop Purchasing Bill," which requires that all enterprises selling shoes, clothes, or textiles to Maine's state government
guarantee that the commodities were not manufactured in sweatshops.

- Recognizing that sprawling suburban development leads to increased
  car dependence and loss of farm land, the state of Oregon requires that
  its cities establish a physical limit to their outward spread ("urban growth
  boundary"), thereby keeping rural land off-limits to urban development.
- Recognizing that corporations, over the long haul, have wrecked havoc
  on many local communities, the city of Arcata, California, has passed a
  new rule mandating that the town council "establish policies and programs that ensure democratic control over corporations conducting
  business within Arcata, in whatever ways are necessary to ensure the
  health and well-being of the community and the environment."
- Recognizing that carelessly designed and constructed buildings result
  in immense waste in the United States, the U.S. Navy has a new rule
  stipulating that all facilities and new construction incorporate sustainable building and design principles.[57]

The above examples are a start; but, ultimately, even more far-reaching
new rules will be necessary to ensure a sustainable world. Here is a sampling of some of the bold new rules now under discussion in the United
States and in other places around the world:

- Transporting food long distances causes pollution and is wasteful of energy resources. Rather than subsidizing fuel and using tax dollars to
  build more roads, how about a transportation tax on all nonlocal food
  products coming into a region? The revenues from such a new tax
  could be used to promote local food systems.
- Rather than having rules that require industrial smokestacks to be
  placed high above ground, thereby ensuring that polluting emissions
  are carried to other areas, the stacks could be lowered and placed next
  to the executive suites of company owners. This new rule would help
  ensure the enactment of zero-emission production technologies.[58]
- And what if progressive building designers were rewarded for the energy they saved in the creation of ecologically sound buildings? For example, incentives could be developed so that designers received, as a
  bonus, a proportion of the new building's energy savings over the first
  five years of operation.[59]
- Or how about a new tariff called "Most Sustainable Nation" to replace
  the "Most Favored Nation" status? This new rule would grant low or

no tariffs to countries that practice fair labor practices and that refrain from despoiling their environment.[60]

- Finally, how about a new rule that holds corporations publicly accountable to society? In this vein, Rabbi Michael Lerner has proposed a social responsibility amendment to the United States Constitution. The proposed amendment reads:

> Every corporation with annual revenues of $20 million or more operating within these United States must receive a new corporate charter every twenty years. To receive this charter, the corporation must prove that it serves the common good, gives its workers substantial power to shape their own conditions of work, and has a history of social responsibility to the communities in which it operates, sells goods and/or advertises. . . .[61]

This amendment might also prescribe "capital punishment" for a corporation found guilty of a crime that causes the death of humans. Under this prescription, a corporation would be dissolved, with its assets directed to the public good, if that corporation (a) marketed products known to be harmful to human health or (b) exposed its employees to conditions that seriously compromised their health. The idea of a social responsibility amendment is based on the often forgotten premise that citizens give corporations charters because they expect them first and foremost to serve a broader public purpose.

These and other visionary "new rule" initiatives will come to pass, provided that citizens come to see democracy not as a passive birthright, but as the active practice of seeking common good.

New Guiding Principles for a Sustainable World

The revitalization of U.S. democracy that the sustainability revolution calls for extends to a critical examination of the U.S. Constitution. Although remarkable for its endurance, the Constitution is a blunt tool for solving the economic, political, and social challenges that the United States now faces. As surprising as it might sound on first hearing, some scholars have actually suggested that the time has come to call for a Constitutional convention for the purpose of drafting a new Constitution. As part of this process, Americans might articulate not a new Bill of Rights, but their first ever Bill of Responsibilities—a set of shared principles and values that citizens pledge to abide by.[62] There is an international-level precedent for just this type of declaration: the newly created Earth Charter.

The Earth Charter traces its origins to a recommendation in the 1987 *Bruntland Commission Report* to the United Nations, namely, to engage in a worldwide effort to articulate the fundamental principles for building a just, sustainable, and peaceful global society in the twenty-first century.

The creation of the Earth Charter was a daunting task. It is the result of a decade-long cross-cultural conversation on humanity's shared ideals, values, and goals, involving thousands of people from countries all over the world. The Charter begins:

> We stand at a critical moment in Earth's history, a time when humanity must choose its future. . . . The choice is ours: form a global partnership to care for Earth and one another or risk the destruction of ourselves and the diversity of life. Fundamental changes are needed in our values, institutions, and ways of living. We must realize that when basic needs have been met, human development is primarily about *being* more, not *having* more. . . . Our environmental, economic, political, social, and spiritual challenges are interconnected, and together we can forge inclusive solutions.[63] (My emphasis.)

What follows is a courageous document that stresses economic justice, global interdependence, and an ethic of respect and care for all of creation. There is no sugarcoating. The Charter makes it clear that, in the final analysis, the problems that the world faces are ethical ones. The imagination and idealism that are desperately needed in our times ring forth in the Earth Charter's conclusion:

> Let ours be a time remembered for the awakening of a new reverence for life, the firm resolve to achieve sustainability, the quickening of the struggle for justice and peace, and the joyful celebration of life.[64]

The Earth Charter serves as a kind of international declaration of sustainability. At the national level, Sweden is now in the process of organizing its entire society around the preconditions (or rules) necessary for a sustainable society. Swedes recognize that it is the environment that ultimately makes the rules. Sweden's program "The Natural Step" was begun by a medical researcher, Karl-Henrik Robert, who reached out to fellow scientists (ecologists, chemists, physicists, physicians) in an effort to reach a consensus on the conditions essential for life to flourish on Earth. It took twenty-one iterations on Robert's initial document to arrive at a final consensus. In the end, the scientists agreed on the basic conditions for a sustainable society—for example, never remove a resource from the earth more rapidly than that resource can replenish itself through natural processes. The Natural Step consensus statement, in the form of a book-let and an audio cassette, has been sent to all 4.3 million households in Sweden.[65] Sweden's Natural Step is powerful because it establishes a shared mental model for the conditions necessary for sustainability. This stimulates creativity because people can engage in a much smarter dia-logue if they have a shared framework for their goals. Already in Sweden, companies are adopting the Natural Step guidelines as preconditions for sustainable operations. For example, Scandic Hotels has produced the 97 percent–recyclable "eco-room."[66]

If it is to happen at all, the creation of a comprehensively sustainable world—one grounded in interdependence and mutual respect—will re-quire that citizens expand their locus of concern beyond the private realm and into the public realm. This could unleash vast stores of social capital and collective intelligence, which in turn could serve as catalysts in the cre-ation of new rules, charters, and programs—milestones on the path to a sus-tainable world.

Reflection

Whenever I hear the admonition, "America: love it or leave it!" I men-tally substitute it with, "Democracy: practice it or lose it!" I yearn for a patriotism born not of a blind love of country, but of resolute commit-ment to democratic ideals and practices. Just because we have been born into a democracy does not mean that we will die in one. Like everything else, democracy is ephemeral.

When the topic of democracy comes up in the classroom, I challenge my students to consider if we, in the United States, even live in a democracy. I begin by offering the classical definition of a democracy: "rule of the many." In a true democracy, it is the people who make the important political and social decisions.[67]

I then query, "In the United States, do 'we, the people,' make the big decisions?" When prompted in this way, students realize that (a) the important private-sector decisions—those regarding work, technology, production—are mostly made by corporations; and that (b) the important government decisions—on military actions, taxes, exploitation of natural resources, toxic-substance regulation—are generally made behind closed doors with little (if any) citizen participation.

This being the case, I prod my students further by asking: "Why do you think that we live in a democracy? What makes the United States a democ-racy?"

Invariably someone responds that "We have free speech."

"True," I agree. "We can stand up and complain in public; we can write scorching letters to our local newspapers. But, again, democracy is 'the rule of the many'—it is actual decision-making power in the hands of the people—not simply the right to complain about how we don't have any power. So, do we really have a democracy?"[68]

At this point, someone usually volunteers, "We have the right to vote and this makes us a democracy. If we don't like what our representatives are doing we can vote them out of office."

I invite students to examine this assumption by asking some more questions: "Does this right to vote give you real power? Do you really have choices in elections, given that generally only rich people or those supported by rich people have the means to run for office? And what does it say about our 'democracy' when the majority of citizens choose not to vote?"

These are uncomfortable questions. Some might even say that asking such questions borders on treason. For example, some citizens regard those who question the wisdom of the United States' war on terrorism as "traitors." This prompts a final question: "What kind of democracy do Americans have when they silence those who raise uncomfortable questions with accusations of treason?"

Questions for Reflection

- What are democratic ideals?
- In what ways do you practice democracy in your community, state, and nation?

Practice: Citizenship and Personal Sustainability

During the time that I was writing this book, I lived in Seattle. My house was located a block from a little corner grocery store. The store used to be a residence, and it still retains a homey feel. As you enter, it has a sofa, some chairs, and a coffee table to the left; shelves with books are situated next to the sofa; paintings by neighborhood artists adorn the walls; a collage of snapshots of local customers is affixed to the check-out counter; and a little journal written by some of the local kids is sometimes for sale by the newspaper rack. Sipping coffee by the window, I watch the morning parade of customers: a mom with her kids, two construction workers, a businessman, a homeless woman, a teacher, a social worker, three students. Then, a girl pops in and puts up a flier saying that she is available for baby- and pet-sitting. The store owner knows them all by name. Conversations bubble up—local real estate prices, local political candidates, local weather, local birds—ideas are generated; plans are hatched.

As I sit there I realize that people come to this little store for more than the odds and ends that they purchase. They stop in for connection; the store is a node in a network; it is a generator of "social capital."

Making a morning visit to this store when I lived in Seattle was a kind of practice for me. It was a way for me to participate in the civic life of my neighborhood. Most communities have these network nodes—libraries, barbershops, coffee houses, and urban parks—where people gather to talk about things that they care about.

The foundation for citizenry is care. When we care about our communities, we are motivated to act on their behalf. One way to identify and activate our care is simply to take a spin around the neighborhood, paying careful attention to the things that attract our concern. Brian Stanfield does this as he bikes to work each day:

> Even on my familiar ride, I realize how many things I care about. At a red light at the first intersection a car stops on the pedestrian crossing, making it more dangerous for people to walk across the street. Care. On the other side of the street, I notice the new supermarket going up. I'm wondering if the new one will be better laid out than the old one. Care. . . . Off now into the railway underpass which needs a good painting. Care. On the other side of the tracks, another supermarket is being built. Three supermarkets within 300 yards! Is this overkill? Care. . . . I am my care. What I do about it is another matter.[69]

As this passage reveals, cares are often felt and expressed in the form of concerns. We can practice citizenship by coupling our concerns about the conditions in our communities with a commitment to make things better. Stanfield calls this the "yes stance." Two specific practices, "appreciative inquiry" and "open space," allow people to operate from the "yes stance"; and in so doing, they allow citizens to build social capital and tap collective intelligence.

Appreciative Inquiry

Appreciative inquiry (AI) involves asking questions that focus on what is good in an organization or community, and it is based on the premise that human organizations grow toward what they persistently ask questions about and concern themselves with. AI is positive and sees potential; it is neither negative nor cynical. The focus is not on problems but on what is working and how to make things better. Hence, instead of asking, "What's the problem here?" or "What's not working?" AI starts with, "Tell me about a time when things were really working well here," followed by, "What changes need to be made to allow those desirable conditions to prevail?" For example, rather than trying to "fix the problem" of poor gender relations in a company by going on a witch-hunt, an AI practioner would begin by asking the people in the organization to tell stories about specific instances when gender relations were healthy and affirming. These stories represent the company's accumulated knowledge of *good* gender relations; and, as such, they form the basis for imagining *excellent* gender relations within the organization. Appreciative inquiry is enlivening and stirs the imagination, often giving rise to powerful innovations that transform organizations.[70]

Just the simple process of focusing on community or organizational assets, instead of shortcomings, can be transformative. For example, in an attempt to revitalize its economy, the town of Greenfield, Iowa, invited citizens to an evening meeting at the town hall to discuss what might be done. They were given three ground rules:

1. Don't say, "That will never happen here."
2. Don't say, "We tried that before and it failed."
3. Don't say, "That will cost too much money."

Rather than complaining about problems, citizens were asked to name the good things (assets) about their community. By the end of the evening, the town residents had listed 110 community assets on the wall. And within a year, more than a quarter of the residents in this town of two thousand people were involved in initiatives to further improve the town—for example, establishing roadside plantings, restoring the hotel, and creating a farmer's market.

Appreciative inquiry can be practiced at any scale. I heard a story about a basketball coach who used AI with her struggling team. On the Monday after a dismal defeat, this coach sat with her players to watch film footage from the game. But rather than pointing out all the errors that her players made, she emphasized and celebrated all the moments of good play. The message was: "This is good. This is what we need to do more of. . . ." After the film, the coach discussed adjustments that the team could make so that their "good play" might extend throughout an entire game.

Appreciative inquiry could even be practiced within an individual household. Imagine five people living together in a rental house. Household "governance" issues invariably arise in such situations. Usually, a lot of time and energy are spent thinking and talking about what isn't working. But AI would call on housemates to stop complaining and, instead, to start focusing on affirmation and appreciation. So, if the problem centered on certain people's not doing their fair share of the work, the discussion would begin with the housemates telling stories of specific instances when each person was doing his or her part. The housemates' accumulated experience of shared responsibility would serve as a launch pad for a discussion aimed at how to make life better in the household.

Open Space

Open space is another methodology for accessing collective intelligence and building social capital in human communities. The methodology was inspired by and modeled on the creative self-organizing properties of natural ecosystems. Specifically, open space is a way of bringing together large numbers of people around issues of shared concern, all the while ensuring that social interactions are as creative and productive as possible.[71] This is a tall order!

At an open space gathering, the attending individuals can include, for example, people from a neighborhood, a church congregation, business merchants, residents of an apartment complex, labor union members, and teachers. Those who are present are invited, if they feel so inclined, to host a discussion or activity of their choosing, as long as it is related to the central theme of the gathering. Anybody can step forward and describe their offering to the group. For example, when I attended an open space gathering of 150 people on sustainability, thirty-odd people came forward to describe their sustainability-related topic, saying things like:

> My name is Marie, and I am concerned about America's heavy dependency on automobiles to the exclusion of more sustainable transportation options like biking and light rail. I would like to meet to discuss the political tools available to individuals and communities for influencing local transportation policy.

> Hi, I'm Bill, and I have developed an effective approach to backyard composting using worms. I have brought the materials for making worm composting bins with me today, and I would like to share this with anybody who might be interested.

All the offerings were then quickly posted on a wall designated for this purpose. For example, Marie wrote "Local Transportation Policy; 10:00–11:30 A.M.; Room D." Next, came the "marketplace." Everybody was invited to walk along the wall (the market) and "shop" among the dozens of offerings for the day. All the sessions were an hour and a half long, and they were in blocks: two morning blocks and two afternoon blocks, each with five to ten choices. After fifteen minutes of "shopping," everyone had their day planned, and the first sessions began. I found this process to be democratic and highly effective.

Citizens don't have to wait for someone from the "outside" to organize open space gatherings; we can do it ourselves. Imagine, for example, that residents in a community are concerned about a new tax or that students on a college campus are concerned about a tuition increase. Rather than individually complaining, they might come together around their issue of concern: in the morning, an open space meeting to explore ideas and perspectives; in the afternoon, a "council" to consider collective actions.

In sum, the sustainability revolution reminds us that we all have a public life. As we tap into our collective intelligence, we can create a genuine democracy, participating fully in the determination of our collective fate.

CONCLUSION

> The main distinction between those who participate fully in their communities and those who withdraw into private life doesn't rest in the active citizens' grasp of complex issues, or their innate moral strength. Instead, those who get involved . . . have learned . . . that they don't need to wait for the perfect circumstances, the perfect cause, or the perfect level of knowledge to take a stand . . . that they can proceed step by step, so that they don't get overwhelmed before they start.
>
> —Paul Loeb[72]

Many people imagine that our environmental problems can be fixed with a little tinkering here and there; but, in fact, something very different is now required of us. Our situation calls to mind the story of the entrance exam for an insane asylum. Candidates are led into a cement-lined room with a row of faucets on one wall. The faucets are all gushing water. There are buckets and mops next to the faucets. The insane see the buckets and mops, and they run for them. Of course, the sane walk over and turn off the faucets. Systems scientist and essayist Donella Meadows told this story and then observed that we modern people are certifiably crazy because we seem to invariably opt for "mop-and-bucket" solutions to the problems facing us. For example, we spend money on new prisons without seriously addressing the root causes of crime. We build more and more new roads, failing to understand that more roads encourage more driving, more cars, more accidents, and more greenhouse-gas emissions. We attack pollution, too, with mops and buckets, putting, as Meadows said, "diapers on smokestacks" to catch nasty emissions rather than designing pollution-free manufacturing process at the front end.[73]

Now, as the sustainability revolution gets under way, humankind is being challenged to grow up—to move beyond cultural adolescence to full adulthood. The earth's biota and natural ecosystems, with hundreds of millions of years of research and development on what works, provide powerful models and templates for the design of a sustainable world. As organizational consultant Margaret Wheatley observes: "If nature uses certain principles to create her infinite diversity and her well-organized systems, it is highly probable that those principles apply to human life and organizations as well. There is no reason to think we'd be the exception."[74]

As explained in this chapter, the tools now exist in the technological, economic, and political realms to build a genuinely sustainable society. For example, new design tools permit the construction of machines, vehicles, and buildings that are ten times more efficient than status quo constructions.[75] In addition, economic tools such as full-cost accounting and tax shifting help minimize the discrepancy between price and cost, and monitoring tools such as the genuine progress indicator measure well-being in holistic ways. And, finally, when political tools such as "new rules" and organizational practices such as open space and appreciative inquiry are implemented, they help revitalize democracy.

But these tools, by themselves, won't change things; there won't be a sustainability revolution if no one comes. The tools have to be used. Participation is essential. Everywhere in the natural world, we observe the centrality of participation. Everything from the subatomic particle to the blue whale exists in a participatory network of interconnections. So it is with us: The essence of our lives is in relationship; the essence of democracy is in participation.

NOTES

1. Sim Van der Ryn and Staurt Cowan, *Ecological Design* (Washington, D.C.: Island Press, 1996), 17.

2. Van der Ryn and Cowan, *Ecological Design,* 17.

3. David Orr, *Earth in Mind* (Washington, D.C.: Island Press, 1994).

4. William McDonough and Michael Braungart, "The Next Industrial Revolution," *Atlantic Monthly* 282 (October 1998): 82–92.

5. McDonough and Braungart, "The Next Industrial Revolution," 88.

6. McDonough and Braungart, "The Next Industrial Revolution."

7. Paul Hawken, *The Ecology of Commerce* (New York: Harper Business, 1993).

8. Mary Guterson, "Living Machines," *In Context*, no. 35 (1993): 37–38.

9. Guterson, "Living Machines."

10. Van der Ryn and Cowan, *Ecological Design*.

11. David Salvesen, "Making Industrial Parks Sustainable," *Urban Land*, February 1996, 29–32, quoted in Timothy Beatley and Kristy Manning, *The Ecology of Place* (Washington, D.C.: Island Press, 1997), 142.

12. Janine Benyus, "Mother Nature's School of Design," *Yes! A Journal of Positive Futures*, Fall 2001, 16–20.

13. Benyus, "Mother Nature's School of Design."

14. Andrew Beattie and Paul Ehrlich, *Wild Solutions* (New Haven, Conn.: Yale University Press, 2001).

15. Beattie and Ehrlich, *Wild Solutions*, 198.

16. Denis Hayes, *The Official Earthday Guide to Planet Repair* (Washington, D.C.: Island Press, 2000).

17. Hayes, *The Official Earthday Guide*; John C. Ryan, *Seven Wonders: Everyday Things for a Healthier Planet* (San Francisco: Sierra Club Books, 1999).

18. Hayes, *The Official Earthday Guide*.

19. Michael Brower and Warren Leon, *The Consumer's Guide to Effective Environmental Choices* (New York: Three Rivers Press, 1999).

20. Brower and Leon, *The Consumer's Guide*.

21. www.HomeEnergySaver.lbl.gov.

22. Hayes, *The Official Earthday Guide*; Ryan, *Seven Wonders*.

23. Brower and Leon, *The Consumer's Guide*.

24. Brower and Leon, *The Consumer's Guide*

25. Hayes, *The Official Earthday Guide*.

26. Hayes, *The Official Earthday Guide*.

27. Brower and Leon, *The Consumer's Guide*.

28. Bill McKibben, "Small Change," *Orion*, January/February 2003, 80.

29. Lester Brown, *EcoEconomy* (New York: W. W. Norton, 2001).

30. McKibben, "Small Change," 81.

31. See chapters 5, 6, and 7 for an elaboration of this point.

32. Hawken, *The Ecology of Commerce*.

33. Staurt Cowan, "Investing in a Watershed Economy," *Stewardship Quarterly* 4, no. 3 (1997).

34. David Korten, "The Post Corporate World," *Yes! A Journal of Positive Futures*, Spring 1999, 12–18.

35. Brown, *EcoEconomy*, 258.

36. Barbara Brandt, *Whole Life Economics* (Gabriola Island, British Columbia: New Society, 1995).

37. Clifford Cobb, Ted Halstead, and Jonathan Rowe, "If the GDP Is Up, Why Is America Down?" *Resurgence Magazine*, no. 175 (March–April 1996): 7–12.

38. This graph is adapted from www.redefiningprogress.org.

39. Christopher Uhl, "Process and Practice: Creating the Sustainable University," in *Strategies for Sustainability: Stories from the Ivory Tower*, edited by Peggy Barlett and Geoffrey Chase (Cambridge, Mass.: MIT Press, forthcoming).

40. Joel Ohringer, "Our Beloved Cars—What a Price We Pay," *In Context*, no. 33 (Fall 1992): 8–9.

41. Brown, *EcoEconomy*, 236.

42. Brown, *EcoEconomy*.

43. Donella Meadows, "CSA Farms Can Help Our Health, Our Land, and Our Farmers," *The Global Citizen*, April 16, 1998 (accessed at www.sustainabilityinstitute.org).

44. Meadows, "CSA Farms Can Help Our Health."

45. Joel Dominguez and Vicki Robin, *Your Money or Your Life* (New York: Penguin Books, 1992).

46. Dominguez and Robin, *Your Money or Your Life*.

47. Vicki Robin and Joel Dominguez, "What Is Money?" accessed at www.newroadmap.org.

48. Robin and Dominguez, "What Is Money?"

49. Robert McChesney, "Corporate Media and the Threat to Democracy," broadcast on Alternative Radio (Boulder, CO), September 14, 1998.

50. Frances M. Lappé and Paul DuBois, *The Quickening of America: Rebuilding Our Nation, Remaking Our Lives* (San Francisco: Jossey Bass, 1994), 21.

51. Benjamin Barber, "McWorld and the Free Market's Threat," broadcast on Alternative Radio (Boulder, CO), October 7, 1998.

52. Robert D. Putnam, *Making Democracy Work: Civic Traditions in Modern Italy* (Princeton, N.J.: Princeton University Press, 1993), 167.

53. David Korten, *When Corporations Rule the World* (West Hartford, Conn.: Kumarian Press, 1995).

54. Lappe and DuBois, *The Quickening of America*.

55. Tom Atlee, "A Compact Vision of Co-intelligence," accessed at www.co-intelligence.org.

56. Atlee, "A Compact Vision of Co-intelligence."

57. This section on "new rules" was inspired by the work of David Morris and colleagues (accessed at www.newrules.org).

58. David Morris, *The New Rules of Localism* (Minneapolis, Minn.: Institute of Local Self-Reliance, 1996). Contact information: Institute of Local Self-Reliance, 1313 5th Street SE, Minneapolis, MN 55414; www.newrules.org.

59. Paul Hawken, Amory Lovins, L. Hunter Lovins, *Natural Capitalism: Creating the Next Industrial Revolution* (Boston: Little, Brown, 1999).

60. Hawken, *The Ecology of Commerce*.

61. Michael Lerner, "Social Responsibility Amendment," *Tikkun* 12 (July/August 1997): 33–35.

62. Sharif Abdullah, *Creating a World That Works for All* (San Francisco: Berrett-Koehler, 1999).

63. Accessed at www.earthcharter.org.

64. Accessed at www.earthcharter.org.

65. Karl H. Robert, "Educating a Nation: The Natural Step," *In Context*, no. 28 (1991): 10–13.

66. Brian Nattrass and Maru Altomare, *The Natural Step for Business* (Gabriola Island, British Columbia: New Society, 1999).

67. Robert McChesney, "Corporate Media and the Threat to Democracy," broadcast on Alternative Radio (Boulder, CO), September 14, 1998.

68. McChesney, "Corporate Media and the Threat to Democracy."

69. R. Brian Stanfield, *The Courage to Lead* (Gabriola Island, British Columbia: New Society, 2000), 17–18.

70. David L. Cooperrider and Diana Whitney, *Appreciative Inquiry* (San Francisco: Berrett-Koehler, 1999).

71. Tom Atlee, "The Politics of Understanding," *Yes! A Journal of Positive Futures*, Fall 1996, 37–39.

72. Paul Loeb, *Soul of a Citizen* (New York: St. Martin's Griffin, 1999), 8–9.

73. Donella Meadows, "Mop-and-Bucket Solutions Keep Us Forever Cleaning Up," *Global Citizen*, September 6, 1995 (accessed at www.sustainabilityinstitute.org).

74. Margaret J. Wheatley, *Leadership and the New Science* (San Francisco: Berrett-Koehler, 1999), 162.

75. Hawken, Lovins, and Lovins, *Natural Capitalism*.

10

EMPOWERMENT: TRANSFORMING SELF, TRANSFORMING SOCIETY

It is my deep conviction that the only option is a change in the sphere of the spirit, in the sphere of human conscience. It's not enough to invent new machines, new regulations, new institutions. We must develop a new understanding of the true purpose of our existence on earth. Only by making such a fundamental shift will we be able to create new models of behavior and a new set of values for the planet.

—Vaclav Havel, President of Czech Republic[1]

In the early stages of the industrial revolution, people knew they were living in a time of great change. The situation is similar today. We live in a time of transition and tumult—a time when we feel a great hunger for meaning and purpose.

The sustainability revolution has come forth out of this turmoil. This revolution challenges society to redefine its bottom line—which, of course, is what makes "sustainability" such a necessary, as well as radical, concept. The shift away from a culture based on exploitation, profit, militarism, and separation to one based on sustainability—that is, one grounded in stewardship, interdependence, social justice, and peace—will not come easily.

In order for this new paradigm to take root in society, three conditions must be satisfied. First, we must have a compelling vision of the new, life-sustaining world that lies ahead (foundation 10.1). Second, citizens must believe that the radical changes required to create this life-sustaining society are possible (foundation 10.2). Finally, major social transformations, such as a those implicit in a sustainability revolution, depend on an active and educated citizenry, skilled in the creative use of power (foundation 10.3). The practices that accompany these three foundations explore the role of vision, activism, and insight in personal empowerment and transformation.

FOUNDATION 10.1: THE ROLE OF VISION IN SOCIAL TRANSFORMATION

As humans, we have the ability to envision the future that we would like for ourselves. In the embrace of a compelling personal vision, our lives are filled with purposefulness. Likewise, at the national level, a noble vision engenders the allegiance of citizens and can result in purposeful action toward societal goals.

The Wellsprings of Vision

Powerful visions aren't just born of thin air. For example, in Colonial America the idea of creating a democracy didn't just arise spontaneously. The colonists had never experienced anything like democratic governance; they came from nations ruled by monarchs (e.g., England, France). Hence, when the Founding Fathers began to ponder how they might unite thirteen separate and sovereign states into one unified nation, they must have been deeply challenged. By good fortune, though, they found themselves living in the midst of a loosely organized nation-state, the Iroquois Confederation. Several hundred years before the colonists arrived, five Indian nations—the Mohawk, Onondaga, Seneca, Oneida, and Caygua—occupied the lands from New England to the Mississippi. They united to form the Iroquois Confederation, which was governed by the "Great Law of Peace."[2] European settlers were aware of this confederation to varying degrees. Benjamin Franklin became intimately familiar with the Iroquois system of governance when he served as the Indian commissioner for the colony of Pennsylvania in the 1750s; he suggested that the Iroquois Confederation might offer a template for the European settlers as they struggled to govern themselves. In this context the Iroquois model of governance may have been one of the sources of inspiration for the Albany Plan of Union drafted in the middle eighteenth century. And, although the drafters of the U.S. Constitution were most heavily influenced by the philosophers of the French Enlightenment (e.g., Locke, Rousseau, Montesquieu), it may be that some of the provisions placed in the Constitution were inspired by the Iroquois Confederation. For example, Iroquois leadership was not based on heredity but on selection; and Iroquois leaders could be impeached. Although these concepts of "selection" and "impeachment" ran counter to European traditions, they were adopted nonetheless. Anthropologist Jack Weatherford believes that the Euro-Americans may also have been influenced by the civil processes the Iroquois Confederation used in arriving at decisions:

> Europeans were accustomed to shouting down any speaker who displeased them; in some cases they might even stone him or inflict worse damage. The [Iroquois Confederation] permitted no interruptions or shouting. They even imposed a short period of silence at the end of each oration in case the speaker had forgotten some point or wished to elaborate or change something he had said.[3]

It may be that the relicts of these Iroquois decision-making sensibilities are reflected today in the practices of civil debate and compromise.

The Iroquois Confederation also offered Euro-American women a compelling vision of equality of the sexes. Married women in Colonial America had no legal existence. They had no right to vote, no authority to sign contracts, no right to legal custody of their children, and no right to own property. In sum, they were legally dead.[4] In light of these conditions, it is hard to imagine how a small cadre of nineteenth-century American women's rights activists dared to dream their revolutionary dream. It seems likely that:

> They caught a glimpse of the possibility of freedom because they knew women who lived liberated lives, women who had always possessed rights beyond the suffragists' wildest imagination—Iroquois women.[5]

Indeed, in the 1800s, communal exchange and trade between Native Americans and Euro-Americans was not uncommon. And while the majority of Euro-American women may have looked on native women with disdain, early feminists such as Elizabeth Cady Stanton, Lucretia Mott, and Matilda Joslyn Gage had personal contact with Iroquois women; and they did not fail to notice that, in Iroquois society, not only was family lineage through mothers but women had a full voice in political affairs as well. It was the Iroquois clan mothers who nominated chiefs and could remove them from power. "To Stanton, Gage, Mott, and their feminist contemporaries, the Native American conception of everyday decency, nonviolence and gender justice must have seemed like the promise land."[6] In this vein, it bears noting that European women captured by Indians often refused repatriation when it was offered—in part because of the disdain that would have been heaped upon them by European men, but also because they enjoyed a status in Iroquois society that was unheard of in the society of their birth.

As these historical footnotes illustrate, the seed for social transformation sometimes consists of simply seeing, in the flesh, a radical alternative to the status quo. Just as the gender norms and governance model of the Iroquois helped shape (to some degree) the Euro-American vision of democracy and gender equality, so it is that in our day the myriad technical, economic, and political innovations in the realm of sustainability now spreading throughout the world provide an impetus for reimagining humanity's future (see chapter 9).

The Power of Imagination

The capacity for vision is usually linked to one's power of imagination. That imagination is essential for creative problem solving is illustrated in the true story about the poet Robert Desnos, who was a prisoner in a Nazi concentration camp. One day Desnos and others were taken away from their barracks. The prisoners rode on the back of a flatbed truck, knowing that the truck was going to the gas chamber. No one spoke. Soon they arrived, and the guards ordered them off the truck. As they began to move toward the gas chamber, Desnos suddenly jumped out of line and grabbed the hand of the woman in front of him. He was animated, and he began to read her palm. The forecast was good: a long life, many grandchildren, abundant joy. A person nearby offered his palm to Desnos. Here, too, Desnos foresaw a long life filled with happiness and success. The other prisoners came to life, eagerly thrusting their palms toward Desnos, and, in each case, he foresaw a long and joyous life. The guards became visibly disoriented. Minutes before, they were on a routine mission, the outcome of which seemed inevitable; but now they became tentative in their movements. Desnos was so effective in creating a new reality that the guards were unable to go through with the executions. They ordered the prisoners back onto the truck and took them back to the barracks. Desnos never was executed.[7]

Robert Desnos was famous for his belief in imagination; he believed it could transform society. Desnos is not alone. The people at a U.S. organization called Wild Earth see a day when half of America will be returned to its original wild state. For reference, less than 2 percent of the lower forty-eight states are presently designated as "wilderness."

When David Foreman, the head of Wild Earth, came to central Pennsylvania to share his vision, hundreds of citizens showed up.[8] Foreman began by asking people to imagine woodland bison, mountain lions, and wolves once again inhabiting the Appalachian highlands:

Imagine visiting Pennsylvania's forests with the expectation of encountering elk, fishers, bobcats, spotted salamanders, flying squirrels, lynx, and gray fox. Imagine ambling along valley streams and regularly spotting river otters, bog turtles, brook trout, freshwater mussels, American shad, mink, and beaver. Imagine looking up to the tree tops and sky and regularly spotting pine martins, peregrine falcons, henslow sparrows, pileated woodpeckers, and bald eagles.

The simple act of clearly articulating a compelling vision is an essential first step in its manifestation. When the Wild Earth vision to "rewild" America was first put forth in the 1980s, it was regarded as impossible; but in recent years, this "rewilding" vision has gained force. Now state governments in Florida, Ohio, Minnesota, and Arizona, among other places, have begun to consider and, in some cases, activate plans to "rewild" their states (figure 10.1).

Rewilding usually begins with measures to expand the size of already existing conservation areas, such as national parks, wilderness areas, and wildlife refuges. Next come efforts to link all these isolated conservation areas via

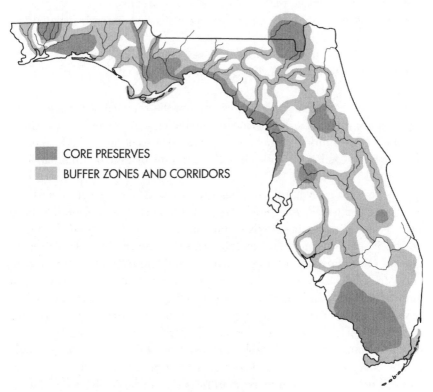

CORE PRESERVES
BUFFER ZONES AND CORRIDORS

Figure 10.1.   A blueprint for the rewilding of Florida?

wildlife corridors into an interconnected wildlands network. The final element in rewilding is the reintroduction of carnivores—for example, eastern cougar and wolf in the case of eastern North America.

Ecosystems without their carnivores can be likened to bicycles without chains—the various ecosystem parts fail to fully engage. The animating role of carnivores in ecosystems was illustrated when wolves were reintroduced into Yellowstone. In the absence of wolves, the elk at Yellowstone were behaving more like domesticated cattle than wild ungulates. With the return of wolves, the elk snapped to attention; the ecosystem was electrified, the "chain" engaged.

The Wild Earth vision has power because it reinforces a universal human desire to heal wounds. Perhaps this is why Forman took time early in his Pennsylvania workshop to invite participants to publicly identify the wounds that have been inflicted on the Pennsylvania landscape. Slowly, citizens stood and, one by one, spoke of the wounds:

- roads that separate animal populations and act as death traps;
- acid-mine drainage that poisons lakes and streams;
- housing developments and shopping malls that entomb meadows in concrete;
- clearcuts that transform verdant forests into eroding hillsides;
- dams that turn free flowing rivers into stagnant lakes.

Conservation easements, stream restoration, the removal of certain roads, the establishment of urban growth boundaries—strategies to heal these wounds are as limitless as human ingenuity. Yet sometimes the healing of old wounds paradoxically requires the creation of new wounds, at least temporarily (see box).

This vision for rewilding America extends to America's very heartland, the Great Plains. Imagine, for a moment, immense herds of buffalo once again roaming across hundreds of millions of acres of restored shortgrass prairie in eastern Montana and the Dakotas; down through Nebraska, Kansas, Oklahoma; and on into North Texas. Dare to think of this region, north to south, as a grand "buffalo commons"—restored to prairie with most of its fences and roads removed. Although this vision may sound preposterous, we do have compelling ecological reasons for taking it seriously.

Agriculture is clearly not the long-term vocation of the Great Plains. This region's rainfall is unpredictable, and its irrigation water is increasingly scarce. Given the region's capricious climate, farming has always been risky and will probably become more so in the years ahead. However, if the re-

REMOVING DAMS: AN ACT OF REWILDING

In the 1990s when Bruce Babbitt was secretary of the interior he spoke matter-of-factly about the importance of blowing up certain dams. The idea of tearing down public works like dams was met with outrage by many citizens. After all, dams provide hydropower and at the same time create lakes for recreation. But if the natural world—pulsating with life and complexity—becomes our point of reference, then it is the act of building a dam that will seem like the "outrage." Free-flowing rivers are a complex mix of riffles, pools, eddies, and swifts; stream substrates vary from fine silts and coarse sands, to pebble and cobble. Each combination of depth, water velocity, and substrate creates a unique aquatic habitat. Most aquatic organisms have specific habitat requirements. Some prefer the pools, others the swifts; some occupy the water column, others the distinct substrate types. A dam obliterates this matrix of habitats and thus leads to biotic impoverishment. There are seventy-five thousand dams in the United States; less than 1 percent of U.S. waterways are undammed. This is the context within which Babbitt spoke of removing certain dams.

gion were restored to native prairie, it might become an "American Serengeti," supporting wild buffalo, deer, elk, and antelope in significant numbers in perpetuity.[10]

We also have a compelling economic rationale for considering this buffalo commons vision. A county-by-county demographic characterization of the western Great Plains—by two geographers from Rutgers University, Drs. Frank and Deborah Popper—reveals that significant portions of the region are in economic decline. Present-day human populations are often less than four per square mile in many counties (compared to the U.S. average of sixty-eight), and young people have been abandoning this region as evidenced by the fact that the median age in many counties is greater than thirty-five years (vs. the U.S. average of thirty).[11]

Whether a buffalo commons will come to pass is anyone's guess. The point is that the sustainability revolution is serving as an incubator and nursery for hundreds of visions like that of the buffalo commons. Compelling visions and technical approaches now exist for reinventing virtually all segments of American life along sustainable lines.

In this time of turmoil, Americans are questing, perhaps as never before, for a vision of a full and dignified life. According to historian Jacob Needleman, "The cultural hero of the present age is no longer the Warrior or the Savior or the Adventurer, the Lover, or the Wise Man. It is the Seeker."[12] Both individually and collectively, Americans are a people in search of a vision worthy of their allegiance. The sustainability revolution has been born of this search. The central metaphor these days is no longer the isolated machine but the interconnected community; the bottom line is no longer profit but well-being; the goal is no longer personal salvation but mutual indwelling; life is no longer a struggle to endure but a mystery to embrace. If we are to create a sustainable society, we will need foresight like that practiced by Benjamin Franklin and Lucretia Mott; we will need imagination like that demonstrated by Robert Desnos; and we will need vision like that offered by David Forman and his colleagues at Wild Earth. Theologian Walter Wink puts it this way:

> The future belongs to whoever can envision a new and desirable possibility, which faith then fixes upon as inevitable. This is the politics of hope. Hope envisages its future and then acts as if that future is now irresistible, thus helping to create the reality for which it longs.[13]

Reflection

A man I barely knew carefully wrote the following string of letters on a piece of paper and asked me what it said:

opportunityisnowhere

When confronted with this puzzle, many people read "Opportunity is no where," but this string of letters could also be read "Opportunity is now here." It just depends on how we choose to see it. Likewise, if we choose to see the world in a hopeless downward spiral, then opportunity is indeed "no where"; but if we catch a glimpse of the emerging sustainability revolution, then opportunity is "now here."

Creating a sustainable society involves enormous imagination and vision; and it requires an ability to think "outside of the box." A popular puzzle, the problem of the nine dots, epitomizes the challenge before us.

•   •   •

•   •   •

•   •   •

To solve the puzzle, connect all nine dots by making four straight lines, but without lifting your pencil from the paper and without retracing any line.

If you are like most people, you will begin by drawing lines around the edges of the square; but then you will recognize that the dot in the middle will be left untouched. You might make various attempts to include the center dot, but nothing works until you begin to "think outside of the box," as defined by the nine dots. The solution lies in extending lines outside of the grid.

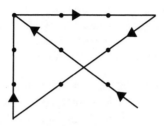

Creating a sustainable society is like solving the problem of the nine dots, over and over again. It requires that that we use our imagination and vision to create an exciting and life-affirming world outside of the "box."

Questions for Reflection

- If you had to write a job description for your life, what would it look like?
- In what ways might your personal vision, if realized, contribute to the well-being of others—those alive today and those in future generations?

Practice: Empowerment through Vision

Many young people have been captivated by Joseph Campbell's famous admonition: "Follow your bliss." If you bound out of bed in the morning, delighted to be spending your days in exciting work in the service of life, then you are most likely "following your bliss." Likewise, if you receive your paycheck and think to yourself, "I can't believe they are paying me to do this," then you are surely in touch with your bliss. On the other hand, if you don't want to get out of bed in the morning, then you have left your bliss behind. Likewise, when your work—what you bring to the world—is shabby and uninspired, then you have lost contact with your bliss. Nor do you have much bliss when your mind is constantly engaged in worry, fear, grasping, and judgment.

Bliss is born of clarity of purpose; it issues forth when one has a clear vision for his or her life. When my daughter, Genny, turned sixteen, she decided that she was going to go off to the mountains for three days by herself. She called it her "solo." I thought of it as her first "vision quest." In many aboriginal cultures, young people went on "vision quests" to learn their place in the world. Teacher Mark Burch describes the vision quest as

> a prolonged period of special ceremonies, secret teachings, physical trials, meditation, and fasting, culminating in a solitary ordeal that did not end until the initiates met a spirit guide who revealed to them their divinely ordained role in the tribe and became their life-long companion in pursuing it.[14]

In our day, we can also engage in a vision quest of sorts by actively searching for meaning and purpose—questing to find our unique calling in the grand scheme of things. Insights do not come forth if we sequester ourselves in comfort. We must stick our necks out into the wind of risk.

Articulating a vision for one's life is not a one-time act; it is an ongoing process of discernment. At age fifty-three, my friend Richard was in search of a vision for the next phase of his life. He saw it as a search for passion. Richard engaged in a multistep practice to discern where his passions met the earth's callings. First, he spent several hours each day, over a period of weeks, making a list of the things that had made him feel most alive and engaged in life. He left no stone unturned, exploring all periods of his life and all aspects: relationships, work, play, spirituality, education. Richard's list spanned almost thirty pages. His second step was to study the list, looking for patterns. By studying this "map," Richard was able to articulate the values, qualities, and types of activities that he wanted his future work to embody (as well as those things that were toxic to his spirit). Finally, when he felt like he had taken the process as far as he could on his own, he invited a group of nine friends to his home for an evening. He began by summarizing what he had learned from his quest, and then he invited the group to help him think in concrete terms about how he might bring his talents, principles, and energies into the world of work. I was among those invited, and I offered my thoughts and ideas; but mostly what I remember from the evening was being a witness to a remarkably open process of personal discernment. A life vision generally issues forth from a place of spaciousness and trust. Richard created space by quitting his job and trusting that a new way would open.

Many people find that engaging in artistic endeavors—music, art, dance—creates space that softens them, ventilates their imagination, and cultivates vision. The arts are not something just for professionals; they are for all of us. It has taken me a long time to really believe this is so.

Engaging in artistic endeavors is a form of play, and "play is the spring-head of discovery, the heart of art, the exploratory urge made manifest."[15] For many of us, it has been a long time since we last played, so we are out of practice; our "play muscles" have atrophied. But this need not stop us. Grab a hunk of clay and experience the sensual delight of working with it for a morning; or pick up some paints and experiment with them as you sit by a pond for the afternoon; or choose an emotion, like joy, or a value, like courage, and spend an evening experimenting with how you might express it with your entire being—pulsating, swaying, rolling, howling.[16]

Different forms of artistic expression awaken different parts of ourselves. Dance and movement bring awareness to our bodies and how we hold ourselves in the world; painting calls forth our deep seeing; music calls forth our soulfulness. In all cases, the arts can help us gain access to hidden parts of ourselves and thereby bring us to a fullness of being.[17]

As we use art to fathom our depths, we are like sculptors or carvers trying to bring forth what is hidden. An interesting practice is to take a block of soapstone and do as my daughter was told to do when she attended a Waldorf school: find out what is in there by carving it free. After two weeks of carving, Genny discovered a goddess, perhaps a muse in her life's journey.

Crafting a personal mission statement can also help in defining and articulating a vision for your life.[18] One way to arrive at a mission statement is to take a piece of paper and write down the things that are most important and meaningful in your life right now. Then, on a second piece of paper, list the things that you still yearn for to make your life more complete. Finally, pick the three or four most important things in each of these two lists and compose a one-hundred-word declaration summarizing your most cherished values and deepest longings. Write this statement in the first person with present-tense, declarative sentences—for example, "I am a peacemaker in my family and community." Next, trim your statement to fifty words, then to twenty-five words. In this way, you will be attempting to arrive at your most dearly held values, beliefs, commitments, and longings.

The final statement can be transcribed to a small card to be placed in your wallet. Jazz up the card with an artistic boarder. This statement illustrates what you are all about in the world. When the phone rings and someone asks you to go to the mall, watch television, or join a book group, you can run the invitation through the filter of your mission statement, asking yourself, "Is going to the mall with a friend part of what I am about in life?" If the answer is "yes," you can go in the full knowledge that what you are doing is congruent with your life's mission. If the answer is "no," you will be able to decline the invitation, knowing that you are actually saying "yes" to your life. Using a mission statement is a way of grounding yourself in your life; it helps you resist the tendency to constantly conform to outside pressures.

Many people find that they can reduce their mission statement to a single word. Often it is a word that has resonated with them ever since they were young and that calls forth their essence and purpose in the world. Perhaps the word is "beauty" or "joy" or "truth" or "compassion." When they hear their word or see manifestations of it, they are profoundly enlivened. Once we are clear on our "life word," we can endeavor to make our lives an expression of that word.

Ultimately, we will be held accountable (by ourselves) to the inner yearnings for wholeness and integrity arising from our hearts. If we don't discover and live by our own vision and calling, the messages of others will insinuate themselves into our consciousness and subdue our passions. We will become fearful, rather than forceful, as we succumb to our culture's insistent pleas:

- Be careful!
- Avoid unnecessary risks!
- Control your passions!
- Feel less!
- Play it safe!
- Conform!

When we acquiesce to these messages—thereby sacrificing our dreams for a saccharine security—the light of spirit and consciousness goes out within us.

FOUNDATION 10.2: UNDERSTANDING THE SOCIAL TRANSFORMATION PROCESS

The sustainability revolution is, most fundamentally, a massive social change movement, entailing a complete reorganization and reimagination of governance, commerce, science, education, civic life, and values. Just as we have difficulty imagining how Americans in the past could have tolerated slavery, child labor, and the overt oppression of women; people one hundred years from now may find it unimaginable that we contemporary Americans toxified our life-support systems, dropped bombs on other people's cities, knowingly allowed millions of children throughout the world to starve to death, and sexually abused our own children.

At present, the sustainability revolution is advancing on three fronts. The first is what Joanna Macy calls "holding actions." When Julia Butterfly Hill spent two years in the top of a tree to slow the cutting of the redwoods in California, that was a holding action; when the U.S. government passes legislation to reduce emissions from car exhausts, that is a holding action; when a citizen refuses to invest his savings in companies that build weapons of mass destruction, that is also a holding action. The goal of holding actions is to slow down the process of destruction of the biosphere in the hopes of protecting, to the degree possible, Earth's vitality for the life-sustaining society that is slowly emerging.[19]

But holding actions are not enough. We also need new structures and designs, such as those showcased in the previous chapter—for example, new ways to gauge progress, new rules, green buildings, and zero-waste manufacturing processes. There has probably never been a time in human history when so many new life-sustaining ways of doing things—from farming to building to business to education to medicine—have appeared in so short a time.

But even these new structures, along with the holding actions, will not be enough to ensure the creation of a life-sustaining world. The sustainability revolution also calls on humanity to cultivate "ecological consciousness"— that is, to see ourselves not as separate from but as part of the living earth; not above and outside, but seamlessly embedded in life's matrix. All three things—the holding actions, the creation of new forms, and the embrace of new consciousness—are equally important in bringing forth a sustainable society.

A Conceptual Model of Social Transformation

We have a tendency to imagine that the so-called Founding Fathers created American democracy in one sitting, but, in fact, many of our cherished freedoms have been the result of major social change movements.[20] One of the best-kept secrets in America is that citizen-based social movements are the centerpiece of democracy. For example, in the early days of our Republic, fewer than 10 percent of the population voted. That privilege was restricted to white males who owned land; no one else was permitted to vote. By virtue of a string of social movements over the last two hundred years, this democratic "given" has been extended to those without property, to women, and to all ethnic groups.

Now, in the first decades of the twenty-first century, we are witnessing active social movements in many sectors of society. Citizens are concerned about access to health care, possible harmful effects of genetically modified foods, gender- and race-based discrimination, unjust labor practices, the erosion of basic civil rights like free speech, militarization, corporate control of media . . . the list is long. On the surface, it appears like a cacophony of separate causes—voices saying "no" to this and "no" to that—but under the surface those voices share a common longing for a life-sustaining world imbued with social, environmental, and economic justice.

An understanding of the dynamics of social change movements is essential to our understanding the unfolding of the sustainability revolution and the roles that citizens play in societal transformation. Bill Moyer, who studies citizens movements, breaks the process of social change into eight stages, which I have abbreviated to six.[21] First, comes the sleep stage: everyone is asleep. For example, the local river is becoming polluted; but politicians are looking the other way, and the public is preoccupied with other things as well. At this initial stage, things go on as usual, even though a problem clearly exists.

Stage 2 begins when the public starts to wake up. It might be that some boaters notice that fish are beginning to die in the river. At this stage, the authorities usually assure citizens that they have no need to worry, saying perhaps that the fish die-off is "just a coincidence" or that it's "part of a normal cycle." But some people are not convinced, and they begin to band together to challenge the official government position.

In stage 3, citizens begin to meet; they build alliances and organize education campaigns. They conduct marches and news conferences: articles appear in the newspaper, and the issue begins to make its way into conversations.

Things really take off in stage 4. This stage is often initiated by a "trigger event"—a dramatic happening that focuses public attention on the problem. Some triggers are unplanned. For example, the Three Mile Island incident galvanized popular opinion against nuclear power. Other trigger events are orchestrated. A group might hang a banner from the Statue of Liberty in hopes of focusing public attention on the loss of American civil liberties.

Stage 5 involves winning over the majority. Broad coalitions become possible at this stage, as mainstream institutions expand their programs to include the issue. The issue also begins to show up in electoral campaigns. At this stage, a vast audience becomes ready to think about alternatives to the existing situation; but there can also be a mistaken perception of failure because concrete gains have yet to be realized.

The final stage is reached when the struggle shifts from opposing present policies to creating a dialogue about possible alternatives. At this stage, the power holders are ready to accept change; it is a time of negotiation and compromise. Concrete improvements are finally instituted. For example, some significant measures may be initally adopted to clean up the polluted river, but more remains to be done. Later the issue will surface again; the "bar" will be raised higher. This six-phase process will repeat, and further advances will be achieved.[22]

Moyer notes that the public has to be won over three times in any social change movement. First, they have to be convinced that a problem really exists; then, they have to be persuaded that the government's present policies are not going to resolve the problem; and, finally, they have to be convinced that those calling for change have a viable solution.[23]

Those in power usually offer strong resistance to change by sowing fear and misinformation. For example, even after U.S. citizens were convinced that nuclear power was problematic and after power holders conceded that it had major faults, citizens were still warned that without nuclear power, "the economy and jobs would collapse, and the country would lose its superpower status."[24]

In sum, Moyer's social change model provides a way to understand society's current struggle between maintaining the status quo and adopting a whole new paradigm based on sustainability. It appears that the sustainability movement as a whole is somewhere around stage 4 in Moyer's scheme. Most people acknowledge that there is a problem; participants have initiated a massive education effort; and natural trigger events continue to snap us to attention—like ozone thinning, climate change, endocrine disruption, songbird declines, corporate scandals, terrorist attacks, and financial meltdowns.

Dr. Everett Rogers and colleagues at the University of Southern California have proposed that when 5 percent of society accepts a powerful, positive social change idea, it becomes "embedded." After this, it will still take a lot of networking and educating before 20 percent of the population comes to embrace the idea; beyond this point, the idea generally becomes unstoppable.[25]

Sustainability as a unitive social change movement has great momentum because it is based on widely held universal values, such as respect for life, conservation of resources, cooperation, and equity. The unique power of the sustainability revolution lies in its power to draw forth idealism, creativity, and integrity, as well as entrepreneurialism.

The Role of Citizens in Social Transformation

All of us, whether we know it or not, play roles in social change movements; some roles are just not as obvious as others. Social activist Alan AtKisson has developed a model that illustrates our various roles. He uses a giant amoeba (the single-celled animal that creeps along engulfing its food) as a symbol for human culture (figure 10.2). The molecules that make up AtKisson's amoeba represent people in society.

The amoeba moves by sticking out a pseudopod (foot) into new territory. Its pseudopod is out on the edge testing new environments, never happy with the status quo. In human society, the pseudopod is the domain of "innovators," people who are always seeking to improve things—the ones offering creative solutions to societal problems. But an innovative idea or creative response by itself won't change things. The bulk of the amoeba will not slosh forward in the direction of the pseudopod unless forces come to bear on it (social pressures).

The bulk of the amoeba (culture) consists of "mainstreamers," and these mainstreamers come in three forms. "Middle-of-the-roaders" occupy the space at the middle of the mainstream. These are the rank-and-file members of society; things seem "okay" to them just as they are. Thus, they lack motivation to seek change. "Laggards," located at the tail end of the mainstream, are particularly complacent. Beyond the laggards are the "reac-

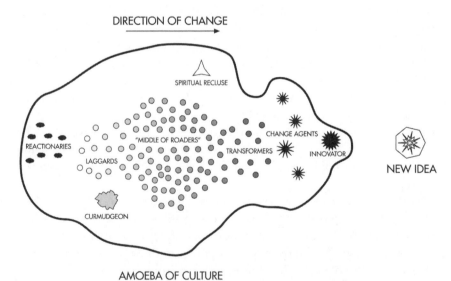

AMOEBA OF CULTURE
Figure 10.2.   The "amoeba of culture" model of social change[26]

tionaries"; they lie outside the mainstream and actively resist change because it threatens their vested interests. Perhaps they are land developers and don't want agricultural land to be protected from development, or maybe they manufacture chemicals and are wary about having too many safety restrictions or warnings on their products.

Given these formidable forces, how might new ideas (innovations) or creative responses to problems gain support? The situation is not entirely hopeless. Among the "mainstreamers" resides a third subgroup that AtKisson refers to as "transformers." Think of them as the molecules located close to the base of the pseudopod. These are regular folks who happen, by disposition, to be particularly curious and open to new ideas. Things don't seem as "okay" to them; they are willing to risk change in the hope of improvement. Perhaps they would like to preserve farmland, or perhaps they believe that strict restrictions should be put on the manufacture of chemicals. In short, they are concerned and awake. If they are given a well-reasoned argument and concrete suggestions, they are ready to act.

The catalysts in this amoeba model are called "change agents"; these are the people who act as organizers and networkers. They might not have the innovative ideas, but they know what to do with a good idea. The change agents form the bridge between the innovators and the transformers (at the forward end of the mainstream). The transformers, in turn, form the bridge to the mass of the mainstream (i.e., to the middle-of-the-roaders). By virtue of these bridges, the mass of the amoeba can begin to move in the direction of the pseudopod—the innovation (see box).[27]

AtKisson's "amoeba of culture" provides a sort of cartoon of the social transformation process. It is best not to take his labels literally (or personally!). For example, I may be a "laggard" on a certain issue, but this doesn't mean that I *am* a laggard. The role that we assume on any given social change issue is usually the result of our level of awareness. If we are behaving as "laggards" or "reactionaries," it may be that we simply have not had the opportunity to openly and deeply explore more expansive ways of thinking about certain issues.

Chaos as a Prelude to Transformative Change

We could easily look at all the change, upheaval, and disturbance occurring throughout the world today and conclude that things are falling apart and that humankind is headed for disaster. However, scientists have learned that great turmoil often precedes major shifts in the natural world. For example, Ilya Prigogine won the Nobel prize in 1977 by demonstrating that

THE AMOEBA OF CULTURE ROLE PLAY

A school class or a group of conscientious citizens can engage in what is called an "amoeba of culture" role play.[28] I did this recently with my students. Our role play was centered on car-based transportation. The middle-of-the-roaders owned cars and didn't see much of a problem with the way things were; the laggards absolutely loved their cars and couldn't imagine living without them; the reactionaries owned gas stations and car dealerships, and they were prepared to actively oppose any movement away from heavy dependence on cars. But there was an innovator who had an exciting idea: the establishment of a convenient light-rail system in combination with an extensive system of bike trails. Change agents—tired of traffic jams, pollution, and car payments—were attracted to this idea and were anxious to learn more.

To get things started, I gave each of the twenty-five students an index card describing their role. For example, the middle-of-the-roader card said: "You live in the suburbs and have two cars; you rely on your car for work and shopping and vacations. You sometimes think that it would be nice to get rid of one of your cars because this would save money, but you can't imagine how your family could manage with just one car."

Because the majority of people in society are in the mainstreamer categories (i.e., middle-of-the-roaders, laggards, or transformers), 70 percent of the students received cards in these categories; the remaining cards went to change agents and reactionaries (10 percent each), with the last cards going to an innovator, a curmudgeon, an iconoclast, and a wise elder.

After participants received their cards, I gave them a few minutes to get into character. Once everyone was ready, I set the scene, a downtown sidewalk cafe, and invited everyone to enter. Initially, the participants didn't know each other's roles, but as they talked to one another, the different orientations become apparent. Pretty soon the innovator and the change-oriented folks had found one another and began to hatch a plan for attracting others to the light-rail idea.

The role play lasted for about forty minutes. During the debriefing, the students who played middle-of-the-roaders talked about how comfortable this role was for them; laggards and reactionaries also observed that they found it easy to be against things and to shout down new ideas. The change-oriented people had a harder time, but eventually they hit on the idea of approaching people, one by one, and inviting them to a free concert in the town hall, which would be followed by a "fun" slide presentation on light-rail transportation. Every time someone agreed to attend, the change agents placed a pink sticky-note on the attendee's shirt, which read "the rail rocks." Pretty soon most people in the room were sporting these pink stickies, and the group experienced a palpable shift, a softening, toward the idea of light rail. Even in the absence of dramatic shifts, people gain an understanding of the social change process by simply engaging in this role play.

certain chemical systems actually shift to greater order after they are disturbed. In other words, disorder can actually act as an ally, cajoling a system to self-organize into a more ordered state, one better attuned to the demands of its changed environment.[29]

Research in the field of chaos theory has shown how "strange attractors" can appear in systems that are in a chaotic state—that is, the state just shy of the point where the system plunges into random behavior without any order. At the present time, "sustainability" might be acting as a "strange attractor," weaving order out of chaos in our society.

The idea that growth and greater order emerge out of states of chaos appears counterintuitive until we reflect on our own life experience. The times of hardship and suffering—when we personally descend into chaos—invariably lead to our personal growth and the expansion of consciousness. So it is with social systems: disruption, confusion, and chaos are necessary to awaken creativity. Seen in this light, the very unraveling of our old worldview—a view based on separation from nature, economic exploitation of Earth's body, and dominance relations among people—is providing the creativity and energy necessary to fuel the sustainability revolution, with its emphasis on human–nature unity, sustainable economics, and the sharing of power. As Margaret Wheatley, author of *Leadership and the New Science*, astutely observes:

> In the dream of dominion over all nature, we believed we could eliminate chaos from life. We believed there were straight lines to the top. If we set a goal or claimed a vision, we would get there, never looking back, never forced to descend into confusion or despair. These beliefs led us far from life, far from the processes by which newness is created. And it is only now, as modern life grows ever more turbulent and control slips away, that we are willing again to contemplate chaos. . . . The destruction created by chaos is necessary for the creation of anything new.[30]

In sum, the United States has been a laboratory for social change. We have access to clearly articulated models that explain the dynamics of change, and our culture is rich with examples of successful social change movements: All of this engenders in Americans an abiding sense that big change, such as that being called forth by the sustainability revolution, is possible.

Reflections

Some time ago I began to ask my students what words came to mind when they heard the word "activist." Many of the words they offered—"demonstrator," "hippie," "environmentalist," "agitator," "malcontent"—had a negative shading. I was surprised that activists were not seen simply as citizens practicing democracy.

I believe that it is important for young Americans to get a taste of activism and to experience and celebrate the freedoms guaranteed us in the Bill of Rights. Thus, I took a class of fifteen students to Washington, D.C., for a teach-in and demonstration scheduled to coincide with the International Monetary Fund (IMF)–World Bank meetings in April 2000. We arrived in the city late in the afternoon on a Friday, two days before the big meetings were set to begin. The students spent their first evening listening to economists, political scientists, and ecologists describe the history and operations of the IMF and World Bank. The next day more information was exchanged, including classes on the practice of nonviolent direct action. On Sunday, the day the meetings were to begin, the students had a variety of options. Some, who had trained and felt ready, decided to join the demonstrators who were going to block the streets leading to the IMF–World Bank meeting area; other students decided to participate in the officially sanctioned march; some from this second group volunteered to serve as peacekeepers during the march; and several students decided to stand apart as observers and simply witness the day's events.

Driving home that evening, I listened as the students told their stories of the day. Whether they opted to observe, serve as peacekeepers, march, or participate in a blockade didn't seem to much matter; they all had a taste of being in the midst of a social change movement—engaged in the machinations of democracy.

Questions for Reflection

- Which, if any, of Joanna Macy's three social change realms—holding actions, building new structures, creating ecological consciousness— do you feel drawn to? Why might this be so?
- Given your skills, values, and beliefs, can you imagine a way that you might participate in the sustainability revolution; and, if so, what might your participation look like?

Practice: Empowerment through Activism

In the face of the monumental social and environmental problems now confronting humanity, many people feel overwhelmed and powerless. It seems that we have all been conditioned, to varying degrees, to believe that we can do nothing about the world's problems. Our situation is similar to that created by psychologist Martin Seligman in his famous conditioning experiment involving three groups of dogs. The dogs in group 1 were subjected to mild shocks, which they could avoid by depressing a panel with their noses. The dogs in group 2 received the same shocks as those in group 1, but they were unable to seek any relief. The dogs in the third group received no shocks. After the dogs were "conditioned," they were placed in boxes containing two compartments divided by a low wall. The same mild shocks were then administered to dogs in all three groups. The dogs in groups 1 and 3 quickly learned to jump over the wall to avoid these shocks. However, the dogs in group 2 laid down and whimpered, making no attempt to jump over the wall. These dogs could have easily escaped the shocks, but they had learned in the first experiment that they could do nothing to escape the shocks—they had been conditioned to believe that they were powerless. Seligman was later able to decondition the dogs in group 2 out of their learned helplessness. We, too, can be deconditioned from modes of thinking and beliefs that have left us feeling disempowered.[31]

One reason that so many feel disempowered is that they are caught in the trap of the "perfect standard." Paul Loeb, who studies social activism, describes the trap this way:

> Before we allow ourselves to take action on an issue, we must be convinced not only that the issue is the world's most important, but that we have perfect understanding of it, perfect moral consistency in our character, and that we will be able to express our views with perfect eloquence.[32]

Holding ourselves to this "perfect standard" ensures that we will never act.

There is power in our personal presence. We can avoid the trap of the "perfect standard" if we recognize that we are all authorities where our own lives are concerned. I have learned this lesson from my sister, Ann Marie. For many years she has been in the habit of making pilgrimages to Washington, D.C., to talk to her congressional representatives and their aides. Ann Marie is a schoolteacher, not a political analyst, so one day I asked her how she mustered the courage and authority to lobby on the Hill. She admitted that at first it was scary, but then one day it hit her that these "officials" were ordinary people, just like her, who just happened to find themselves in unusual circumstances. When I asked her what she actually said when she visited her congressional representatives, she responded that she simply shares with them her vision for America and the pain she feels when our leaders fail to live up to our country's highest principles and values. Ann Marie doesn't buy into any "perfect standard"; her "standard" is personal authenticity. An essential practice for cultivating activism is to simply believe in the legitimacy of our individual life experience and to bring our voice forward.

There is power in the language of the heart. Often the most powerful language for effecting change comes not from the head, but from the heart. We know this "heart language" when we hear it. For example, imagine that your local government officials have convened to discuss a proposal to widen a road to accommodate more traffic. Representatives from the police department, the fire department, and the transportation commission give formal, computer-aided presentations explaining why it is necessary to widen the road. The issue appears to be settled, but then, in the back of the room, a mother with an infant on her shoulder rises and speaks of what a wider road, with more cars and faster speeds, will mean for the safety of the children in her neighborhood. This woman is real; she talks from the heart; her facts come from her own experience. Suddenly, widening the road is no longer an exercise in abstractions. There is a shift in the room. The council hesitates. What seemed to be a mundane matter is suddenly imbued with life and care.

One practice for cultivating "heart language" is to distill political discourse down to its bare essence. You can engage in this practice by starting with a piece of classic bureaucratese, such as the 1999 warning from the World Commission on Forests and Sustainable Development:

> There has been a clear global trend toward a massive loss of forested areas. . . . The current trends are toward an acceleration of the loss of forested area, the loss of residual primary forests, and progressive reduction in the internal quality of residual forest stands. . . . Much of the forest that remains is being progressively impoverished, and all of it is threatened.

With a red pencil in hand, have a go at whittling this statement down to something like:

> Worldwide, forests are disappearing at an ever-increasing rate and those forests that remain are becoming more and more degraded; no forests are secure.

Finally, really chop away until you get it down to just five words:

> We are destroying our forests.

The commission's original statement is lifeless; the five-word summary stirs the heart.[33]

The use of "heart language" allows you to penetrate to the heart of issues. For example, take the issue of substance abuse. As a society, we spend an immense amount of time discussing how to regulate cigarettes, alcohol, and other substances. Instead, we could go to the core of the issue and ask: Why is it that so many people in our culture fall into substance abuse in the first place? Might it be that we reach for drugs in their various guises because we suffer a hole in the middle of our being that we are trying to fill—a hole that results from (among other things) our being brought up in a society that systematically teaches us to be unhappy with who we are. We feel a collective sense of relief when someone finally says, in clear language, what everyone already knows to be true but is afraid to say. When we are grounded in the heart and speak the truth, we unleash great power.

There is power in seeking common ground. Almost all political discourse in America is highly polarized. If you doubt this, try this experiment: Go to the library and sit down with some magazines representing viewpoints both conservative *(Commentary, National Review)* and liberal *(The Progressive, The Nation)*. These magazines cover many of the same topics, and the articles are generally carefully researched and well written. However, the conclusions on any given issue will be very different, depending on whether you have chosen a liberal or a conservative magazine. Neither side is fully right or fully wrong. Both have part of the truth, and they milk that piece for all that it's worth. By reading only the magazines that reinforce your views, you easily fall into the trap of self-righteous moral indignation, imagining that your side holds the whole truth; but ordered righteousness is born of fear—it constricts life.[34] I tell you this from my personal experience; it has been one of the most difficult lessons I have had to learn in life.

We can escape the tyranny of half-truths if we cultivate the practice of critically examining our positions on issues—that is, by asking what is true about our stand and what is untrue about it. For example, if you are pro-choice, you might ask, "What is right about abortion?" followed by "What is wrong about abortion?" The idea is to overcome the strong tendency we all have to attach to our "positions" and in so doing to mistake our opinions and positions for who we are. Instead, we need to learn to stop siding with ourselves. This practice is essential if we are to move beyond polarized politics to creative problem solving. What makes this practice so challenging is that we come to an issue, in the first place, because our "heart" tugs us there. We care, and we want our view to prevail. But, again, ours is not the only heart in the room; nor is ours the only head at the table.

Once we recognize that all sides of an issue have wisdom, a wonderful thing invariably happens: We become curious and open, and in the process, we discover that people on all sides of any issue have essentially the same needs. For example, the protagonists and antagonists in the birth control debate both have a need to protect and nurture life; and the "hawks" and the "doves" in the military-spending debate both have a need for safety and security. Where people differ is in their strategies to meet basic needs. The trust and goodwill necessary for genuine communication open up once we understand that our "adversary" almost invariably wants the same thing as we want. This common ground is a source of power.

Seeking common ground, then, is a way of reframing controversy. Experiment with this concept by picking an issue that is highly polarized in your community. For example, take the issue of a logging company that is destroying old-growth forest. Rather than staging a protest, imagine going to the owner of the company and asking, "What would it take for you to stop cutting down the old-growth trees?"

> This question lets the other person create the path for change. . . . The question is an invitation to the [company owner] to co-create options for the future of his business with the community.[35]

In sum, "activism" is citizenship in action: a manifestation of our caring for ourselves, our communities, and our institutions; it is a lifelong practice involving a profound commitment to personal authenticity, truth seeking, and the common good. Gandhi spoke to the budding activist in all of us when he said: "Almost anything you do will seem insignificant, but it is very important that you do it."[36] In the end, it will be the little acts done in a spirit of love that will be important because everyone influences everyone else: If your life is exemplary, you influence your family. If your family is exemplary, your family influences the community. If your community is exemplary, your community influences the nation. If your nation is a exemplary, your nation influences the world.[37]

FOUNDATION 10.3: UNDERSTANDING POWER

A vision, like that for a sustainable society, will only become reality if it can call forth the allegiance and active support of a nation's citizenry. But enormous forces are often at work that conspire to undercut and marginalize new ideas that threaten existing power structures. Hence, any time a bold new idea or a whole new paradigm, like sustainability, is struggling to emerge, a power struggle ensues between those desiring change and those preferring to maintain the status quo.

It is curious that the most important dynamic governing human societal relations—namely, the types and uses of power—is seldom discussed in our schools. This tendency may be part of the reason why most people hold the mistaken belief that *only* those in positions of authority have power.

"Power Over"

Those "in power" usually use "power over" to maintain their power. We see "power over" almost everywhere we look in U.S. culture—in corporations, government agencies, public schools, and private households. Most people aren't aware of the pervasive nature and unsettling character of "power over" until they happen, one day, to question the authority of a teacher, boss, police officer, administrator, or parent. Questioning authority—that is, upsetting established power relationships—is not encouraged in societies based on "power over" relations.[38]

In submitting to "power over," citizens often fail to realize that those in power need what citizens agree to give them so that they can remain in power—namely, labor, resources, and acquiescence. Indeed, "orders, to be effective, must be obeyed."[39]

"Power With"

A type of power capable of subduing "power over" is "power with." "Power with" is the power of solidarity; it is the "power of the people." It becomes manifest when people bond as equals, work as a team, and struggle collectively. The legendary struggles to extend rights to children, labor, women, minorities, and the natural world have been central to the flowering and evolution of democracy in the United States. History is "made" when people organize and discover that "power with" can free them from "power over."

This power of solidarity is potentially so great that power holders often put enormous energy into splitting up those they exercise authority over. In industry, for example, it is common practice to create distinctions between job classifications so that workers will see one another as superiors and inferiors, rather than as sharing a common lot.[40]

Power from Within

A third type of power, "power from within," is also closely associated with social change. Individuals experience "power from within" when they bring their outward actions into alignment with their inner convictions. This power gives an individual the strength to stand up alone and say "no" to injustice; it is the power that people are referring to when they say that "one person can make a difference" (see box).

THE WOMAN WHO REFUSED TO SELL RACIST FRUIT

In her book *In the Tiger's Mouth*, Katrina Shields tells a story about an Irish woman who was less than twenty, with no political background, who was working as a clerk in a Dublin supermarket. The young woman had read in the papers about the apartheid regime in South Africa, and she was aware that the black people there had called on consumers not to buy South African products. She also knew that her store carried South African fruit, and this fact troubled her. So one day she decided to refuse to sell it. She dutifully rang up all the other goods, but she refused to sell the products from South Africa. Customers complained to the manager, and in short order, the young woman was fired. But by the end of the day, all the women clerks in the store also were refusing to sell the South African fruit. And soon workers in other supermarkets around Dublin were coming out in support of the women who refused to sell the "racist" fruit. There were threats of mass shutdowns, and the media jumped on the story. Eventually, according to Shields, "The company agreed to stop selling South African fruit, the young woman got her job back, and, in the process, the whole community had learnt about the power of one person's commitment to act in solidarity with those who are oppressed."[41]

The story about the woman who refused to sell racist fruit refutes the popular myth that there is only a limited amount of power to go around—that is, the more power you have, the less power I have. By discovering her personal power, the Irish woman brought forth a rich vein of untapped power in the community.

Power in the Context of the Sustainability Revolution

Our conceptualization of power is intimately linked with our conceptualization of reality. If we believe, for example, that reality is made up of discrete and separate entities, then power will be understood as one thing's acting on another thing—for example, one thing's pushing on another thing to make it behave. In this mind-set, we will act aggressively in the world to get what we want; and we will build defenses around ourselves and our belongings to keep safe from outside threats. This scenario, of course, describes the "power over" mind-set that leads to domination, control, and subordination, which is so widespread in the world today.[42]

But over the last hundred years, our understanding of the nature of reality has changed. We now know that "things" are not separate and immutable. Matter is no longer understood in terms of "stuff that abides," but instead, as "patterns that perpetuate"—whirlpools in flow. Moreover, from the study of living systems, we now know that development and evolution in the natural world occur not by erecting walls of defense, but by interacting and opening up to currents of matter, energy, and information:

> As life forms evolve in complexity and intelligence, they shed their armor, grow sensitive, vulnerable protuberances—like lips, tongues, ears, eyeballs, noses, fingertips—the better to sense and respond, the better to connect in the web of life and weave it further.[43]

In sum, the true nature of reality is change. Interconnection, relationship, and flexibility are emerging metaphors of the sustainability revolution (table 10.1). This new understanding of reality, grounded in modern science, is forging a new understanding of power. We now know, for example, that a unifying power dynamic in the natural world is "power with" (i.e., synergy). As a case in point, consider the situation of one of your neurons. If that neuron were to "assume" that its powers were its own personal property and seek to keep itself "safe" from other neurons by isolating itself behind a defensive wall, it would shrivel up and die. The neuron's "health and its power lie in opening itself to the charge, letting the signals through. Only then can the larger systems of which it is a part learn to respond and think."[44]

The human nervous system, composed of billions of neurons, offers a metaphor for human society. To function properly, the brain requires the open flow of information from neurons; likewise, to function effectively, human social organizations (e.g., governments) require the open and free circulation of information among their citizens. When feedback is blocked through cover-ups, censored reports, misinformation, and deception, it falls on citizens to open blocked channels of information and give feedback.

In sum, revolutions are always about power; but in the case of the sustainability revolution, the aim isn't to gain control over society, but instead, to change the very way that power is conceptualized—that is, as a process that one engages in, rather than a discrete object that one possesses.

Table 10.1.   The Contrasting Worldviews of "Power Over" and "Power With"[45]

"Power Over" *(Industrial Revolution)*	*"Power With"* *(Sustainability Revolution)*
MECHANISTIC: Earth is like a big machine; things work best when central control and specialization are exercised.	ORGANIC: Earth is alive, animated from within; things work best when interdependence, creativity, and cooperation are encouraged.
STATIC: Everything in the world is self-contained and self-explanatory; nothing ever really changes; it is important to control things.	DYNAMIC: Everything in the world is mutually interdependent; change is the essence of reality; it is important to get into the "flow."
SIMPLE: Keeping things simple through standardization is the best way of ensuring control.	COMPLEX: Life is innately complex; nurturing diversity is the best way of ensuring a healthy world.
CLOSED: Everything outside is inferior and represents a threat; it is important to be surrounded with defensive structures.	OPEN: Much of what is required for survival and meaning comes from outside; seeks new ways to interact with others; welcomes others' views and opinions.
SEPARATE: Life is a problem that can be solved abstractly at a distance. Domination is necessary.	RELATED: Life is a process to be engaged in with feelings and intuitions, as well as critical thinking. Trustworthy relationships must be cultivated.
PRODUCT: Life is about production; the end product is what is important; progress depends on order, predictability, and accountability.	PROCESS: Life is about relationship; interaction is what is important; progress depends on flexibility, openness, and adaptability.
PHYSICAL: Only objects exist in the world; materiality is the essence of life; humans are the end-product of evolution.	SPIRITUAL: Everything has interiority; mystery is the essence of life; humans are a present-time expression of the creative unfolding of the universe.

Embracing the "Other"

The sustainability revolution will depend, ultimately, on the ability of humans to love one another. People have been struggling with the dictum "love thy neighbor" for two thousand years. Now, at the outset of the third millennium, humankind is seeing more clearly than ever before that oppression and warfare are not effective means of achieving either enduring security or well-being for any of us. As we fully grasp the comprehensive nature of our interconnectedness, we come to see that when we do harm to other people or to natural ecosystems, we literally violate ourselves.

The sustainability revolution is grounded in acceptance, not blame. It recognizes that it is in the nature of the human to embody both kindness and selfishness, honesty and deceit, courage and cowardice; this is the dialectic that we all carry within us. Our capacity to fully accept ourselves, warts and all, is the starting place for a culture grounded in well-being.

Self-acceptance is not easy. Many of us have been taught since childhood to be dissatisfied with who we are—to see our flaws before our strengths. It is well known that what we cannot accept in ourselves, we try to push away from ourselves by projecting onto others and then rejecting them. In order to accept the "Other," then, we must first learn to love ourselves—not in an indulgent, juvenile way; but in a forgiving, empathetic, adult way. As we do this, we come to see that each of us is the Other for someone. Hence, "the key to changing your relationship to the 'Other' is to expand your notion of Self to include everyone you see" (see box).[46]

Embracing the Other represents a dramatic departure from our historical modus operandi. The story line of history has been to project blame onto the Other—to scapegoat—to always say that someone else is the problem. So

---

EMPATHY FOR THE OTHER

Angie O'Gorman tells the story of how she was awakened late one night by an intruder standing over her bed. No one else was in the house. Somehow, amidst her fear, she realized that there was a connection between herself and this Other—either they would both be damaged by what ensued, or they would both emerge safely, with their integrity intact. This insight allowed O'Gorman to act with empathy for both herself and the intruder.

She spoke calmly into the darkness, asking him what time it was.

"Two-thirty," he replied.

She expressed concern that his watch might be broken because the clock on her night stand read 2:45. After a pause, she asked how he had entered the house. He said he broke a window. She said this was a problem for her because she didn't have the money to fix it. He said that he was also having money problems.

When she felt it was safe, O'Gorman told him, respectfully, that he would have to leave. He said he didn't want to . . . had no place to go. Lacking the force to make him leave and seeing a person without a home, she said that she would give him a set of sheets but he would have to make his own bed downstairs. The man went downstairs, and O'Gorman sat up in bed for the rest of the night. The next morning they had breakfast together. Then the man left. By cultivating empathy for the Other, O'Gorman entered a new plane—a place where no one wins unless everyone wins, because if someone loses, everyone loses.[47]

much of human history is the story of who killed whom—a tale of vengeance. Much of the carnage has been what anthropologists call "sacred violence"— violence justified in the service of a holy cause. But, in our time, the idea of redemptive violence, finally, is being seen for what it is—a misguided attempt to avoid our own pain and brokenness. As theologian Richard Rohr points out, "If you don't transform your pain, you will always transmit it. . . . You will always find someone who is worthy of your hatred . . . and then, of course, you become the evil you despise."[48]

Ultimately, we are at a moment in history when we are striving to make a shift from a consciousness based on exclusivity to one based on inclusivity. This is a monumental shift. Inclusivity is grounded in connection, whereas exclusivity is based on separation. A consciousness grounded in inclusivity generates trust; one moored in exclusivity spreads fear—especially fear of the Other. When our goal is exclusivity, we silence those with whom we disagree; when inclusivity becomes our goal, we invite everyone to the table because our goal is to create a world that works for all.[49]

In sum, the use of power carries with it moral responsibility. "Power over" is an exercise of force that subjugates the Other. "Power with," to be transformative, must extend beyond oppositional politics—for example, the People versus the System—and it must move to an understanding that there is no "enemy."

Reflection

I am fascinated by the "coming out" process—the act of publicly standing up and speaking one's personal truth, like the young woman from Dublin did. When we come out, we say "yes" to our lives. The strength of our personal convictions overcomes our fears of rejection, embarrassment, and reprisal. We declare that our primary allegiance is to our own personhood.

Ideally, our schools should be training grounds for the "coming out" process. I dream of a day

- when teachers applaud students who have the courage to stand by their convictions, especially when these convictions run counter to status quo values and beliefs;
- when valedictorians publicly "come out," declaring that their "mission" in life is to be fully alive, aware, and awake;
- when college professors declare their readiness to commit civil disobedience, if necessary, in order to honor personal convictions that they hold as dearly as life itself;

- when university presidents make it clear to faculty and students that higher education is, first and foremost, about developing a philosophy of life and finding a calling that is noble and life-affirming.

We are all candles waiting to be lit. May we come to light. May we come out.

Questions for Reflection

- What does "coming out" mean to you?
- What have you been silent about but might one day be willing to "come out" for?

Practice: Empowerment through Insight

Since this is the final chapter of this book, it's an appropriate place for me to introduce a practice that can offer insight throughout life—one that can aid our coming to know ourselves more fully. "To know ourselves" means, most fundamentally, to understand and to be fully aware of the workings of our mind—and thereby to have a clear understanding of things as they really are, without prejudice or illusion. We gain an understanding of our mind by studying it.[50]

Mindfulness Meditation

One formal practice of mind study is called "mindfulness meditation." This practice helps bring a person's awareness into the present moment, and variations of this practice are now being taught to people of all sorts—students, corporate executives, professional athletes—in order to increase their concentration, efficiency, and overall awareness. Both Eastern and Western wisdom traditions recognize the value of meditation in expanding human consciousness.

Mindfulness meditation is, in effect, the practice of retraining the mind so that we might one day live with unbroken awareness in the present. Because this is not easy, we need to start small. The breathing practice introduced in chapter 5 (p. 150) is the entry point to mindfulness meditation. It consists of sitting quietly and placing full attention on the breath. Don't verbalize or conceptualize; simply notice the incoming and outgoing breath. The idea is to extricate yourself from the morass of mental images that usually engulf you, by bringing attention to the here

CHAPTER 10

and now. Form an intention to remain present for a single inhalation, then an exhalation, then a second inhalation, and so forth.

Although it sounds easy, to quiet the mind is quite challenging, especially at first. Often the mind introduces a juicy thought before we have even completed one in-breath; or if we do manage to complete a single breathing cycle, the mind jumps in to congratulate us. The difficulty that you experience in keeping your attention on your breath offers an insight into the workings of your mind. You discover, through your direct experience, that the mind is very busy—it wanders all over the place, and you have little control over it. The mind wanders about because it has been conditioned to do so. Meditation practice brings awareness to the mind, and with this awareness comes understanding and, ultimately, genuine freedom.

Another insight that is gained early in meditation practice concerns the content of our thoughts. For example, when you get distracted in thought during meditation, you can simply note the type of thought that carried you away from your breath. Experiment with this aspect of meditation by assuming the role of the detached observer and by simply watching your thoughts go by without getting caught up in them. Welcome each new thought, acknowledge it, and then let it pass away. For example, if you get caught up in fretting, simply smile to yourself, noting "worry mind." Do the same for "planning mind," "judging mind," "fearful mind," and so forth. "The practice of naming is like becoming a biologist of the mind, classifying the flora and fauna that come into view and noticing how they are related to each other in the ecosystem. You are not specifically looking for anything, but simply noting what occurs and which species predominate."[51]

As you begin to experiment with mindfulness meditation, it may help to think of yourself as a rock resting by a riverbank and to think of your thoughts as the water flowing down the river; just let the thoughts flow by. Simply by noting the general content of your mind's chatter, you can gain clear insight into the mind states that most occupy you, how these mind states arise, and what their effect and duration mean. Normally, you would engage with the passing thoughts and get caught up in their emotional charge—that is, you would become the thoughts. "In our so-called 'normal' state of consciousness we are therefore continually lost in the drama of our lives, unaware of how the process that creates our reality is taking place" (see box).[52]

As you begin to practice mindfulness meditation, you may find that much of your mental energy is absorbed by "wanting mind"—always wanting things to be different from what they are. Venerable Henepola Gunaratana, the abbot of the Bhavana Society monastery in West Virginia, calls this the "if only" syndrome: "If only I had a higher salary, a different boss, a second

AN ANALOGY TO EXPLAIN MINDFULNESS MEDITATION

Meditation teacher Wes Nisker uses a movie theater analogy to explain how mindfulness meditation works:

When we are watching the screen, we are absorbed in the momentum of the story, our thoughts and emotions manipulated by the images we are seeing. But if just for a moment we were to turn around and look toward the back of the theater at the projector, we would see how these images are being produced We would recognize that what we are lost in is nothing more than flickering beams of light. Although we might be able to turn back and lose ourselves once again in the movie, its power over us would be diminished. The illusion-maker has been seen.

Similarly, in mindfulness mediation, we look deeply into our own movie-making process. We see the mechanics of how our personal story gets created, and how we project that story onto everything we see, hear, taste, smell, think, and do.[53]

Understanding how the mind works to create our life "dramas" frees us from taking our "movie" so seriously.

---

car, a different spouse, a better computer, curly hair, a smaller nose . . . then everything would be okay." Exploring "if-only mind" in meditation leads us to another insight—namely, that the mind is very good at creating desires. It is always seeking a different reality, always wanting things to be different from what they are, always fleeing from pain.[54]

This insight reveals that reality is not what has to be fixed (reality is just what it is!); rather, the problem to be fixed is how we have chosen to see the world. Rather than blaming others or blaming ourselves for our unsettled mind states, we can exercise an alternative strategy—to simply be with and accept things as they are. Instead of putting up resistance through mental judgment and emotional negativity, we can yield to the flow of life. This is the essence of all wisdom traditions.

A simple example will help illustrate how the mind constantly creates problems for us. Suppose you are sitting down to write a letter, and you hear the sound of a dog barking outside. Your mind classifies the sound as "dog barking" and judges it as annoying. This happens automatically—unconsciously. The mind creates problems like this "because it holds the unconscious belief that its resistance, which you experience as negativity or unhappiness in some form, will somehow dissolve the undesirable condition."[55] This thinking, of course, is wrong; the resistance that the mind creates is far more distressing than the actual cause that provoked the resistance. Rather than resisting, you could allow the sound of the barking dog, or whatever else may cause a negative reaction, to simply flow right through you. The point here is not that we should become passive. We still act in the world, but our energy is no longer wasted in resisting what already is.[56]

As one develops a mindfulness meditation practice, she becomes adept at simply accepting what is, which includes developing the capacity to be with painful feelings—loneliness, anger, and despair. Once the practioner allows herself the space to experience her pain, she notices that all feelings are ephemeral. They come and go. Meditation teachers call this flux the "weather of the mind." One may experience sun one hour (joy) and clouds the next hour (sadness), which may be followed by mist (melancholy) and so forth. When you can simply be with your painful mind states, you are no longer controlled by them. Thus, you come to realize that the most an emotion can do is make you feel it—if you are willing to feel it, that is!

As a minipractice, you can pick an emotion that is very active in your life right now—maybe it is jealousy or greed or anger. Just pick one, and when it comes up, be willing to really feel it in your body. Don't get caught up in the story—the content—behind the emotion; simply experience the physical sensation of this emotion in your body. If you do this practice for a time, you will become more and more fearless; you will be able to face the emotions that previously caused you great pain because you were in constant fear of feeling them.

So, mindfulness meditation is a practice that allows us to deal skillfully with the things that are unwanted in our lives. We break the conditioned response of running from such things, and instead, we open to them. Likewise, rather than fiercely grasping after that which brings us happiness, we learn to become less attached to pleasure. Yes, we still enjoy things, but we don't grasp after them. By doing so, we can free ourselves of both fears (that which we normally run from) and desires (that which we normally grasp after).[57]

The Power of Now

Over time, meditation practice enables one to become ever-more grounded in the present moment. Indeed, one learns that the "Now" is all there is. As spiritual teacher Eckhart Tolle observes, "Life is now. There was never a time when your life was not now, nor will there ever be."[58] The present moment holds the key to liberation from the ego because the ego is very much grounded in the past and future. For the egoistic mind, the past harbors one's identity, and the future offers the promise of fulfillment. The mind and the ego are not able to get any traction in the "Now." The present, to the extent that it exists for the egoistic mind, is either perceived through the eyes of the past or seen simply as a means to some future end.[59] Read this paragraph again. Take it in sentence by sentence, and really allow Tolle's point to settle in.

As one gains the capacity to live more fully in the Now, egoistic mind states associated with the future (anxiety, stress, tension, fear) and the past (guilt, sadness, bitterness, resentment) begin to diminish. In the crucible of the Now, problems melt away. If this seems mysterious, try it: Stop and bring your attention fully to the Now, and your "problems" will begin to dissolve away. It cannot be otherwise.[60]

The mind cannot function and remain in control without time. Hence, by living in the Now—the eternal present—one's identity is no longer anchored in the mind. This is not to say that the mind isn't important. The mind is an impressive asset if used rightly, but most of us don't use our minds; our minds use us. We are not in control. If we were in control, we could be free of our mind's thought traffic whenever we wanted; but most of us live at the mercy of our minds.[61] Tolle estimates that 80 to 90 percent of most people's thinking is repetitive; he also notes that many of our thoughts are negative and therefore harmful.[62]

Loving Kindness

The insights gained through mindfulness meditation awaken one to the great capacity that humans have for love and kindness. Just the simple act of paying attention to another being reveals our connection and draws us into intimacy. One specific meditation practice consists of releasing thoughts of hatred and greed by actively cultivating thoughts of "loving kindness." The practice begins by allowing thoughts of loving kindness to flow toward oneself:

> May I be well, happy, and peaceful. May no harm come to me. May no difficulties come to me. May no problems come to me. . . . May I also have patience, courage, understanding, and determination to meet and overcome inevitable difficulties, problems, and failures in life.[63]

Next, the phrasing is changed slightly so that thoughts of loving kindness can be extended to one's family: "May my parents and brothers and sisters be well, happy, and peaceful. May no harm . . . etc." Then, the circle of attention is widened to include friends, strangers, "enemies," and finally all beings.

For many the idea of trying to consciously evoke feelings of loving kindness may seem artificial or false. We have been led to believe that feelings are supposed to arise spontaneously, but those who experiment with this practice discover that by setting an intention to extend loving kindness, they are able to widen their circle of understanding and compassion.[64]

Because our thoughts are often transformed into speech and action, many find it helpful to do this practice at the beginning of the day. If you begin the day extending loving kindness to yourself and others, you in effect set an intention to be open to everyone—friend and foe—during the day. It is especially important to include your "enemies" in this practice. Indeed, it is only when your enemies are free of danger and it is only when they know joy and peace of mind that you yourself will experience ease of well-being.

These meditation practices are, in essence, about the art of being present. In his book *Stepping Lightly*, Mark Burch offers a vivid vision of what living mindfully can mean:

> Imagine yourself for a moment living in a more mindful and attentive way. See yourself moving more slowly through life, taking time to notice things, time to find your balance as you walk, time to notice how things look and smell around you. See yourself looking deeply into other people's eyes as you talk with them, studying their faces with attention and sensitivity. See yourself deeply enjoying the pleasure of love-making, a fresh salad, a starry evening sky, walking barefoot in wet grass. Imagine yourself gazing steadily inward, knowing and accepting yourself, your feelings, longings, spiritual intuitions, dreams. See yourself so clear, so centered and strong, that you can calmly remain with those in crisis, listening, loving, understanding, giving.[65]

In sum, the practice of mindfulness meditation is a lifelong discipline that leads to ever-greater presence and ever-widening consciousness. With practice, it provides an understanding of the patterns of the mind that block the flow of life. This practice is important because, as Tolle observes, "The pollution of the planet is only an outward reflection of an inner psychic pollution: millions of unconscious individuals not taking responsibility for their inner space. . . . If humans clear inner pollution, then they will also cease to create outer pollution."[66]

CONCLUSION

> The true value of a human being is determined primarily by the measure and the sense in which he has attained liberation from the self.
>
> —Albert Einstein[67]

Sometimes, when we are having a difficult time figuring out how to get from point A to point B, someone will offer us a map. This is what the three foundations of this chapter have offered. Point A is our present predicament; point B is the open, just, and sustainable society that is waiting to be born. The "map" comes to life on a parchment infused with imagination and vision. Humankind needs, as never before, a vision that is worthy of its best efforts and that honors the integrity and splendor of the living earth.

Examining the map closely, we see the many steps involved in the creation of a life-sustaining world. This isn't a quick day trip; it will take time for us to awaken and summon our courage to act as innovators, change agents, and transformers in the service of life.

Like any other journey, this new revolutionary saga needs a power source. "Power over" is destabilizing and violates the natural order. It needs to be replaced by "power with"—that is, power based on openness, trust, and cooperation.

All journeys have their exterior and interior dimensions. This journey toward sustainability is as much about personal transformation as it is about societal transformation. Indeed, the two are inextricable intertwined. As we become empowered as individuals, the impossible becomes possible within society.

NOTES

1. Vaclav Havel, "Introductory Essay," in *Creating a World that Works for All*, by Sharif Abdullah (San Francisco: Berrett-Koehler, 1999), viii.

2. Jacob Needleman, "Founding America," *Yes! A Journal of Positive Futures*, Spring 2002, 12–17.

3. Jack Weatherford, *Indian Givers* (Minneapolis, Minn.: Crown, 1988), quoted from Needleman, "Founding America," 15.

4. Sally R. Wagner, "American Woman," *Yes! A Journal of Positive Futures*, Spring 2002, 18–19.

5. Wagner, "American Woman," 18.

6. Wagner, "American Woman," 19.

7. Susan Griffin, "Can Imagination Save Us?" *Utne Reader*, July–August 1996, 43–45.

8. The Wildlands Project (www.wildlandsproject.org).

9. This map is based on Reed F. Noss, "Protecting Natural Areas in Fragmented Landscapes," *The Natural Areas Journal* 7, no. 1 (1987): 2–13.

10. Ernest Callenbach, *Bring Back the Buffalo!* (Washington, D.C.: Island Press, 1996).

11. Callenbach, *Bring Back the Buffalo!*

12. Needleman, "Founding America," 17.

13. Walter Wink, *The Powers That Be* (New York: Galilee Doubleday, 1998), 185.

14. Mark A. Burch, *Stepping Lightly* (Gabriola Island, British Columbia: New Society, 2000), 144.

15. Gregg Levoy, *Callings: Finding and Following an Authentic Life* (New York: Three Rivers Press, 1997), 122.

16. Levoy, *Callings*.

17. Levoy, *Callings*.

18. This discussion of personal mission statements was inspired by David Tait.

19. Joanna Macy, *Coming Back to Life* (Gabriola Island, British Columbia: New Society, 1998).

20. Bill Moyer, *Doing Democracy* (Gabriola Island, British Columbia: New Society, 2001).

21. Moyer, *Doing Democracy*.

22. Moyer, *Doing Democracy*.

23. Moyer, *Doing Democracy*.

24. Moyer, *Doing Democracy*, 84–85.

25. Katrina Shields, *In the Tiger's Mouth* (Gabriola Island, British Columbia: New Society, 1994).

26. This figure is adapted from Alan AtKisson, *Believing Cassandra* (White River Junction, Vt.: Chelsea Green, 1999), 182.

27. AtKisson, *Believing Cassandra*.

28. AtKisson, *Believing Cassandra*.

29. Margaret Wheatley, *Leadership and the New Science* (San Francisco: Berrett-Koehler, 1999).

30. Wheatley, *Leadership and the New Science*, 119.

31. Jeremy W. Hayward, *Letters to Vanessa* (Boston: Shambhala, 1997).

32. Paul R. Loeb, *Soul of a Citizen* (New York: St. Martin's Griffin, 1999), 38.

33. This example is drawn from Donella Meadows, "Saying It Straight," *The Global Citizen*, December 16, 1999 (accessed at www.sustainabilityinstitutue.org).

34. A. B. Schmookler, "Creating a Culture of Meaning." Talk given on February 27, 1997, at Foundation for Global Community, Palo Alto, CA.

35. Fran Peavey, *Strategic Questioning* (San Francisco: Crabgrass Organization, 2001), 10. Contact information: Crabgrass, 3181 Mission Street, San Francisco, CA 94110.

36. Shields, *In the Tiger's Mouth*, 15.

37. Shields, *In the Tiger's Mouth*.

38. Shields, *In the Tiger's Mouth*.

39. Priscilla Boucher, "Reclaiming the Power of Community," in *Putting Power in Its Place*, edited by Judith Plant and Christopher Plant (Gabriola Island, British Columbia: New Society, 1992), 47.

40. Shields, *In the Tiger's Mouth*.

41. Shields, *In the Tiger's Mouth*, 12.

42. Macy, *Coming Back to Life*.

43. Macy, *Coming Back to Life*, 53.

44. Macy, *Coming Back to Life*.

45. This table was inspired by Diarmuid O'Murchu, *Our World in Transition* (New York: Crossword, 1992).

46. Sharif Abdullah, *Creating a World That Works for All* (San Francisco: Berrett-Koehler, 1999), 80.

47. This story was related in Wink, *The Powers That Be*. The original story appeared in Angie O'Gorman, "Defense through Disarmament: Nonviolence and Personal Assault," in *The Universe Bends Toward Justice*, edited by Angie O'Gorman (Gabriola Island, British Columbia: New Society, 1990), 242–246.

48. Richard Rohr, *New Great Themes of Scripture* (Cincinnati, Ohio: St. Anthony Messenger Press, 1999).

49. Abdullah, *Creating a World*.

50. Venerable Henepola Gunaratana, *Mindfulness in Plain English* (Boston: Wisdom, 1991).

51. Wes Nisker, *Buddha's Nature* (New York: Bantam Books, 1998), 195.

52. Nisker, *Buddha's Nature*, 26.

53. Nisker, *Buddha's Nature*, 26.

54. Gunaratana, *Mindfulness in Plain English*.

55. Eckhart Tolle, *The Power of Now* (Novato, Calif.: New World Library, 1999), 159–160.

56. Tolle, *The Power of Now*.

57. Pema Chodran, *Awakening Compassion*, audio cassette recording (Boulder, Colo.: Sounds True, 1995).

58. Tolle, *The Power of Now*, 41.

59. Tolle, *The Power of Now*.

60. Tolle, *The Power of Now*.

61. Tolle, *The Power of Now*.

62. Tolle, *The Power of Now*.

63. Gunaratana, *Mindfulness in Plain English*, 101.

64. Nisker, *Buddha's Nature*.

65. Burch, *Stepping Lightly*, 106.

66. Tolle, *The Power of Now*, 65.

67. Albert Einstein, *Ideas and Opinions* (New York: Three Rivers Press, 1954), 12.

Part III: Summary

HEALING OURSELVES, HEALING EARTH

Let the beauty we love be what we do.

—Rumi

We live in an epic time—"a time between times." Western civilization's old "story"—based on objectification, separation, exclusivity, and control—no longer fits with our culture's new understanding of reality. The old story depicted the world as dead. We now know this objectification of reality is mistaken; it contradicts the leading-edge discoveries in physics, biology, and systems science; it also violates people's deepest intuitions. Thus, the old story is dying. It flickers out one person at a time.

Look into the eyes of fellow citizens: Some are clinging desperately to the old story; their eyes are pinched into slits; they are afraid. Continue looking, and you will encounter others who have rejected the old story but have found nothing to replace it with; they are confused and prone to cynicism and addictive behaviors—their eyes are agitated. Look some more, and you will find those whose eyes shine with meaning and purpose because they live in a world that is alive, not dead. Where others see separation, they see connection. These are the people who are hospicing out the old reality and giving birth to the sustainability revolution.

Across the country and the world, tens of millions of people are summoning the courage to question economism, mindless consumption, and the domination ("power over") mentality. Increasingly, people are embracing sustainability principles as the only pathway to a sane future. This is a silent revolution, little mentioned in the popular press, simmering just below the surface of contemporary life. But it is real. The pieces are slowly coming together: a new company that is offering solar technologies; farmers who are committed to sustainable farming practices; builders who create highly efficient structures that require little energy to heat and cool; cities that are discouraging car transit in favor of bike and rail; landholders who sustainably manage their forests; businesses that are more concerned with treating their employees justly and with producing high-quality goods than with growth and profit maximization; towns that use indicators based on sustainability to gauge their progress . . . the list goes on.

A caterpillar becoming a butterfly offers a wonderful analogy for how the sustainability revolution is now spreading throughout modern culture. As the caterpillar approaches the time of metamorphosis, certain cells begin the process of building the various parts of the new butterfly. Biologists call these "imaginal" cells. Now, think of all the sustainability initiatives bubbling up around us today as "imaginal" cells, growing and interweaving to make a new just and sustainable society (what a beautiful butterfly indeed!).[1]

Although many people are clearly working hard to bring forth a life-sustaining society, their efforts may not be enough. We would be naïve in the extreme to ignore the power that humans now have (and, therefore, the potential we have) to irreparably harm all life on Earth. That the sustainability revolution will flicker out is quite possible; the needed changes may not come in time.

Who, then, is holding things back? Who is standing in the way of the life-sustaining society? Businessman and corporate consultant Robert Greenleaf has a sobering answer:

Not evil people. Not stupid people. Not apathetic people. Not the "system." Not the . . . reactionaries. The better society will come, if it comes, with plenty of evil, stupid, apathetic people around and with an imperfect, ponderous, inefficient "system" as the vehicle for change. Liquidate the offending people, radically alter or destroy the system, and in less than a generation they will all be back.

The real enemy is fuzzy thinking on the part of good, intelligent, vital people, and their failure to lead. Too many of us settle for being critics and experts. There is too much intellectual wheel spinning, too much retreating into research, too little preparation for and willingness to undertake the hard and high risk tasks of building better institutions in an imperfect world, too little disposition to see the problem as residing "in here" and not "out there." In short, the enemy is good people who have the potential to lead but do not lead. They suffer; society suffers.[2]

Creating a sustainable society hinges on whether people who have met their material needs can now craft a simpler, more satisfying new way of life. Healing and wisdom will ultimately come through the reforging of connections—namely, our connections to our bodies, the natural world, and our home places. Sustainability is, ultimately, about loving—loving the earth and loving life. So, that's it: We are all here on Earth to learn to be good lovers! May we awaken!

NOTES

1. Sarah van Gelder, "The Transformation of Business: An Interview with Willis Harman," *Yes! A Journal of Positive Futures*, no. 41 (Summer 1995): 52–55.
2. Robert K. Greenfield, "The Institution as Servant" (Indianapolis: R. K. Greenleaf Center). Contact information: R. K. Greenleaf Center, 921 East 86th Street, Suite 200, Indianapolis, IN 46240; www.Greenleaf.org.

BIBLIOGRAPHY

Abdullah, Sharif. *Creating a World that Works for All.* San Francisco: Berrett–Koehler, 1999.

Armstrong, Thomas. *Awakening Your Child's Natural Genius.* New York: G. P. Putnam's Sons, 1991.

AtKisson, Alan. *Believing Cassandra.* White River Junction, Vt.: Chelsea Green, 1999.

Ball, Phillip. *A Biography of Water: Life's Matrix.* Berkeley: University of California Press, 2001.

Beattie, Andrew, and Paul Ehrlich. *Wild Solutions.* New Haven, Conn.: Yale University Press, 2001.

Berry, Thomas. *The Dream of the Earth.* San Francisco: Sierra Club Books, 1988.

———. *The Great Work.* New York: Bell Tower, 1999.

Blood, Casey. *Science, Sense and Soul.* Los Angeles: Renaissance Books, 2001.

Bormann, Herbert F., Diana Balmori, and Gordon T. Geballe. *Redesigning the American Lawn.* New Haven, Conn.: Yale University Press, 1993.

Brandt, Barbara. *Whole Life Economics.* Gabriola Island, British Columbia: New Society, 1995.

Brower, Michael, and Warren Leon. *The Consumer's Guide to Effective Environmental Choices.* New York: Three Rivers Press, 1999.

Brown, Lester. *EcoEconomy.* New York: W. W. Norton, 2001.

Brown, Molly. *Growing Whole.* Center City, Minn.: Hazelden Educational Materials, 1993.

Burch, Mark. *Stepping Lightly.* Gabriola Island, British Columbia: New Society, 2000.

Callenbach, Ernest. *Bring Back the Buffalo!* Washington, D.C.: Island Press, 1996.

Capra, Fritjof. *The Hidden Connections.* New York: Doubleday, 2002.

Chambers, Nicky, Craig Simmons, and Mathis Wackernagel. *Sharing Nature's Interest.* London: Earthscan, 2000.

Chiras, Daniel. *Lessons from Nature.* Washington, D.C.: Island Press, 1992.

Cohen, Joel. *How Many People Can the Earth Support?* New York: W. W. Norton, 1995.

Colburn, Theo, Diane Dumanoski, and J. Peter Myers. *Our Stolen Future.* New York: Dutton, 1996.

Cooperrider, Daniel, and Diana Whitney. *Appreciative Inquiry.* San Francisco: Berrett–Koehler, 1999.

Dominguez, Joel, and Vicki Robin. *Your Money or Your Life.* New York: Penguin Books, 1992.

Duncan, David James. *My Story as Told by Water.* San Francisco: Sierra Club Books, 2001.

Durning, Alan T. *How Much is Enough: The Consumer Society and the Future of the Earth.* New York: W. W. Norton, 1992.

Ehrlich, Paul, and Anne Ehrlich. *Extinction.* New York: Random House, 1981.

Eisenberg, Evan. *The Ecology of Eden.* New York: Alfred A. Knopf, 1998.

Eisenstein, Charles. *The Yoga of Eating.* Washington, D.C.: New Trends, 2003.

Eisler, Riane. *The Chalice and the Blade.* New York: HarperCollins, 1987.

Flannery, Tim. *The Eternal Frontier.* New York: Atlantic Monthly Press, 2001.

Fox, Matthew. *The Coming of the Cosmic Christ.* San Francisco: Harper San Francisco, 1988.

Goldberg, Natalie. *Writing Down the Bones.* Boston: Shambhala, 1986.

Grace, Eric. *The World of the Monarch Butterfly.* San Francisco: Sierra Club Books, 1997.

Gunaratana, Henepola. *Mindfulness in Plain English.* Boston: Wisdom, 1991.

Hanh, Thich Nhat. *Peace is Every Step.* New York: Bantam Books, 1991.

———. *The Heart of Understanding.* Berkeley, Calif.: Paralax Press, 1988.

Hardin, Jesse Wolf. *Coming Home: ReBecoming Native, Recovering Sense of Place.* Reserve, N.Mex.: The Earthen Spirituality Project.

———. *Kindred Spirits: Sacred Earth Wisdom.* Columbus, N.C.: Swan-Raven, 2001.

Hawken, Paul, Amory Lovins, and L. Hunter Lovins. *Natural Capitalism: Creating the Next Industrial Revolution.* Boston: Little, Brown, 1999.

Hawken, Paul. *The Ecology of Commerce.* New York: Harper Business, 1993.

Hayes, Denis. *The Official Earthday Guide to Planet Repair.* Washington, D.C.: Island Press, 2000.

Hayward, Jeremy. *Letters to Vanessa.* Boston: Shambhala, 1997.

Heinrich, Bernd. *The Trees in My Forest.* New York: HarperCollins, 1997.

Hertsgaard, Mark. *Earth Odyssey.* New York: Broadway Books, 1998.

Hubbard, Barbara. *Conscious Evolution: Awakening the Power of our Social Potential.* Novato, Calif.: New World Library, 1998.

Jensen, Derrick. *A Language Older than Words.* New York: Context Books, 2000.

Kabat-Zinn, Jon. *Full Catastrophe Living.* New York: Dell, 1990.

Kaza, Stephanie. *The Attentive Heart.* Boston: Shambhala, 1996.

Kohl, Judith, and Herbert Kohl. *The View from the Oak.* New York: The New York Press, 1997.

Korten, David. *When Corporations Rule the World.* West Hartford, Conn.: Kumarian Press, 1995.

Krafel, Paul. *Seeing Nature.* White River Junction, Vt.: Chelsea Green, 1999.

La Chapelle, David. *Navigating the Tides of Change.* Gabriola Island, British Columbia: New Society, 2001.

Langer, Ellen. *Mindfulness.* Cambridge, Mass.: Perseus Books, 1989.

Lappe, Frances M., and Paul DuBois. *The Quickening of America: Rebuilding Our Nation, Remaking Our Lives.* San Francisco: Jossey-Bass, 1994.

Leopold, Aldo. *A Sand County Almanac.* Oxford: Oxford University Press, 1949.

Lerner, Michael. *Spirit Matters.* Charlottesville, Va.: Hampton Roads, 2000.

Levoy, Greg. *Callings: Finding and Following an Authentic Life.* New York: Three Rivers Press, 1997.

Liebes, Sidney, Elisabet Sahtouris, and Brian Swimme. *A Walk Through Time.* New York: John Wiley and Sons, 1998.

Loeb, Paul. *Soul of a Citizen.* New York: St. Martin's Griffin, 1999.

Lovelock, James. *Gaia: A New Look at Life.* Oxford: Oxford University Press, 1979.

Macy, Joanna. *Coming Back to Life.* Gabriola Island, British Columbia: New Society, 1998.

Mander, Jerry. *In the Absence of the Sacred.* San Francisco: Sierra Club Books, 1991.

Margulis, Lynn, and Dorion Sagan. *Microcosmos.* Berkeley: University of California Press, 1997.

———. *What is Life?* Berkeley: University of California Press, 1995.

Margulis, Lynn. *Symbiotic Planet.* New York: Basic Books, 1998.

McKibben, Bill. *The End of Nature.* New York: Anchor Books, 1989.

McKinney, Michael L., and Robert. M. Schoch. *Environmental Science: Systems and Solutions.* New York: West, 1996.

Meadows, Donella, Dennis Meadows, and Jorgen Randers. *Beyond the Limits: Confronting Global Collapse, Envisioning a Sustainable Future.* Post Mills, Vt.: Chelsea Green, 1992.

Mello, Robert A. *Last Stand of the Red Spruce.* Washington, D.C.: Island Press, 1987.

Moore, Thomas. *Care of the Soul.* New York: HarperCollins, 1992.

Moyer, Bill. *Doing Democracy.* Gabriola Island, British Columbia: New Society, 2001.

Nattrass, Brian, and Maru Altomare. *The Natural Step for Business.* Gabriola Island, British Columbia: New Society, 1999.

Nisker, Wes. *Buddha's Nature.* New York: Bantam Books, 1998.

———. *The Essential Crazy Wisdom.* Berkeley, Calif.: Ten Speed Press, 2001.

O'Murchu, Diarmuid. *Our World in Transition.* New York: Crossword, 1992.

Orr, David. *Earth in Mind.* Washington, D.C.: Island Press, 1994.

Palmer, Parker. *The Courage to Teach.* San Francisco: Jossey-Bass, 1998.

Parks, Sharon. *Big Questions, Worthy Dreams.* San Francisco: Jossey-Bass, 2000.

Peavey, Fran. *Strategic Questioning.* San Francisco: Crabgrass Organization, 2001.

Pimm, Stuart. *The World According to Pimm: A Scientist Audits the Earth.* New York: McGraw-Hill, 2001.

Putnam, Robert. *Making Democracy Work: Civic Traditions in Modern Italy.* Princeton, N.J.: Princeton University Press, 1993.

Quammen, David. *Song of the Dodo.* New York: Simon & Schuster, 1996.

Quinn, Daniel. *Ishmael.* New York: Bantam, Turner, 1992.

————. *The Story of B.* New York: Bantam Books, 1996.

Rathje, William, and Cullen Murphy. *Rubbish.* Tucson: University of Arizona Press, 2001.

Raymo, Chet. *Skeptics and True Believers.* New York: Walker, 1998.

————. *An Intimate Look at the Night Sky.* New York: Walker, 2001.

————. *Natural Prayers.* St. Paul, Minn.: Hungry Mind Press, 1999.

Rezendes, Paul. *The Wild Within.* New York: Penguin Putnam, 1998.

Robbins, John. *Diet for a New America.* Walpole, N.H.: Stillpoint, 1987.

Rosenberg, Marshall. *Nonviolent Communication: A Language of Compassion.* Encinitas, Calif.: Puddle Dancer Press, 1999.

Ryan, John, and Alan Durning. *Stuff: The Secret Lives of Everyday Things.* Seattle, Wash.: Northwest Environmental Watch, 1997.

Ryan, John. *Seven Wonders: Everyday Things for a Healthier Planet.* San Francisco: Sierra Club Books, 1999.

Sardello, Robert. *Freeing the Soul from Fear.* New York: Riverhead Books, 1999.

Seeds, Michael. *Horizons: Exploring the Universe.* Pacific Grove, Calif.: Brooks/Cole, 2002.

Segal, Jerome. *Graceful Simplicity: Toward a Philosophy and Politics of Simple Living.* New York: Henry Holt, 2000.

Seidel, Peter. *Invisible Walls.* Amherst, N.Y.: Prometheus Books, 1998.

Settle, Robert, and Pamela Alreck. *Why They Buy: American Consumers Inside and Out.* New York: John Wiley and Sons, 1986.

Shepard, Paul. *The Others: How Animals Made Us Human.* Washington, D.C.: Island Press, 1996.

Shields, Katrina. *In the Tiger's Mouth.* Gabriola Island, British Columbia: New Society, 1994.

Spretnak, Charlene. *The Resurgence of the Real.* New York: Routledge, 1999.

Stanfield, R. Brian. *The Courage to Lead.* Gabriola Island, British Columbia: New Society, 2000.

Stella, Thomas. *The God Instinct.* Notre Dame, Ind.: Sorin Books, 2001.

Strauss, Susan. *The Passionate Fact.* Golden, Colo.: North American Press, 1996.

Suzuki, David. *The Sacred Balance.* Amherst, N.Y.: Prometheus Books, 1998.

Swimme, Brian, and Thomas Berry. *The Universe Story.* San Francisco: Harper San Francisco, 1992.

Swimme, Brian. *The Hidden Heart of the Cosmos.* Maryknoll, N.Y.: Orbis Books, 1996.

———. *The Universe is a Green Dragon.* Santa Fe, N.Mex.: Bear, 1984.

Thomashow, Mitchell. *Bringing the Biosphere Home.* Cambridge, Mass.: MIT Press, 2002.

Tolle, Eckhart. *The Power of Now.* Novato, Calif.: New World Library, 1999.

Van der Ryn, Sim, and Staurt Cowan. *Ecological Design.* Washington, D.C.: Island Press, 1996.

VanMatre, Steve. *Earth Education.* Greenville, W.Va.: The Institute for Earth Education, 1990.

Volk, Tyler. *Gaia's Body.* New York: Copernicus/Springer-Verlag, 1998.

Wackernagel, Mathis, and William Rees. *Our Ecological Footprint.* Gabriola Island, British Columbia: New Society, 1996.

Wheatley, Margaret. *Leadership and the New Science.* San Francisco: Berrett–Koehler, 1999.

Whyte, David. *Crossing the Unknown Sea: Work as a Pilgrimage of Identity.* New York: Riverhead Books, 2001.

Wilber, Ken. *Grace and Grit.* Boston: Shambhala, 1991.

———. *A Brief History of Everything.* Boston: Shambhala, 1996.

Wink, Walter. *The Powers that Be.* New York: Galilee Doubleday, 1998.

Zimmerman, Michael. *Science, Nonscience, and Nonsense: Approaching Environmental Literacy.* Baltimore, Md.: Johns Hopkins University Press, 1995.

INDEX

ABOUT THE AUTHOR

Christopher Uhl is professor of biology at Pennsylvania State University. He teaches ecology and environmental science and conducts research in the realm of human ecology.